Bernhard Kienzler, Volker Metz, Alba Valls (eds.)

Final Workshop Proceedings of the Collaborative Project „Fast / Instant Release of Safety Relevant Radionuclidesfrom Spent Nuclear Fuel"

(7th EC FP CP FIRST-Nuclides)

Antwerp 05 – 07 November 2013

Karlsruhe Institute of Technology
KIT SCIENTIFIC REPORTS 7716

Final Workshop Proceedings of the Collaborative Project „Fast / Instant Release of Safety Relevant Radionuclides from Spent Nuclear Fuel“

(7th EC FP CP FIRST-Nuclides)

Karlsruhe 01 - 02 September 2014

Edited by
Bernhard Kienzler, Volker Metz, Alba Valls

KIT Scientific Publishing

Report-Nr. KIT-SR 7716

Karlsruher Institut für Technologie (KIT)
Institut für Nucleare Entsorgung

Impressum

KIT Scientific Publishing

Karlsruher Institut für Technologie (KIT)
KIT Scientific Publishing
Straße am Forum 2
D-76131 Karlsruhe

KIT Scientific Publishing is a registered trademark of Karlsruhe Institute of Technology. Reprint using the book cover is not allowed.

www.ksp.kit.edu

Print on Demand 2016

ISSN 1869-9669
ISBN 978-3-7315-0525-9
DOI: 10.5445/KSP/1000054569

Foreword

The present document contains the proceedings of the Final Workshop (FW) of the EURATOM FP7 Collaborative Project FIRST-Nuclides (Fast / Instant Release of Safety Relevant Radionuclides from Spent Nuclear Fuel). The electronic version of these proceedings is available from the webpage of the project (http://www.firstnuclides.eu/). The Final Workshop was hosted by KIT/INE and held in Karlsruhe (Germany) 1st – 2nd September 2014. The project started in January 2012 and had duration of three years. The consortium consisted of 10 beneficiaries and 12 associated groups. Most of the members of the consortium participated in the Final Workshop, furthermore interested scientists from NRC, PNNL, and University of Western Ontario were present.

These proceedings serve several purposes. The key purpose is to document and make available the outcome of the FIRST-Nuclides project to a broad scientific community. For this purpose, a considerable part of the project activity is reported by the proceedings, together with the scientific-technical contributions containing details of the work and the "Topical Sessions" dealing with different subjects of interest for the development of the project. In the Final Workshop of FIRST-Nuclides, the topical session focused on X-ray Absorption Fine Structure (XAFS) investigations of HLW waste glass and spent nuclear fuel. The project facilitated training mobility measures for students, and the outcome of these measures was also reported during the workshop. Additional purposes of the proceedings are to ensure on-going documentation of the project outcome, promote systematic scientific-technical development throughout the project and to allow thorough review of the project progress.

All Scientific and Technical papers (S&T) included in the proceedings have been reviewed by the EUG (End-User-Group). The EUG is specifically set up within the project representing the interest of the national waste management and/or national regulatory organizations who will use the results of the project for their Safety Cases. For this reason, this EUG had a strong involvement on guiding the priorities of the activities of the project.

The proceedings give only very brief information about the project structure and the different activities around the project, such as training measures and dissemination of knowledge. This type of information can be found in detail under http://www.firstnuclides.eu/.

The editors of the proceedings thank all contributors to the project, especially those submitting Scientific and Technical contributions, and the workpackage leaders who provided the summary of the different workpackages for publication in these proceedings. Special thanks are given also to the reviewers, the members of the EUG, whose effort and hard work reflected their commitment and dedication to the project and ensure a proper direction and a high quality of the research within the project FIRST-Nuclides.

Table of contents

Workpackage Overview

S + T Contributions

Training and mobility measures

Additional presentations

The Collaborative Project FIRST-Nuclides

The EURATOM FP7 Collaborative Project "Fast / Instant Release of Safety Relevant Radionuclides from Spent Nuclear Fuel (CP FIRST-Nuclides)" started in January 1, 2012 and extends over 3 years. The European nuclear waste management organisations contributing to the Technology Platform "Implementing Geological Disposal (IGD-TP)" considered the fast / instant release of safety relevant radionuclides from high burn-up spent nuclear fuel as one of the key topics in the deployment plan. For this reason, the CP FIRST-Nuclides deals with understanding the behaviour of high burn-up uranium oxide (UO2) spent nuclear fuels in deep geological repositories.

The project was implemented by a Consortium with ten beneficiaries (Karlsruhe Institute for Technology (KIT) Germany, Amphos 21 Consulting S.L. (AMPHOS21) Spain, Joint Research Centre – Institute for Transuranium Elements (JRC-ITU) European Commission, Forschungszentrum Juelich GmbH (JÜLICH) Germany, Paul Scherrer Institut (PSI) Switzerland, Belgian Nuclear Research Centre- Centre d'Etude de l'Energie Nucleaire (SCK•CEN) Belgium, Centre National de la Recherche Scientifique (CNRS) France, Fundació Centre Tecnològic (CTM) Spain, Magyar Tudományos Akadémia Energiatudományi Kutatóközpont (MTA EK) Hungary, Studsvik Nuclear AB (STUDSVIK) Sweden). Thirteen organisations have contributed to the project without any funding as Associated Groups (AG), i.e. organisations from France (Commissariat à l'énergie atomique et aux énergies alternatives, CEA), United States of America (Los Alamos Natonal Laboratory, SANDIA National Laboratories), United Kingdom (Nuclear Decommissioning Authority (NDA), National Nuclear Laboratory (NNL), University Cambridge, Center for Nuclear Engineering of the Imperial College London, University Lancaster), Finland (Posiva Oy, Teollisuuden Voima (TVO)), Czech Republic (ÚJV Řež, a. s.), Spain (Ciemat) and Germany (Gesellschaft für Anlagen- und Reaktorsicherheit (GRS) mbH). These groups had particular interest in exchange of information.Finally, a group of six implementation and regulatory oriented organizations (SKB (Sweden), NAGRA (Switzerland), ONDRAF/NIRAS (Belgium), ANDRA (France), BfS (Germany), ENRESA (Spain) participate as an "End-User Group (EUG)". This group ensured that end-user interests (waste management organisations and regulators) are reflected in the project work, reviewing the project work and the scientific-technical outcome.

Main events of FIRST-Nuclides were the

- Kick-off meeting held in Barcelona, Spain, February 9-10, 2012, where the partners agreed upon harmonizing the experimental conditions. A standard water composition was selected and it was agreed to use different fuel sample sizes, covering the whole suite of potential size distributions to be expected under disposal conditions. Furthermore, the EUG presented comments and recommendations on the project work program.
- 1st Annual Workshop held in Budapest, Hungary, October 9-11, 2012 (Kienzler et al., 2013)
- 2nd Annual Workshop held in Antwerp, Belgium, November 5-7, 2013 (Kienzler er al., 2014) and the
- Final Workshop held in Karlsruhe, Germany, September 1-2, 2014 (Kienzler et al., 2015). The EUG assessed the scientific-technical progress of the project, and reviewed the scientific-technical contributions, as well as the plan for dissemination of the final results.

The Final Workshop was held in direct connection to the 27th Spent Fuel Workshop, September 03 - 05, 2014 at the same location. This fact allowed to present a summary and 8 detailed papers on FIRST-Nuclide investigations.

With respect to the applications of the results for safety analyses and the Safety Case, the EUG rated the outcome of FIRST-Nuclides as follows:

- The value of the studies performed in the Project appears to be high.
- The subject is highly relevant for all waste management organisations.
- The Instant Release Fraction (IRF) contributes substantially to the peak releases after container breaching and to potential radiological consequences.
- Specifically, the results include:
  - Data from experimental determination of rapid release fractions for moderate and high burn-up UO_2 fuels, including doped fuels.
  - An increased data base for release of Cs and I from high burn-up fuel
  - Comprehensive comparisons of IRF with fission gas release (FGR).
  - The improvement of analytical techniques for some difficult to measure radionuclides such as ^{14}C and ^{79}Se.
  - Improved data on fission product retention of TRISO fuel.
  - Insights into mechanisms related to fission product release.

Indispensable for CP FIRST-Nuclides is the documentation of the scientific and analytical state of knowledge and the dissemination of the generated knowledge not only to the WMO but also to the general scientific community and other interested parties. This dissemination strategy is demonstrated by the high number of presentations at international conferences and the publications in peer-reviewed journals presently under preparation. The aspect of training and education of the next generation of spent nuclear fuel specialists was well pursued in the FIRST-Nuclides.

References

Kienzler, B., Metz, V. Duro, L. Valls, A. (2013). 1st Annual Workshop Proceedings of the 7th EC FP CP FIRST-Nuclides project. KIT SCIENTIFIC Reports 7639.

Kienzler, B., Metz, V. Duro, L. Valls, A. (2014). 2nd Annual Workshop Proceedings of the 7th EC FP CP FIRST-Nuclides project. KIT SCIENTIFIC Reports 7676.

Kienzler, B., Metz, V. Duro, L. Valls, A. (2015). Final Workshop Proceedings of the 7th EC FP CP FIRST-Nuclides Project.

1 Final Workshop

The Final Workshop of the Collaborative Project FIRST-Nuclides is the third and last workshop organized during the project duration. It was held in Karlsruhe (Germany) 1st – 2nd September 2014. The workshop was hosted by KIT/INE. 56 attendees attended the workshop, representing beneficiaries, Associated Groups, the End-User Group and project external organizations. The workshop was organized in two days and it was followed by the Spent Fuel Workshop, held at the same venue on 3rd – 5th September 2014.

The first day of the workshop was focused on the scientific and technical work done during the project life. WP leaders presented summaries of the results obtained during the three years of the project. The details of work carried out by each beneficiary were presented in a poster session.

The second day started with a session conducted by Lawrence Johnson as representative of the End-User-Group (EUG). This session presented the reflections of the EUG members with respect to the work carried out in the frame of the project and the results obtained along the three years of FIRST-Nuclides. An advice for future work relevant for the end-users was also highlighted.

The second day was completed by i) the topical session, in this case focused on X-ray Absorption Fine Structure (XAFS) investigations applied to nuclear waste glass and to spent nuclear fuel, ii) presentations by the applicants of mobility measures who presented their work carried out during their trainee at KIT/INE and iii) presentations by Associated Groups on topics related with the FIRST-Nuclides project.

1.1 Objectives

The Final Workshop combined different activities and meetings with the following objectives:

- Informing about the scientific progress achieved during the three years of the project.
- Informing about the administrative status.
- Informing/agreeing upon forthcoming reporting.
- Discussing various topics of interest for the consortium.
- Discussing upon still open or new questions to be solved in order to work on a forthcoming project proposal.

Emphasis was put on scientific-technical topics with administrative issues kept to the minimum necessary.

1.2 Workshop sessions

WP summaries

Workpackage leaders presented overviews of their respective WP. Results obtained during the project, highlights on specific achievements, open questions or further work were issues WP leaders focused on. The following presentations were given within this session.

V. Metz. Summary of WP1 "Samples and Tools".

D.H. Wegen. Summary of WP2 "Fission Gas Release and Rim and Grain Boundary Diffusion"

K. Lemmens. Summary of WP3 "Dissolution based release"

J. de Pablo. Summary of WP4 "Modelling"

A. Valls. Summary of WP5 "Knowledge, reporting and training"

Poster presentations

The Final Workshop was followed by the 27th Spent Fuel Workshop, held at the same venue on 3rd – 5th September 2014 where the beneficiaries had the opportunity presenting their results to an international audience. Therefore, the poster session aimed on facilitating in depth discussions between the specialists of the beneficiaries, the End Users, the Associated Groups and the external participants. 18 posters were presented during the poster session. Following, a list of the presented poster is provided:

B. Kienzler, V. Metz, L. Duro, A. Valls, V. Montoya. Generic Poster of FIRST-Nuclides project.

B. Kienzler, V. Metz, E. Gonzáles-Robles, M. Herm, O. Roth (KIT/INE) Modelling the Diffusion of Fission Products in Fuel Rods.

E. González-Robles, M. Lagos, E. Bohnert, N. Müller, M. Herm, B. Kienzler, V. Metz (KIT/INE) Leaching experiments with cladded pellet and fragments of a high burn-up nuclear fuel rod segment under Ar/H_2 atmosphere.

M. Pękala, A. Idiart, L. Duro (AMPHOS21) Modelling of spent fuel saturation with water – implications for the instant release fraction (IRF).

L. Aldave de las Heras, O. Beneš, A. Bulgheroni, P. Carbol, I. Marchetti, E. Moore, D. Papaioannou, D. Serrano-Purroy , S. Van Winckel, D. H. Wegen, T. Wiss (JRC-ITU) ITU's contributions towards FIRST-Nuclides CP.

D. Papaioannou, D. H. Wegen, D. Serrano-Purroy , S. Van Winckel, L. Aldave de las Heras, R. Nasyrow, G. Paperini, R. Gretter (JRC-ITU) Post irradiation characterisation of spent nuclear fuels.

P. Carbol, A. Bulgheroni, D. H. Wegen, I. Marchetti, T. Wiss (JRC-ITU) Oxygen and water diffusion in UO_2 under oxidising and reducing conditions.

L. Aldave de las Heras, D. Serrano-Purroy, S. Van Winckel, M. Sandow, S. Millet, R. Sureda Pastor, A. Martinez Torrents, J. de Pablo (JRC-ITU) Development of methods for the determination of pure ß emitters based on sequential injection flow analysis coupled to ICP-MS.

D. Serrano-Purroy, L. Aldave de las Heras, S. Van Winckel , J.-Y. Colle, O. Beneš, R. Sureda Pastor, A. Martinez Torrents, J. de Pablo (JRC-ITU) IRF of two UO_2 BWR Spent Nuclear Fuels.

H. Curtius, N. Lieck, G. Kaiser, M. Güngör, M. Klinkenberg, D. Bosbach (JÜLICH) Leaching of spent UO_2 TRISO coated particles -Instant radionuclide release fraction and microstructure evolution-.

E. Curti, I. Günther-Leopold, H.P. Linder, N. Milosavljevic (PSI) Results of SNF aqueous leaching experiments at PSI.

E. Curti, A. Froideval-Zumbiehl, M. Martin, A. Bullemer, I. Günther-Leopold, A. Puranen, D. Jädernäs, O. Roth, D. Grolimund, C.N. Borca and A. Velea (PSI + STUDSVIK) X-ray absorption spectroscopy of Se on high-burnup UO_2 spent fuel.

Th. Mennecart, C. Cachoir, K. Lemmens (SCK·CEN) Characterisation and leaching tests for the experimental determination of IRF radionuclides from Belgian high-burnup spent nuclear fuel.

Traboulsi, J. Vandenborre, G. Blain, B.Humbert, F. Haddad, M. Fattahi (CNRS) Radiolytic Corrosion of UO_2 by $^4He^{2+}$ irradiation of water.

Casas, A. Espriu, D. Serrano-Purroy, J. de Pablo (CTM) IRF modelling from Spent Nuclear Fuel leaching experiments.

Casas, J. Giménez, A. Espriu, D. Serrano-Purroy, R.M. Sureda, A. Martinez Torrents, J. de Pablo (CTM/UPC) UPC and CTM activities at First Nuclides Project.

E. Slonszki and Z. Hózer (MTA-EK) Isotopes dissolution during wet storage of damaged and leaking VVER fuel.

O. Roth, A. Puranen, C. Askeljung, M. Granfors, D. Cui (STUDSVIK) Characterization and leaching of high burn up spent fuel with and without matrix dopants.

N. Rauff-Nisthar, C. Boxall, R. Wilbraham, D. Hambley, C. Padovani (Lancaster University) Corrosion behaviour of AGR simulated fuels.

Topical session and Associated Group presentations

The main objective of organizing the Topical Sessions is covering the key areas along with the project. One talk was given during the Final Workshop Topical session focused on XAFS analysis on spent fuel and vitrified wastes.

J. Rothe (KIT-INE) XAFS investigations of genuine nuclear waste glass and spent fuel particles at ANKA.

Additional presentations were given on a topic of general interest, especially the context of the present project within the EURATOM FP7 program on geologic disposal. These presentations were given by the associated group.

B. Pastina (POSIVA) FIRST-Nuclides transport modelling from spent fuel to the biosphere and uncertainties propagation.

C. Padovani (RWM), O. Roth (Studsvik), A. Puranen (Studsvik), C. Askeljung (Studsvik), M. Cowper (AMEC) The characteristics and leaching behaviour of AGR spent fuel – scoping studies.

B.D. Hanson, E. Buc, E. Mausolf (PNNL). Non-Congruent Release: What does it mean?

Training mobility measures session

The three applicants for mobility measures gave short talks presenting their work carried out during their trainee at KIT/INE. The titles of the presentations are provided below.

A. Martínez-Torrents, E. González-Robles, E. Bohnert, I. Casas, J. de Pablo, V. Metz (CTM – KIT/INE) α-Radiolysis under alkaline conditions in 0.05 and 5.0 molar NaCl.

P. Szabó, V. Metz, E. González-Robles (MTA-EK – KIT/INE) Mission to the KIT-INE.

D. García, J. Lützenkirchen, N. Finck, M. Huguenel, L. Calmels, V. Petrov, M. Grivé, L. Duro (AMPHOS21 – KIT/INE) Sr sorption onto goethite and zirconium oxide in high ionic strength systems. Experimental and preliminary modelling results.

EUG session

Lawrence Johnson presented the reflections of the EUG members with respect to the work done and the results obtained during the project as well as open questions for future work.

All the scientific-technical contributions submitted by the beneficiaries were reviewed by the EUG members (End-User-Group).

Workpackage Overview

2 Overview WP1: Selection, characterisation and preparation of materials and set-up of tools

Volker Metz

Karlsruhe Institute of Technology (KIT), Institute of Nuclear Waste Disposal (IN)
** Corresponding author: volker.metz@kit.edu*

2.1 Summary

The basic activities of the project FIRST-Nuclides – coordinated in WP1 "Selection, characterisation and preparation of materials and set-up of tools" were to select, provide and prepare spent nuclear fuel (SNF) samples for subsequent experimental investigations. The objectives of this workpackage included the complete characterization of the selected SNF materials with respect to the individual fuel characteristics and irradiation history, achieving permission by the fuel owners for publication of key parameters as well as the installation of experimental and analytical tools. All experimentally working partners of the FIRST-Nuclides project (i.e. KIT, JRC-ITU, JÜLICH, PSI, SCK•CEN, CNRS, CTM, MTA EK and STUDSVIK) contributed to this workpackage. Initially the activities within WP1 were planned to be conducted until June 2013.

Two deliverable have been prepared documenting the available experimental and theoretical data of studied high burnup SNF samples, including the results of characterisation studies and the description of experimental / analytical methodologies and tools to be applied in WP1, WP2 and WP3.

- D1.1 "Characterisation of spent fuel to be used in the project, revealing the relevance for the Safety Case"
- D1.2 "Characterisation of spent nuclear fuel samples and description of methodologies and tools to be applied in FIRST-Nuclides"

Due to delays in the documentation and with respect to sample preparation of high burnup SNF materials used by JRC-ITU and CTM and the delayed permission by the fuel owners (RWE / EON) to publish data (granted at 14.8.2014), it was decided producing an additional Deliverable D1.3 which includes the up-dated information of the characterization of all samples as well as the completed radiochemical analyses of a fuel pellet (SCK•CEN)

3 Overview WP2: Gas release + RIM and GRAIN boundary

Detlef Wegen *

European Commission, Joint Research Centre (JRC), Institute for Transuranium Elements (ITU)
** Corresponding author: Detlef.Wegen@ec.europa.eu*

3.1 Introduction

In the first project year the focus was on setting up experimental facilities, characterisation and preparation of samples. In the second year of the experimental work programme in WP2 first results have been obtained by the partners but also new challenges arose. In the last project year most of the experiments have been finalised and evaluated.

Work package two is divided into the two components "Experimental determination of fission gas release" and "Rim and grain boundary diffusion". In the first component the focus is on the quantification of fission gases and fission gas release in high burn-up (HBU) UO_2 spent nuclear fuels (SNF). Fission gas sampled in the plenum of a fuel rod are analysed as well as the grain boundary inventory and the cross sectional distribution of fission gases and volatile fission products.

The second component "Rim and grain boundary diffusion" deals with investigations on oxygen diffusion in spent UO_2 fuel. The examination of diffusion mechanisms will result in the quantification of water penetration into the fuel (grain and grain boundaries) structures and subsequently couple the diffusion/corrosion phenomena. Furthermore, investigations on irradiated and unirradiated fuel kernels separated from high temperature gas cooled reactor (HTR) fuel are planned which are complementary to those on light water reactor (LWR) fuel.

The experimental part of WP2 started in project month 4 and will end in project month 36 (Wegen et al., 2012; Wegen et al., 2012a; Wegen et al., 2013a; Wegen et al., 2013b; Wegen et al., 2014; Wegen et al., 2014a).

The JOINT RESEARCH CENTRE – INSTITUTE FOR TRANSURANIUM ELEMENTS (JRC-ITU) is the leading organization of WP2. In the first project year the fission gas release (FGR) from a spent fuel rod owned by KIT was measured (Wegen et al., 2013). The determination of the inventory of fission gas and fission products in grain boundaries were measured in the last project year (Wegen et al., 2014).

The Karlsruher Institut für Technologie (KIT) analysed in the first project year fission and activation products in the fission gas sampled at JRC-ITU from the plenum of a fuel rod segment by puncturing (González-Robles et al., 2013; González-Robles et al., 2013a). The development, testing and implementation of analytical methods for fission and activation products have started in the first project year one and were continued in the second. Leaching experiments in which gas and solution analyses were performed started in the first year and continued until project month 33 (Wegen et al., 2014).

STUDSVIK NUCLEAR AB (STUDSVIK) investigates in the frame of WP2, the radial fission gas and volatile fission product distribution (Xe, I, and Cs) by Laser-Ablation Mass Spectroscopy (LA-MS) on HBU boiling water reactor (BWR) SNF (Wegen et al. 2014, Roth et al., 2013; Puranen et al., 2013).

The JOINT RESEARCH CENTRE – INSTITUTE FOR TRANSURANIUM ELEMENTS (JRC-ITU) is the leading organization of WP2. The investigation of diffusion effects started in the first project year with the characterisation and preparation of spent fuel samples, which will be used for corrosion experiments in $H_2{}^{18}O$ water at room temperature. In 2013, it was planned to determine the $^{18}O/^{16}O$ depth profiles using a shielded SIMS (secondary ion mass spectrometry) to quantify the oxygen diffusion into spent nuclear fuel (SNF) (Wegen et al., 2014a).

Forschungszentrum Jülich GmbH (Jülich) is working on spent high temperature reactor fuel. Within the first half of the project the radionuclide inventory in the fuel kernel and in the coatings were determined and compared to calculated values as well. Further on investigations of the microstructure and of the elemental distribution of the fuel kernel and of the coatings are performed before (first half of the project) and after leaching (second half of the project). After cracking of the tight coatings the fission gas release fraction was measured in the first 18 months. Then static leaching experiments with the separated fuel kernels and coatings were started in 2013 end continued in 2014 in order to determine the fast instant radionuclide release fraction (Wegen et al., 2014a; Curtius et al., 2013; Curtius et al., 2013a).

Unirradiated tristructural-isotropic (TRISO) fuel particles are investigated by the CENTRE NATIONAL DE LA RECHERCHE SCIENTIFIQUE (CNRS) at the ARRONAX cyclotron. The particles are irradiated using a He^{2+}-beam in the dose rate range of 0.8 – 4.4 kGy/min. The corrosion of UO_2 TRISO particles is investigated in view of grain boundary effects and secondary phase formation and the influence of hydrogen. The experiments started in 2013 with studies on the role of grain boundaries, followed by investigations of fuel particle corrosion under hydrogen and in varying dose rates (Wegen et al., 2014a; Vandenborre et al., 2013; Vandenborre et al., 2013a; Traboulsi et al., 2013).

3.2 Achievements

Experimental determination of fission gas release

JRC-ITU has examined the release of the fission products ^{88}Sr, ^{87}Rb, ^{137}Cs, and ^{136}Xe from powder of irradiated BWR UO_2 (54 GWd/t_{HM}) originating from the core region of the fuel pin. The sample was heated in a Knudsen cell coupled to a mass spectrometer under vacuum conditions at a rate of 10 K/min until complete vaporization of the fuel at 2,460 K. The measured release profiles indicate two release mechanisms for ^{137}Cs, one with rather low intensity starting just below 1,000 K and a second significant release after 1,500 K. The release of the fission products ^{88}Sr, ^{87}Rb, ^{137}Cs, and ^{136}Xe has been semi-quantified (Colle et al., 2014).

KIT has continued leaching experiments on a well characterised samples of cladded PWR fuel and PWR fuel fragments under Ar/H_2 (p_T = 40 bar; pH_2 = 3 bar) in bicarbonate water (19 mM $NaHCO_3$ + 1 mM NaCl). After a cumulative contact time of 246 days, 15% of the fission gases (Kr+ Xe) were released into the gas phase for the cladded fuel pellet while for fuel fragments 11% of the fission gases (Kr+ Xe) inventory were released after 27 days (González-Robles et al., 2013b; González-Robles et al., 2013c; González-Robles et al., 2014).

STUDSVIK finished the evaluation of Laser Ablation data on cross sections from a standard UO_2 fuel and an Al/Cr-additive fuel. The caesium and iodine profiles measured on both fuels are very similar and follow the radial burn-up profile. Caesium, iodine and to some extent selenium also appear to collect in some fuel cracks. Selenium was tentatively identified by the good agreement of the isotopic ratios of mass 77, 79 and 82 with the calculated inventory. Chromium and especially aluminium are heterogeneously distributed in the Al/Cr-additive fuel (Puranen et al. 2014).

Rim and grain boundary diffusion

Jülich investigates UO_2TRISO coated particles from spent HTR fuel (burn-up ~100 GWd/t_{HM}). UO_2 fuel kernels were exposed to synthetic groundwater (19 mM NaCl and 1 mM $NaHCO_3$) for 276 days under oxic (air) and anoxic/reducing (Ar/H_2) conditions. The U release was below detection limit. During 276 days maximal 0.2% ^{90}Sr and 8% $^{134/137}Cs$ were released from grain boundaries. Depending on the environmental conditions, different release functions were observed. ESEM (environmental scanning electron microscope) investigations show on samples leached in oxic environment, numerous intragranular open pores acting as new accessible leaching sites and white spherical spots containing Mo and Zr. Under anoxic/reducing conditions numerous metallic precipitates (Mo, Tc and Ru) filling the intragranular pores and white spherical spots containing Mo and Zr, were detected (Curtius et al., 2014).

CNRS studies the oxidation of UO_2 in unirradiated HTR fuel kernels by α-radiolysis products of water. Radiolysis products are produced by 66.5 MeV He^{2+} beam irradiation of water in the ARRONAX cyclotron giving a dose rate of 4.37 kGy/min. The obtained results show that gaseous H_2 produced by water radiolysis completely inhibits the UO_2 corrosion by interaction between H_2 and the UO_2 surface. The UO_2 radiolytic corrosion expressed as a function of the absorbed dose show a ten-fold higher dissolution rate for the lower dose rate (0.8 kGy/min) than that found for the higher one (4.4 kGy/min). If OH radicals are produced near the UO_2 surface they interact directly with the UO_2 surface and not with each other to form H_2O_2. In this case also the radiolytic corrosion is higher than in presence of molecular H_2O_2. It is also shown that the UO_2 radiolytic corrosion under He^{2+} radiation occurs essentially on the grain boundaries and not on the crystallized UO_2 grains (Vandenborre et al., 2014; Vandenborre et al. 2014).

JRC-ITU analysed and modelled diffusion profiles measured on UO_2 by parallel diffusion processes. No new experimental results on oxygen diffusion experiments have been obtained because during refurbishment of the autoclave in the hot cell, intended for oxygen diffusion experiments, a Ti-welded tube joint broke. For safety reasons a replacement of the complete autoclave setup is necessary. Despite all efforts undertaken it was not possible to get the set-up back in working conditions because of the delivery time of special valves needed. It was decided to prolong WP2 from 33 to 36 months and to publish the results in an open journal with reference to FIRST Nuclides (Carbol et al., 2013; Carbol et al., 2014).

3.3 References

Carbol, P., Marchetti, I., Wegen, D.H. (2013). WP2: Rim and Grain Boundary Diffusion. 2nd Annual Workshop Proceedings of the 7th EC FP CP FIRST-Nuclides project (eds. Kienzler et al.). KIT SCIENTIFIC Reports 7676, 79-86.

Carbol, P., Marchetti, I., Wegen, D.H., Bulgheroni, A., Wiss, T. (2014). WP2: Grain boundary diffusion in spent fuel and poly-crystalline UO_2. Final Workshop Proceedings of the 7th EC FP CP FIRST-Nuclides Project (eds. Kienzler et al.).

Colle, J.-Y., Beneš, O., Serrano-Purroy D., Sureda Pastor, R., Martinez Torrents, A. (2014). Fission product release from irradiated fuel. Final Workshop Proceedings of the 7th EC FP CP FIRST-Nuclides Project (eds. Kienzler et al.).

Curtius, H., Müller, E., Müskes, H.W., Klinkenberg, M., Bosbach, D. (2013). Selection and Characterisation of HTR Fuel. 1st Annual Workshop Proceedings of the 7th EC FP CP FIRST-Nuclides project (eds. Kienzler et al.). KIT SCIENTIFIC Reports 7639, 41.

Curtius, H., Müskes, H.W., Güngör, M., Liek, N., Bosbach, D. (2013a). First results on instant radionuclide release fraction from spent UO2 TRISO coated particles. 2nd Annual Workshop Proceedings of the 7th EC FP CP FIRST-Nuclides project (eds. Kienzler et al.). KIT SCIENTIFIC Reports 7676, 97-103.

Curtius, H., Lieck, N., Kaiser, G., Güngör, M., Klinkenberg, M., Bosbach, D. (2014). Instant radionuclide release fraction from spent UO_2TRISO coated particles and microstructure evolution. Final Workshop Proceedings of the 7th EC FP CP FIRST-Nuclides Project (eds. Kienzler et al.).

González-Robles, E., Bohnert, E., Loida, A., Müller, N., Metz, V., Kienzler, B. (2013). Fission gas measurements and description of leaching experiments with of KIT's irradiated PWR fuel rod segment (50.4 GWd/t_{HM}). 1st Annual Workshop Proceedings of the 7th EC FP CP FIRST-Nuclides project (eds. Kienzler et al.). KIT SCIENTIFIC Reports 7639, 231.

González-Robles, E., Bohnert, E., Müller, N, Herm, M., Metz, V. (2013a). Determination of the fission gas release in the segment N0204 and gas phase result of anoxic leaching experiment. 2nd Annual Workshop Proceedings of the 7th EC FP CP FIRST-Nuclides project (eds. Kienzler et al.). KIT SCIENTIFIC Reports 7676, 37-44.

González-Robles, E., Wegen, D.H., Papaioannou, D., Kienzler, B., Nasyrow, R., Metz, V. (2013b). Physical characterisation of spent nuclear fuel: First steps to further Instant Release Fractions investigations. 8th EC Conference on the Management of Radioactive Waste, EURADWASTE 2013.

González-Robles, E., Bohnert, E., Metz, V., Wegen, D.H., Papaioannou, D., Kienzler, B. (2013c). Physical characterisation and calculation of the initial and boundary conditions of a commercial UO_2 spent nuclear fuel regarding the radionuclide release. 37th Symposium on the Scientific Basis for Nuclear Waste Management.

González-Robles, E., Lagos, M., Bohnert, E., Müller, N., Herm, M., Metz, V., Kienzler, B. (2014). Leaching experiments with cladded pellet and fragments of a high burn-up nuclear fuel rod segment under argon/H_2 atmosphere. Final Workshop Proceedings of the 7th EC FP CP FIRST-Nuclides Project (eds. Kienzler et al.).

Puranen, A., Roth, O., Granfors, M. (2013). Investigating the radial distribution of potential rapid release radionuclides in irradiated nuclear fuel. Symposium: E No. 2 15, E-MRS 2013 Spring Meeting.

Puranen, A., Granfors, M., Roth, O., (2014). Laser ablation study of irradiated standard UO_2 fuel and Al/Cr doped UO_2 fuel. Final Workshop Proceedings of the 7th EC FP CP FIRST-Nuclides Project (eds. Kienzler et al.).

Roth, O., Puranen, A., Low, J., Granfors, M, Cui, D., Askeljung, C. (2013). Spent fuel leaching experiments and laser ablation studies performed in Studsvik - Status and preliminary results. 2nd Annual Workshop Proceedings of the 7th EC FP CP FIRST-Nuclides project (eds. Kienzler et al.). KIT SCIENTIFIC Reports 7676, 163-174.

Traboulsi, A., Vandenborre, J., Blain, G., Barbet, J., Fattahi, M. (2013). Impact of Water Radiolysis on Uranium dioxide corrosion. Migration 2013.

Traboulsi, A., Vandenborre, J., Blain, G., Humbert, B., Barbet, J., Fattahi, M. (2014). Radiolytic Corrosion of Uranium Dioxide: Role of Molecular Species. The Journal of Physical Chemistry C, 118(2), 1071–1080.

Vandenborre, J., Traboulsi, A., Blain, G., Barbet, J., Fattahi, M. (2013). Radiolytic Corrosion of Grain Boundaries onto the UO_2 TRISO Particle Surface. 1st Annual Workshop Proceedings of the 7th EC FP CP FIRST-Nuclides project (eds. Kienzler et al.). KIT SCIENTIFIC Reports 7639, 161.

Vandenborre, J., Traboulsi, A., Vandenborre, J., Blain, G., Humbert, B., Barbet, J., Fattahi, M. (2013a). Impact of Water Radiolysis on Uranium Dioxide Corrosion. 2nd Annual Workshop Proceedings, 7th EC FP – FIRST-Nuclides, 5th-7th November, Antwerp, Belgium, KIT SCIENTIFIC Reports 7676, 131-138.

Vandenborre, J., Traboulsi, A., Blain, G., Humbert, B., Haddad, F., Fattahi, M. (2014). Radiolytic corrosion of uranium dioxide under He^{2+} irradiation. Final Workshop Proceedings of the 7th EC FP CP FIRST-Nuclides Project (eds. Kienzler et al.).

Wegen, D.H., González-Robles, E., Puranen, A. (2012). DELIVERABLE (D-N: 2.1) - Status of fission gas release studies (12 months), FIRST Nuclides (Contract Number: FP7-295722). European Atomic Energy Community, JRC Scientific and Policy Reports, JRC76116.

Wegen, D.H., Carbol, P., Curtius, H., Vandenborre, J. (2012a). DELIVERABLE (D-N: 2.2) Status of Rim and Grain Boundary Diffusion Experiments (12 months). European Atomic Energy Community, JRC Scientific and Policy Reports, JRC77256.

Wegen, D.H., Papaioannou, D., De Weerd, W., Rondinella, V.V., Glatz, J.-P. (2013). Fission gas release measurement on one 50.4 GWD/t_{HM} PWR fuel segment. 1st Annual Workshop Proceedings of the 7th EC FP CP FIRST-Nuclides project (eds. Kienzler et al.). KIT SCIENTIFIC Reports 7639, 201.

Wegen, D.H., González-Robles, E., Puranen, A. (2013a). DELIVERABLE (D-N: 2.1) - Status of fission gas release studies (24 months), FIRST Nuclides (Contract Number: FP7-295722). European Atomic Energy Community, JRC Scientific and Policy Reports, JRC87028.

Wegen, D.H., Carbol, P., Curtius, H., Vandenborre, J. (2013b). DELIVERABLE (D-N: 2.2) Status of Rim and Grain Boundary Diffusion Experiments (24 months). European Atomic Energy Community, JRC Scientific and Policy Reports, JRC87025.

Wegen, D.H., Beneš, O., González-Robles, E., Puranen, A. (2014). DELIVERABLE (D-N: 2.1) - Status of fission gas release studies (44 months), FIRST Nuclides (Contract Number: FP7-295722). European Atomic Energy Community, JRC Scientific and Policy Reports, JRC92238.

Wegen, D.H., Carbol, P., Curtius, H., Vandenborre, J. (2014a). DELIVERABLE (D-N: 2.2) Status of Rim and Grain Boundary Diffusion Experiments (34 months). European Atomic Energy Community, JRC Scientific and Policy Reports, JRC92264.

4 Overview of WP3 (first-nuclides): Dissolution based release

Karel Lemmens

Belgian Nuclear Research Centre (SCK•CEN) (BE)

4.1 Objectives

The overall objectives of WP3 of FIRST-Nulides were the quantification of the fast release of gaseous and non-gaseous activation and fission products into the aqueous phase during spent nuclear fuel leach tests, and – to the extent possible - the determination of their chemical speciation. The experiments covered high burnup spent nuclear fuels (HBU-SNF) having a burnup mostly in the range of 50 to 60 GWd/t_{HM}, different irradiation histories, reactor types and sample positions in the fuel rods. Special emphasis was given to the determination of IRF values of caesium, iodine, selenium and carbon-14 (from fuel and cladding) as well as to the redox speciation of selenium. Additionally, dissolution rates for relevant isotopes were determined for damaged VVER fuel elements (to be disposed of in Germany, Hungary and Finland). The results from experiments using fuel samples with/without cladding allow evaluation of the effect of the presence of the cladding material.

4.2 Description of work performed in the project

Seven institutes were involved in WP3, i.e. KIT, JRC-ITU, PSI, SCK•CEN, CTM, Studsvik and EK. All of these institutes, except for EK, have performed leach experiments on samples of high burnup fuels in laboratory conditions. EK has contributed by collecting and interpreting the isotopic dissolution data of damaged and leaking VVER fuels. The leach experiments by the various institutes were harmonized as much as possible to allow a better comparison of the leach data. For this reason, a standard leachant (19 mM NaCl and 1 mM $NaHCO_3$) was used by most laboratories, and tests with similarly prepared samples (cladded fuel segments) were added by several laboratories. Nevertheless, there were also differences in experimental approaches, such as the redox conditions (reducing or oxidizing atmosphere), leaching mode (static or pseudo-dynamic), and alternative sample preparations. Both PWR and BWR fuels were tested. One test series with a MOX fuel was added as well. Test durations up to 1 year were reached. The leachates were analyzed for a large number of radionuclides, depending on the techniques available in the participating laboratories. However, most laboratories have measured (or planned to measure) the most critical IRF radionuclides, i.e. Cs and I isotopes. Other isotopes that were measured by several, but not all laboratories are ^{14}C and ^{79}Se. For some isotopes, the analytical methods were developed in the framework of the project. Special efforts were made to characterize selenium in the spent fuel by solid state analyses. More details about the contribution of each institute are given hereunder.

KIT has performed leach experiments on a spent PWR UOX fuel with average burnup of 50.4 GWd/t_{HM}. They have tested in parallel the leaching of cladded fuel segments and fuel fragments. The tests were performed with the standard leachant in autoclaves under argon/hydrogen gas atmosphere (i.e. under reducing conditions). After one day of preleaching, samples of the leachate were taken at regular time intervals, without renewal of the solution, i.e. in static mode. The IRF of relevant radionuclides such as I, Cs, Tc, Sr and U isotopes was determined and reported for test durations up to 245 days. The autoclave set-up also allowed the measurement of the fission gas release during the leach tests. The results are presented in González-Robles et al. (2014).

JRC-ITU has performed leach experiments on a spent BWR UOX fuel with an average burnup of 42.22 GWd/t_{HM}. They have tested in parallel the leaching of cladded fuel segments and fuel powders that were taken separately from the core zone and the rim zone of the fuel. The tests were performed with the standard leachant in glass test tubes under air atmosphere (i.e. under oxidizing conditions). At each sampling, the leachate was completely renewed, so the tests were done in pseudo-dynamic mode. The IRF of Cs, Tc, Sr, Rb, Mo and U isotopes was determined and reported for test durations up to 190 days. The results are presented in Serrano-Puroy et al. (2014).

PSI has performed leach experiments on a spent BWR UOX fuel with an average burnup of 57.5 GWd/t_{HM}, spent PWR UOX fuel with an average burnup of 56.5 GWd/t_{HM}, and spent MOX fuel with an average burnup of 63 GWd/t_{HM}. They have tested in parallel the leaching of cladded fuel segments, the leaching of fuel fragments, and the leaching of separated claddings with some adhering fuel residues. The tests were performed with the standard leachant in glass columns under air atmosphere (i.e. under oxidizing conditions). After 7 days, the leachate was removed completely and then replaced by fresh leaching solution. Afterwards, samples of the leachate were taken at regular time intervals, without renewal of the solution, i.e. in static mode. The IRF of I and Cs isotopes was determined and reported for test durations up to 182 days. In collaboration with Studsvik, samples of the BWR fuel were analysed with micro-XRF and micro-XAS to determine the selenium distribution on the microscale its oxidation state and its structural environment (next-neighbour distances and coordination numbers). The results are presented in Curti et al. (2014a,b).

SCK•CEN has performed leach experiments on a spent PWR UOX fuel with an average burnup of 50.5 GWd/t_{HM}. They have tested in parallel the leaching of cladded fuel segments and opened fuel segments where the cladding and fuel fragments are separated from each other but leached together. The tests were performed with the standard leachant in glass columns under air atmosphere (i.e. under oxidizing conditions), identical to the columns used by PSI. After two complete leachate renewals in the first days, samples of the leachate were taken at regular time intervals, without further renewal of the solution, i.e. in static mode. The IRF of I and Cs isotopes was determined and reported for test durations up to 357 days. Apart from these, many other isotopes were analysed, a.o. ^{14}C, ^{99}Tc and ^{238}U. The results are presented in Mennecart et al. (2014).

CTM has performed leach experiments on a spent BWR UOX fuel with an average burnup of 45 GWd/t_{HM}. They have tested in parallel the leaching of cladded fuel segments and fuel powders that were taken separately from the core zone and the rim zone of the fuel. The tests were performed with the standard leachant in glass test tubes under air atmosphere (i.e. under oxidizing conditions). At each sampling, the leachate was completely renewed, so the tests were done in pseudo-dynamic mode. The IRF of Cs, Tc, Sr, Rb, Mo and U isotopes was determined and reported for test durations up to 190 days. The experiments of CTM were performed in the laboratories of JRC-ITU. The results are presented in Martínez-Torrents et al. (2014).

Studsvik has performed leach experiments on samples of six types of a spent fuel, i.e.

BWR UOX fuels with an average burnup of 50.2 GWd/t_{HM} (test on a cladded segment), 54.8 MWd/t_{HM} (test on a cladded segment), and 57.1 MWd/t_{HM} (test on fuel fragments + separated cladding)

An Al/Cr doped BWR UOX fuel with a burnup of 59.1 MWd/t_{HM} (test on fuel fragments + separated cladding)

A Gd doped PWR UOX fuel with an average burnup of 54.4 MWd/t_{HM} (test on fuel fragments + separated cladding)

A spent PWR UOX fuel with an average burnup of 70.2 MWd/t_{HM} (test on fuel powder)

The tests on cladded fuel segments and fuel fragments + separated cladding were performed with a slightly modified standard leachant (10 mM NaCl and 2 mM $NaHCO_3$) in glass test tubes under air atmosphere (i.e. under oxidizing conditions). At each sampling, the leachate was completely renewed, so the tests were done in pseudo-dynamic mode. The IRF of I, Cs, Tc, Sr, Rb, Mo, Se, U and many other isotopes was determined and reported for test durations up to 364 days. The tests on fuel powder were performed using a simultaneous grinding and leaching method. Laser Ablation Inductively Coupled Plasma Mass Spectroscopy analysis was applied on fuel cross-sections from the BWR fuel with burnup 57.1 MWd/t_{HM} and the Al/Cr doped BWR fuel, to reveal the radial profile of Cs, I and Se. The results are presented in Roth et al. (2014a,b) and Puranen et al. (2014a,b)

EK has collected and interpreted the isotopic dissolution data of damaged and leaking VVER fuels that had been stored in storage pools in the period 2003-2007 (damaged fuel) and 2009-2010 (leaking fuel). The damaged fuels had a burnup in the range 10.1-26.7 GWd/t_{HM}. The leaking fuel had a burnup of 14 GWd/t_{HM}. The leachant was in this case the water of the storage pool. In the case of the damaged fuel, the pH of water was ≈7 in the first 14 days after the incident (15 g boric acid per kg of water) and ≈4-4.5 (21 g boric acid per kg of water) in the remaining period, with measurements up to 1,368 days. In the case of the leaking fuel, measurements up to 369 days are available. During the storage, the water purification system has led to periods of increasing and decreasing radionuclide concentrations. The dissolution rate was calculated in various ways for a series of isotopes, a.o. Cs, I and U isotopes.

The results are presented in Slonszki et al. (2014).

4.3 References

Curti, E., Günther-Leopold, I., Linder, H.-P., Milosavljevic, N. (2014). Results of Leaching Experiments on Spent UO_2 and MOX Nuclear Fuel Carried out at PSI. Final Workshop Proceedings of the 7th EC FP CP FIRST-Nuclides Project (eds. Kienzler et al.).

Curti, E., Froideval-Zumbiehl, A., Martin, M., Bullemer, A., Günther-Leopold, I., Puranen, A., Jädernäs, D., Roth, O., Grolimund, D., Borca, C.N., Velea, A. (2014). X-Ray Absorption Spectroscopy of Selenium in High-Burnup UO_2 Spent Fuel form the Leibstadt and Oskarshamn-3 Reactors. Final Workshop Proceedings of the 7th EC FP CP FIRST-Nuclides Project (eds. Kienzler et al.).

González-Robles, E., Lagos, M., Bohnert, E., Müller, N., Herm, M., Metz, V., Kienzler, B. (2014). Leaching Experiments with Cladded Pellet and Fragments of High Burn-Up Nuclear Fuel Rod Segment under Argon/H_2 Atmosphere. Final Workshop Proceedings of the 7th EC FP CP FIRST-Nuclides Project (eds. Kienzler et al.).

Martínez-Torrents, A., Sureda, R., de Pablo, J., Clarens, F., Serrano-Purroy, D., Aldave de las Heras, L., Casas, I. (2014). Corrosion Test of Commercial UO_2 BWR Spent Nuclear Fuel for Fast/Instant Release Studies. Final Workshop Proceedings of the 7th EC FP CP FIRST-Nuclides Project (eds. Kienzler et al.).

Mennecart, T., Lemmens, K., Cachoir, C. (2014). Characterisation and Leaching Tests for the Experimental Determination of IRF Radionuclides for Belgian High-Burnup Spent Nuclear Fuel. Final Workshop Proceedings of the 7th EC FP CP FIRST-Nuclides Project (eds. Kienzler et al.).

Puranen, A., Granfors, M., Roth, O. (2014). Laser Ablation Study of Irradiated Standard UO_2 Fuel and Al/Cr Doped UO_2 Fuel. Final Workshop Proceedings of the 7th EC FP CP FIRST-Nuclides Project (eds. Kienzler et al.).

Puranen, A., Granfors, M., Roth, O. (2014). Aqueous Leaching of ^{79}Se from Spent Nuclear Fuel. Final Workshop Proceedings of the 7th EC FP CP FIRST-Nuclides Project (eds. Kienzler et al.).

Roth, O., Puranen, A., Askeljung, C., Cui, D. (2014). Selection of Materials and Characterization of Samples Used in Spent Fuel Leaching and Laser Ablation Studies. Final Workshop Proceedings of the 7th EC FP CP FIRST-Nuclides Project (eds. Kienzler et al.).

Roth, O., Askeljung, C., Puranen, A., Granfors, M., Cui, D., Low, J. (2014). Leaching of High Burnup Spent Fuel With and Without Matrix Dopants. Final Workshop Proceedings of the 7th EC FP CP FIRST-Nuclides Project (eds. Kienzler et al.).

Serrano-Purroy, D., Aldave de las Heras, L., Van Winckel, S., Martínez Torrents, A., Sureda, R., Glatz, J.P., Rondinella, V.V. (2014). WP3. Dissolution Based Release, IRF Corrosion Tests of Commercial UO_2 BWR Spent Nuclear Fuel. Final Workshop Proceedings of the 7th EC FP CP FIRST-Nuclides Project (eds. Kienzler et al.).

Slonszki, E., Hózer, Z. (2014). Isotopes Dissolution during Wet Storage of Damaged and Leaking VVER fuel. Final Workshop Proceedings of the 7th EC FP CP FIRST-Nuclides Project (eds. Kienzler et al.).

5 Overview WP4 (3rd year)

Joan de Pablo

UPC

5.1 Objectives

The objectives of WP 4 cover initial speciation of fission products in LWR fuel, and multi-scale modelling of the migration / retention processes of fission products in the HBU spent fuel, in the cladding, and the estimation of the fission product total release through the spent fuel rod.

On the other hand, a semi-empirical model will be developed to predict fission product release to water from gap, grain boundaries and grains.

5.2 Introduction

In this period, KIT was modelling temperature and diffusion coefficients during reactor operation from the center of the pellet to the periphery (rim zone). Modelling is compared to the results obtained by using Laser Ablation performed at Studsvik.

On the other hand, AMPHOS 21 focused the modelling on how an initially "dry" fracture (micro and macro cracks) was invaded by water and the instantaneous radionuclide transfer into the solution upon wetting. In this case, modelling was also compared to leaching spent fuel results obtained with segmented high burn-up fuel samples, good agreement was obtained with some of the radionuclides studied, during the first days of the experiment.

UPC-CTM focused on the initial release (few days), comparing dissolution rates for the different radionuclides taking into account the samples segmented (gap, cracks, fissures) vs. powder (open grain boundaries, bubbles, microcracks).

5.3 Modeling

Behavior of fission products in fuel rods

The central temperature of a fuel rod was calculated using heat conductivity as a function of the burn-up and [1] the heat capacity of UO_2 fuel taken from IAEA TECDOC-949 [2] and Popov (2000) [3]. For calculating the diffusion and distribution of fission products in UO_2 fuel, Popov recommends Lucuta's equation for the heat conductivity of UO_2.

Calculations were performed by using the FlexPDE 6.35 code, which is a flexible solution system for partial differential equations. A standard geometry was applied for the calculations: Fuel zone: 0.00-4.40 mm, rim zone: 4.40-4.60 mm, gap: 4.60-4.65 mm, cladding: 4.65-5.375 mm. The calculation was performed in R, Z cylinder coordinates.

The modelling was performed for an average burn-up of 50 GWd/t_{HM} and power rates of 180 W/cm (KKL), 260 W/cm (KKG) and 340 W/cm (e.g. power ramps or fuel investigations by SCK•CEN from PWR Tihange-1) and Studsvik's fuel nominated as C1 (see Deliverable No: 1.2 "Characterization of spent nuclear fuel samples and description of methodologies and tools to be applied in FIRST-Nuclides").

Main conclusions of this modelling were:

- Diffusion processes depend on the temperature.
- Small gaps between the pellets or between pellet and cladding effect the temperature significantly.
- Radial distribution of I, Xe, Se and Cs in a spent fuel pellet modelled by assuming a temperature-dependent sorption.
- The calculated total concentration in the colder region of the pellet corresponds to Studsvik's measurements.
- Calculated release of volatile radionuclides into the gap regions corroborate with the fission gas release (FGR).
- After discharge from the reactor, diffusion processes of volatiles do not play any role due to the negligible diffusion coefficients at interim storage temperature.

5.4 Water saturation of the fuel

The conceptual model assumes that the rate of release for some radionuclides from a SNF pellet (or a fragment of it) upon contact with water is controlled by the wetting of surfaces of cracks present in the pellet. Calculations were carried out to quantify the mass (moles) of radionuclides released from the crack surfaces and accumulated in the external reservoir as a function of time. This allows direct comparison with data available from laboratory leaching experiments.

Modelling results indicate that wetting of crack surfaces present in a SNF pellet may quantitatively explain radionuclide release patterns observed in laboratory leaching experiments performed on SNF pellets or their fragments. Specifically, the time scales of initial fast radionuclide release and the later slower releases are on a scale comparable with the times calculated for complete saturation of the "macro" and "micro" crack systems. The proposed model is capable of capturing these general trends in a semi-quantitative manner. In order to increase the predictive capacity of the model further data on the statistical properties of cracks and on the distribution of radionuclides on the surfaces of cracks is needed.

5.5 Modeling of radionuclide release to water

Fast/Instant release rate of Tc, Mo, Sr, Rb and Cs as well as Uranium dissolution rate were evaluated for different samples, segmented and powder, by using a semi-empirical model based on the experimental fitting by using three different first-order kinetic equations corresponding to different parts of the fuel.

Leaching fuel experiments of different burn-up's, three PWR (48, 52 and 60 MWd/kg_U) and three BWR (42, 53 and 54 MWd/kg_U) were studied.

Normalized dissolution rate of Cesium is always higher than for the rest of measured radionuclides, especially when looking at results obtained from cladded samples. For powder samples this trend is also observed except for Rb, which shows more similar release rates. This is also true for Sr rates obtained for PWR samples. This is assumed to be due to the degree of matrix segregation. The higher the fraction of a radionuclide is segregated, the higher is its release because it is assumed that this release is practically instantaneous as soon as the water gets in contact with them.

In general, the percentage of radionuclide released is higher for powdered than for cladded samples, except in the case of Cs in PWR samples. It can be assumed that for Cs its content in gaps and fractures is similar than its content in grain boundaries and micro fissures, which is not the case for the rest of radionuclides, that show greater content in grain boundaries due to their lower migration once segregated from the matrix.

When comparing PWR and BWR fuel samples, clear differences in terms of normalized release rates are observed, being the values obtained for PWR fuels larger than for BWR fuels. Once again, such differences become more significant as we look to Cs and Rb and to some extent to Sr, which are the ones expected to be segregated from the matrix in higher degrees. When looking at the percentages dissolved, they seem to be higher for pellets (higher availability to water) than for powders, except for Cs and PWR fuels.

5.6 References

Lucuta, P.G., Matzke, H., Hastings, I.J. (1996). A Pragmatic Approach to Modelling Thermal Conductivity of Irradiated UO_2 fuel: Review and recommendations. Journal of Nuclear Materials, 232(2–3), 166-180.

IAEA (1997). Thermophysical Properties of Materials for Water Cooled Reactors. International Atomic Energy Agency, Vienna.

Popov, S.G., Carbajo, J.J., Ivanov, V.K., Yoder, G.L. (2000). Thermophysical Properties of MOX and UO_2 Fuels Including the Effect of Irradiation. Oak Ridge National Laboratory (ORNL).

6 Overview WP5: Knowledge, reporting and training

Alba Valls

Amphos 21 Consulting S.L. (ES)

6.1 Objectives

Work package 5 (WP5) of the FIRST-Nuclides project is focused on knowledge management and knowledge dissemination, reporting and training. The objectives of this WP are (i) to provide access to all scientific-technical results for all interested parties, (ii) to elaborate a state of the art report on the project and (iii) to organize, within the project, dissemination of the results, training and education for the next generation of spent nuclear fuel specialists.

6.2 Description of work performed in the project

Several dissemination and training activities have been implemented within WP5. A brief description of each one of these activities is presented below.

Dissemination activities

Generic poster: a generic poster of the project which is regularly updated. The last version of the poster is available at the project webpage, under *****.

Newsletters: three newsletters have been edited along the project. The main objective of the newsletter is to communicate to the broader community the progresses achieved every year, as well as presenting the last news and the coming events related with the project. All of them are available at the project webpage, under ****

Deliverables: each WP has several deliverables where it is compiled all the information related with the work done. Those reports are available at the project webpage under ****.

Web portal: A web portal has been created and constantly updated to inform both the consortium members and the broader community. For this reason, the website consists of two different parts; one public and another one restricted to the consortium members. The latter is a way of internal communication.

Annual workshops: every year a workshop have been organized by the coordination team and the host institutions (AMPHOS21, MTA-EK, SCK•CEN and KIT/INE). 1 kick-off meeting and 3 annual workshops took place during the project life.

Presentation at international conferences: Both the coordinator of the project and members of the beneficiary institutions have attended several international conferences giving oral presentations either on the project as a whole or specific work carried out in the frame of the project.

Peer review publications: The coordination team encourages all beneficiaries to publish their work in peer review journals such as *Journal of Nuclear Materials.* During the last workshop, the WP leaders have planned to publishing 4 papers which would cover a review of the scientific achievements within all the technical WP.

Training activities

Lab-Exchange meeting: 17 participants from four partner institutions (KIT-INE, JRC-ITU, Studsvik and PSI) have attended at this meeting organized by PSI on March 2013. The aim of the meeting was to serve as a discussion forum for experimentalists of the project to discuss analytical and technical experimental details.

Organization of a training course: KIT-INE, JRC-ITU and AMPHOS21 organized a training course (9th - 10th July 2013, Karlsruhe). The course focused on the radionuclide release from LWR spent nuclear fuel (SNF), experimental methods available to quantify radionuclide release and relevant characteristics of LWR SNF. Twelve participants attended the course, from six project partners and three associated groups.

Training mobility measures: The project has allocated resources for three training mobility measures, which consider the visit of one student from a participant institution to a different partner organization.

Albert Martínez (PhD student from UPC/CTM) was hosted at KIT-INE to improve experimental abilities in working in an inert gas glovebox, gas sampling in a shielded box (gas-MS) or handling of 0.1 wt.% ^{238}Pu doped $UO_2(s)$ pellets, among other technical issues.

Péter Szabó (PhD student from EK-MTA) was also at KIT-INE, being trained on sample preparation for ^{14}C analyses (LSC), uranium analyses, introduction to geochemical modeling, laser fluorescence measurements, and other skills of interest for his scientific development within the project.

David García (PhD student from AMPHOS21) has benefited from a training mobility also at KIT-INE on development of sorption models.

- *7 travel grants to attend the Final workshop*: 7 PhD students have taken advantage of the travel grants. The grant covered travelling, accommodation and workshop fee expenses in order young people have the opportunity of attending the final workshop of the project.

6.3 State of the Art (SoA)

A scientific-technical state-of-the-art was reported at the beginning of the project. This document is composed of two different sections. The first section aims on providing basic information such as description of the fuel, irradiation induced processes in UO_2 or modeling tools for fuel performance. The second section summarizes the results obtained from more than 100 published experiments using different samples, experimental techniques, solutions, etc.

During the last year of the project, a database compiling all the information and results of the experiments reported at the first SoA report was developed and presented during the Final Workshop of the project. In addition, results obtained during the FIRST-Nuclides project are included at the database.

Database description

The database is developed in Excel format and it contains 4 different types of excel spreadsheets:

A spreadsheet consisting on a *table* that contains a *summary* of the relevant parameters and results of each sample included in the database.

Interactive spreadsheet where the user can select the data to be *plotted* (selection of both the "x" and "y" axis). Data result of the experiments carried out in the frame of FISRT-nuclides project are shown in a different color in order to see if they are in agreement with previous literature data. In addition, an interactive table allows the user to select which of the parameters from the summary table must be shown in each column. This option may be useful for understanding the presence of an outlier point.

Several spreadsheets, one per publication, providing *detailed information* on the samples, materials, experimental conditions, analysis techniques, results, uncertainties, etc..

6.4 References

The database is available at the FIRST-Nuclides project website, under ***.

More information regarding this database is available at Deliverable 5.1 – 3rd update which can be also download from the project website (www.firstnuclides.eu).

S + T Contributions

7 Selection of materials and characterisation of samples used in spent fuel leaching and laser ablation studies

Olivia Roth, Anders Puranen, Charlotta Askeljung and Daqing Cui*

Studsvik Nuclear AB (SE)
** Corresponding author: olivia.roth@studsvik.se*

7.1 Abstract

This paper describes the selection and characterization of six high burn-up SNF samples for leaching experiments and laser ablation studies performed at the Hot Cell laboratory in Studsvik within the FIRST-Nuclides project (European Union's European Atomic Energy Community's (Euratom) Seventh Framework Programme FP7/2007-2011 under grant agreement n° 295722).

The sample preparation performed prior to the experimental studies is also described in this paper.

7.2 Introduction

The aim with the project is to study the fraction of fission and activation products that is fast/instantly released from spent nuclear fuel upon contact with aqueous media. The fraction consists of readily soluble phases in the gap between fuel and cladding, cracks and grain boundaries. Some of these fission and activation products have a long half-life and are for this reason important for the safety assessment of deep repositories for spent nuclear fuel.

At the Studsvik Hot Cell laboratory, spent fuel leaching studies have been conducted since around 1980. During 1990-1996 a comprehensive research program was initiated aiming at mapping the most important parameters influencing the stability of spent nuclear fuel in water (Forsyth, 1997). Since then the program has been extended with leaching experiments of high burn-up fuel and instant release experiments (Johnson et al., 2012; Zwicky et al., 2011).

In these previous experiments, standard UO_2 fuel has been used. Today new fuel types with additives and dopants are taken into use in commercial reactors. The additives and dopants effects properties such as grain size and fission gas release which in turn may affect the instant release behavior of the fuel. The main objective of this study is to investigate how these changes in the fuel matrix affect the instant release process.

Samples from six irradiated fuel rods were selected to be studied at Studsvik within the EURATOM FP7 Collaborative Project “Fast / Instant Release of Safety Relevant Radionuclides from Spent Nuclear Fuel (CP FIRST-Nuclides)”. The studies consist of leaching experiments and laser ablation studies.

Four of the rods have standard UO_2 pellets, one has Al/Cr doped UO_2 pellets and one has Gd-doped pellets. This selection was made in particular to assess the effect of doping on the instant release behavior. One of the standard fuel samples was leached by a simultaneous grinding and leaching method to specifically assess the release of the grain boundary inventory.

This paper describes the characterization of materials and sample preparation. The outcome of the experiments are presented in Puranen et al. (2014a), Puranen et al. (2014b), Roth et al. (2014), Curti et al. (2014).

7.3 Sample selection and characterization

The main focus of the investigation is to explore the effects of additives and dopants on the fast/instant release of fission products such as Cs and I. The selection of material has previously been described in Roth and Puranen (2012). The selected fuels are listed in Table 7.1.

Table 7.1: Fuels selected for investigation at Studsvik.

Sample name	Reactor type	Fuel type	FGR [%]	Calculated BU (rod average) [MWd/kg_U]
D07	BWR	Std UO_2	~1.6	50.2
L04	BWR	Std UO_2	~3.1	54.8
5A2	BWR	Std UO_2	~2.4	57.1
C1	BWR	Al/Cr doped UO_2	~1.4	59.1
VG81	PWR	Gd doped UO_2	~2.2	54.4
AM2K12	PWR	Std UO_2	~4.9	70.2

Further sample data and fuel inventory data are found in Appendix A (samples D07 and L04), Appendix B (samples 5A2 and C1), Appendix C (sample VG81) and in reference Zwicky et al. (2011) (sample AM2K12).

The fuels have previously been characterized by measuring the fission gas release (FGR). All samples have undergone gamma scanning (either gross gamma scan or digital gamma scan) before experimental start-up. The methods for FGR measurements and gamma scanning are described in Roth and Puranen (2012). The calculated burn-up (rod average) for the rods are provided from the core calculations of the power stations that irradiated the rods.

7.4 Sample preparation for leaching experiments

Spent fuel leaching studies are performed using samples from 6 different fuel rods. 5 of the fuel samples were leached as either cladded fuel segments or fuel fragments + separated cladding. The sixth sample (AM2K12) was leached as powdered specimen using simultaneous grinding and leaching technique. The preparation of these different samples is described below.

7.5 1. Leaching of cladded fuel segments and fuel fragments + separated cladding

For the samples leached as cladded fuel segments and fuel fragments + separated cladding, gamma scanning was undertaken prior to the leaching experiment to identify pellet-pellet interfaces, thereafter samples are cut from the segment. The samples are cut at mid-pellet positions as shown in Figure 7.1. Each sample consists of approximately 2 fuel pellets including cladding.

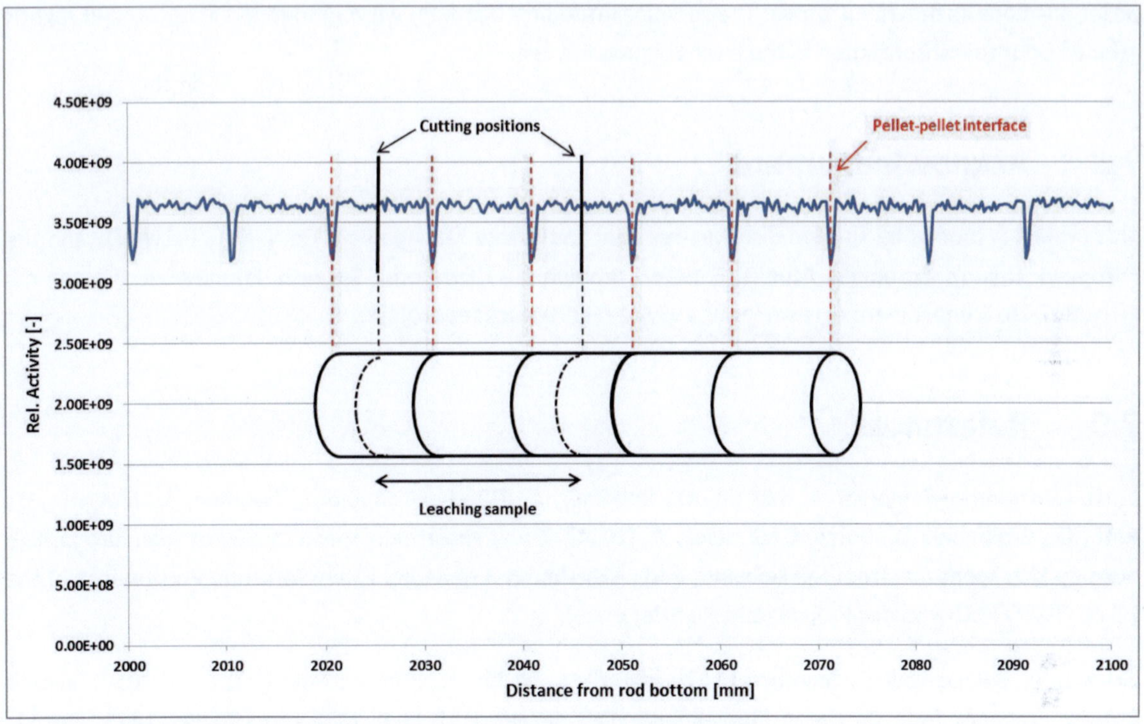

Figure 7.1: *Schematic picture of leaching sample position relative to gamma spectrum.*

The fuel segments are weighed before the start of the leaching experiments. To obtain the samples consisting of fuel fragments + separated cladding, the cladding is cut vertically and bent open which causes the fuel fragments to detach from the cladding (5A2 and C1) or the fuel was slightly crushed using a hammer on the outside of the cladding until the fuel fragment broke away from the cladding (VG81). The fuel fragments are collected and leached together with the cladding (and any remaining fuel still attached to the inside of the cladding).

7.6 Leaching of powdered fuel sample

The AM2K12 sample was leached as powdered specimen using simultaneous grinding and leaching technique. The sample used here has undergone 4 years leaching as fuel fragments + separated cladding, this leaching step is outside the scope of the present study and has been described in Johnson et al. (2012), Ekeroth et al. (2012).

After the 4 year leaching period the experiment was finalized and the fuel fragments stored in Hot Cell. For the powdered leaching study, approximately 8 fuel fragments (1.8 g) of fuel were removed, weighed and transported to a fume hood were the simultaneous grinding and leaching experiment was performed.

7.7 Sample preparation for the laser ablation study

Samples from the fuel 5A2 and C1 were chosen for laser ablation studies. For both fuels the samples were cut out from neighbouring pellets to the samples used for the leaching studies. The samples were cut in mid pellet positions as described above. The samples (including cladding) were embedded in epoxy and slightly grinded prior to transportation to the laser ablation hot cell.

7.8 Acknowledgement

This project is funded by The Swedish Nuclear Fuel and Waste Management Co. (SKB) , Posiva Oy and the European Union's European Atomic Energy Community's (Euratom) Seventh Framework Programme FP7/2007-2011 under grant agreement n° 295722 (FIRST-Nuclides project).

7.9 References

Curti, E., Froideval-Zumbiehl, A., Martin, M., Bullemer, A., Günther-Leopold, I., Puranen, A., Jädernäs, D., Roth, O., Grolimund, D., Borca, C.N., Velea, A. (2014). X-Ray absorption spectroscopy of selenium in high burn-up UO_2 spent fuel from the Leibstadt and Oskarshamn-3 reactors. Final Workshop Proceedings of the 7th EC FP CP FIRST-Nuclides Project (eds. Kienzler et al.).

Ekeroth, E., Cui, D., Low, J., Granfors, M., Zwicky, H.-U., Spahiu, K., Zetterström Evins, Lena (2012). Instant Release Fractions from Corrosion Studies with High Burnup LWR Fuel. MRS Proceedings, 1475, imrc11-1475-nw35-o36, doi:10.1557/opl.2012.565.

Forsyth, R. (1997). The SKB Spent Fuel Corrosion Program – An Evaluation of Results from the Experimental Program Performed in the Studsvik Hot Cell Laboratory. SKB Technical Report, TR 97-25.

Johnson, L., Günther-Leopold, I., Kobler Waldis, J., Linder, H.P., Low, J., Cui, D., Ekeroth, E., Spahiu, K., Evins, L.Z. (2012). Rapid Aqueous Release of Fission Products from High Burn-Up LWR Fuel: Experimental Results and Correlations with Fission Gas Release. Journal of Nuclear Materials, 420, 54-62.

Puranen, A., Granfors, M., Roth, O. (2014). Laser Ablation Study of Irradiated Standard UO_2 Fuel and Al/Cr Doped UO_2 Fuel. Final Workshop Proceedings of the 7th EC FP CP FIRST-Nuclides Project (eds. Kienzler et al.).

Puranen, A., Granfors, M., Roth, O. (2014). Aqueous Leaching of ^{79}Se from Spent Nuclear Fuel. Final Workshop Proceedings of the 7th EC FP CP FIRST-Nuclides Project (eds. Kienzler et al.).

Roth, O., Puranen, A. (2012). Selection of Materials, Preparations and Experimental Set-Up. 1st Annual Workshop Proceedings of the 7th EC FP CP FIRST-Nuclides project (eds. Kienzler et al.). KIT Scientific Reports 7639, 137-144.

Roth, O., Askeljung, C., Puranen, A., Granfors, M., Cui, D., Low, J. (2014). Leaching of High-Burn-Up Spent Fuel with and Without Matrix Dopants. Final Workshop Proceedings of the 7th EC FP CP FIRST-Nuclides Project (eds. Kienzler et al.).

Zwicky, H.-U., Low, J., Ekeroth, E. (2011). Corrosion Studies with High Burnup Light Water Reactor Fuel; Release of Nuclides into Simulated Groundwater during Accumulated Contact Time of up to Two Years. SKB Technical Report, TR-11-03.

7.10 Apendix

Apendix 1

	D07	L04
Pellet type	Standard UO_2	Standard UO_2
Initial enrichment [%]	4.25	4.25
Reactor	Olkiluoto 1	Olkiluoto 1
Years in reactor		
Irradiation time	06.06.2003 –13.05.2008	06.06.2003–13.05.2008
Burnup (rod average) physics calc. [MWd/kg_U]	50.2	54.8
Burnup (rod average) gamma scan [MWd/kg_U]	N.A.	53.4
Released Kr [%]	1.52	3.07
Released Xe [%]	1.57	3.06
Average LHGR [kW/m]	14.3	14.6
Sample position leaching [mm from rod bottom]	785	760
Local BU [MWd/kg_U]	59.3	65.7
Local LHGR [kW/m]	14.5	16.2

N.A. = Not Available

	D07 µg/g_U(initial), at EOB	L04 µg/g_U(initial), at EOB
^{85}Rb	1.74E+02	1.88E+02
^{87}Rb	4.07E+02	4.39E+02
^{90}Sr	8.81E+02	9.44E+02
^{90}Zr	6.55E+01	7.46E+01
^{91}Zr	9.85E+02	1.07E+03
^{92}Zr	1.10E+03	1.19E+03
^{93}Zr	1.21E+03	1.31E+03
^{94}Zr	1.34E+03	1.47E+03

	D07 µg/g$_U$(initial), at EOB	L04 µg/g$_U$(initial), at EOB
^{95}Zr	4.13E+01	4.12E+01
^{96}Zr	1.40E+03	1.53E+03
^{99}Tc	1.31E+03	1.41E+03
^{100}Mo	1.65E+03	1.84E+03
^{100}Ru	2.46E+02	3.14E+02
^{101}Ru	1.37E+03	1.51E+03
^{102}Ru	1.43E+03	1.62E+03
^{103}Ru	2.94E+01	3.19E+01
^{104}Ru	1.03E+03	1.19E+03
^{106}Ru	1.93E+02	2.24E+02
^{104}Pd	5.61E+02	7.19E+02
^{105}Pd	7.24E+02	8.20E+02
^{106}Pd	5.04E+02	6.23E+02
^{107}Pd	4.15E+02	4.93E+02
^{108}Pd	2.81E+02	3.38E+02
^{110}Pd	8.73E+02	1.05E+02
^{127}I	8.05E+01	9.25E+01
^{129}I	2.59E+02	2.86E+02
^{133}Cs	1.78E+03	1.90E+03
^{134}Cs	2.50E+02	2.95E+02
^{135}Cs	6.02E+02	5.80E+02
^{137}Cs	2.08E+03	2.30E+03
^{134}Ba	1.76E+02	2.21E+02
^{135}Ba	1.70E+00	2.47E+00
^{136}Ba	4.40E+01	4.90E+01
^{137}Ba	1.31E+02	1.49E+02
^{138}Ba	2.29E+03	2.53E+03
^{139}La	2.09E+03	2.29E+03
^{140}Ce	2.15E+03	2.39E+03
^{141}Pr	1.90E+03	2.08E+03
^{142}Nd	6.33E+01	8.78E+01
^{143}Nd	9.85E+02	8.88E+02
^{144}Nd	2.37E+03	2.81E+03
^{145}Nd	1.11E+03	1.18E+03
^{146}Nd	1.29E+03	1.48E+03
^{148}Nd	6.61E+02	7.35E+02

	D07 $\mu g/g_U$(initial), at EOB	L04 $\mu g/g_U$(initial), at EOB
^{150}Nd	3.18E+02	3.56E+02
^{153}Eu	2.25E+02	2.52E+02
^{243}Cm	7.80E-01	8.50E-01
^{244}Cm	1.18E+02	1.87E+02
^{237}Np	6.42E+02	6.66E+02
^{238}U	9.20E+05	9.15E+05
^{238}Pu	3.22E+02	3.58E+02
^{239}Pu	3.79E+03	3.44E+03
^{85}Kr	3.92E+01	4.20E+01
^{88}Sr	5.66E+02	6.10E+02
^{91}Y	2.26E+01	2.11E+01
^{144}Ce	3.06E+02	3.13E+02

Apendix 2

	O3C1	5A2
Pellet type	Cr_2O_3/Al_2O_3 doped UO_2	Standard UO_2
Initial enrichment [%]	4.1	3.5
Reactor	Oskarshamn 3	Oskarshamn 3
Years in reactor	7	7
Irradiation time	2000-08-10 – 2008-10-02	2000-08-10 – 2008-10-02
Burnup (rod average) physics calc. [MWd/kg_U]	59.1	57.1
Burnup (rod average) gamma scan [MWd/kg_U]	56.0	N.A.
Released Kr [%]	1.34	2.48
Released Xe [%]	1.46	2.44
Average LHGR [kW/m]	~19	~17
Sample position leaching [mm from rod bottom]	2680	2663
Sample position leaching 2 [mm from rod bottom]	2725	2690
Sample position LA [mm from rod bottom]	2660	2643
Local BU [MWd/kg_U] at sample position	61.1 (calc from gammascan)	60.5 (est from gamma scan)
Average LHGR [kW/m] at sample position	N.A.	N.A.

N.A. = Not Available

	O3C1 µg/g_U(initial), at EOB	5A2 µg/g_U(initial), at EOB
^{85}Rb	1.88E+02	1.73E+02
^{87}Rb	4.46E+02	4.08E+02
^{90}Sr	9.10E+02	8.24E+02
^{90}Zr	1.37E+02	1.26E+02
^{91}Zr	1.04E+03	9.44E+02
^{92}Zr	1.20E+03	1.11E+03
^{93}Zr	1.28E+03	1.19E+03
^{94}Zr	1.47E+03	1.38E+03
^{95}Zr	4.92E+01	4.77E+01
^{96}Zr	1.52E+03	1.42E+03
^{99}Tc-	1.38E+03	1.30E+03
^{100}Mo	1.82E+03	1.73E+03
^{100}Ru	3.35E+02	3.29E+02
^{101}Ru	1.47E+03	1.40E+03
^{102}Ru	1.51E+03	1.46E+03
^{103}Ru	4.35E+01	4.38E+01
^{104}Ru	1.18E+03	1.18E+03
^{106}Ru	2.31E+02	2.37E+02
^{104}Pd	7.77E+02	7.81E+02
^{105}Pd	8.36E+02	8.43E+02
^{106}Pd	5.54E+02	5.75E+02
^{107}Pd	5.07E+02	5.27E+02
^{108}Pd	3.50E+02	3.65E+02
^{110}Pd	1.18E+02	1.23E+02
^{127}I	8.82E+01	8.71E+01
^{129}I	3.01E+02	2.95E+02
^{133}Cs	1.96E+03	1.85E+03
^{134}Cs	2.74E+02	2.66E+02
^{135}Cs	7.08E+02	6.39E+02
^{137}Cs	2.29E+03	2.18E+03
^{134}Ba	2.17E+02	2.12E+02
^{135}Ba	3.58E+00	3.62E+00
^{136}Ba	7.56E+01	7.33E+01
^{137}Ba	1.72E+02	1.63E+02
^{138}Ba	2.56E+03	2.43E+03
^{139}La	2.34E+03	2.22E+03

	O3C1 µg/g_U(initial), at EOB	5A2 µg/g_U(initial), at EOB
^{140}Ce	2.37E+03	2.24E+03
^{141}Pr	2.05E+03	1.93E+03
^{142}Nd	9.33E+01	9.25E+01
^{143}Nd	8.73E+02	7.77E+02
^{144}Nd	2.79E+03	2.64E+03
^{145}Nd	1.13E+03	1.06E+03
^{146}Nd	1.49E+03	1.41E+03
^{148}Nd	6.99E+02	6.65E+02
^{150}Nd	3.56E+02	3.46E+02
^{153}Eu	2.25E+02	2.18E+02
^{237}Np	5.76E+02	5.07E+02
^{238}U	9.12E+05	9.11E+05
^{239}Pu	3.81E+03	3.69E+03
^{243}Cm	8.28E-01	8.23E-01
^{244}Cm	2.42E+02	2.71E+02
^{238}Pu	3.45E+02	3.15E+02
^{239}Pu	3.81E+03	3.69E+03
^{85}Kr	3.81E+01	3.47E+01
^{88}Sr	6.15E+02	5.62E+02
^{91}Y	2.56E+01	2.41E+01
^{144}Ce	2.94E+02	2.79E+02

Apendix 3

	VG81
Pellet type	8% Gd_2O_3
Initial enrichment [%]	2.8
Reactor	Vandellos II
Years in reactor	
Irradiation time	2000-10-10 to 2007-05-05
Burnup (rod average) physics calc. [MWd/kg_U]	54.4
Burnup (rod average) gamma scan [MWd/kg_U]	N.A.
Released Kr [%]	1.92
Released Xe [%]	2.30
Average LHGR [kW/m]	N.A.
Sample position leaching [mm from rod bottom]	875

	VG81
Local BU [MWd/kg$_U$]	55.7
Local LHGR [kW/m]	13.6

N.A. = Not Available

	VG81 µg/g$_U$(initial), at EOB
^{85}Rb	1.30E+02
^{87}Rb	3.23E+02
^{90}Sr	7.14E+02
^{90}Zr	1.57E+01
^{91}Zr	6.90E+02
^{92}Zr	8.93E+02
^{93}Zr	1.01E+03
^{94}Zr	1.12E+03
^{95}Zr	1.90E+02
^{96}Zr	1.23E+03
^{99}Tc	1.18E+03
^{100}Mo	1.53E+03
^{100}Ru	2.45E+02
^{101}Ru	1.25E+03
^{102}Ru	1.49E+03
^{103}Ru	1.54E+02
^{104}Ru	1.13E+03
^{106}Ru	4.98E+02
^{104}Pd	5.02E+02
^{105}Pd	7.21E+02
^{106}Pd	4.90E+02
^{107}Pd	5.15E+02
^{108}Pd	3.53E+02
^{110}Pd	1.18E+02
^{127}I	8.42E+01
^{129}I	2.54E+02
^{133}Cs	1.62E+03
^{134}Cs	3.45E+02
^{135}Cs	3.84E+02
^{137}Cs	2.03E+03

	VG81 µg/g_U(initial), at EOB
^{134}Ba	6.22E+01
^{135}Ba	7.35E-01
^{136}Ba	3.49E+01
^{137}Ba	3.62E+01
^{138}Ba	2.06E+03
^{139}La	1.91E+03
^{140}Ce	1.88E+03
^{141}Pr	1.57E+03
^{142}Nd	4.55E+01
^{143}Nd	9.74E+02
^{144}Nd	1.34E+03
^{145}Nd	9.32E+02
^{146}Nd	1.19E+03
^{148}Nd	6.25E+02
^{150}Nd	3.17E+02
^{153}Eu	1.97E+02
^{243}Cm	8.65E-01
^{244}Cm	2.83E+02
^{237}Np	6.19E+02
^{238}U	9.01E+05
^{238}Pu	3.20E+02
^{239}Pu	6.20E+03
^{85}Kr	3.35E+01
^{88}Sr	5.12E+02
^{91}Y	1.04E+02
^{144}Ce	8.29E+02

8 Fission Product release from irradiated fuel

*Jean-Yves Colle[1], Ondřej Beneš[*1], Daniel Serrano-Purroy[1], Rosa Sureda Pastor[2] and Albert Martinez Torrents[2]*

[1] *European Commission, Joint Research Centre, Institute for Transuranium Elements (EC)*
[2] *Fundació CTM Centre Tecnològic (ES)*
* *Corresponding author: ondrej.benes@ec.europa.eu*

8.1 Abstract

This study reports the release of Sr, Rb, Cs, Xe fission products from an irradiated UO_2 sample using the Knudsen effusion mass spectrometer. The aim of the ongoing study is to compare the release of selected fission products from the fuel as taken from the reactor with the release from same fuel after a leaching experiment in order to identify the instant release fraction. In this paper results for a fuel sample taken from the reactor are presented.

8.2 Introduction

Knudsen Effusion Mass Spectrometry (KEMS) is a powerful technique to study the high temperature thermodynamic and kinetic behavior of materials (Drowart et al., 2005; Hilpert, 1991). It has been used at the Institute for Transuranium Elements (ITU) of the Joint Research Centre (JRC) of the European Commission for more than 20 years to generate fundamental data and understanding for nuclear fuel safety studies with standard measurements of partial vapor pressure and vaporization behavior (Gotcu-Freis et al., 2011a; Gotcu-Freis, 2011b; Gotcu-Freis, 2011c; Hiernaut, 2004), ionization efficiency curves and ionization and dissociation energies (Capone, 1999) on radioactive materials and actinide materials. It has been also intensively used for kinetic studies like fission products and actinide release behavior from irradiated fuel (Hiernaut, 2008a; Capone, 1996; Colle, 2006; Hiernaut, 2001; Hiernaut, 2008b; Hiernaut, 2008c; Hiernaut, 2009; Hiernaut, 2005; Ronchi, 2004; Wiss, 2007; Wiss, 2004; Wiss, 2006). All those applications have some particularities in common: specific requirements related to the radioactivity of the samples and the very high temperatures to vaporize actinide compounds, especially for oxides. The first point requires a setup built in a glove box and thus it is necessary to avoid complex design and it is also necessary to minimize the waste with a highly reliable system. The second point requires a very high temperature furnace for the Knudsen cell.

8.3 Experimental – KEMS set-up

Figure 8.1 shows the actual setup of the KEMS system installed at ITU. It is equipped with quadrupole mass spectrometer {7} (QMG422 from Pfeiffer Vacuum) which offers a mass range of 1 to 512 atomic mass unit (amu). It has a cross beam electron bombardment ion source, an axial Faraday cup and a secondary electron multiplier (SEM) located at 90° to the filter axis used in ion current measurement mode. The furnace is heated by a resistance coil {2} and shielded by seven W and Ta shields {8}. The furnace can be heated with temperature ramps such as 10 or 30 K/min or at constant temperature. The Knudsen cell can be equipped with a gas inlet {10} allowing work with up to 100 Pa pressure in the cell {17}, thus simulating oxidizing conditions by varying the oxygen partial pressure. The system is equipped with a single blend between the ion source chamber and the furnace chamber and requires calibration for each measurement usually done with mass loss technique of a silver sample. A chopper {3} allows making the background measurement and a liquid nitrogen (LN2) cold trap {5} significantly improves the vacuum in the ion source chamber and thus the mass spectra background. The vacuum is generated with a standard turbo pump backed up with a dry primary pump. The vessel (out of the turbo pumps) can be baked up to 80°C by heating up the cooling water. The complete system is placed in an alpha tight glove box surrounded by movable 5 cm lead gamma shields.

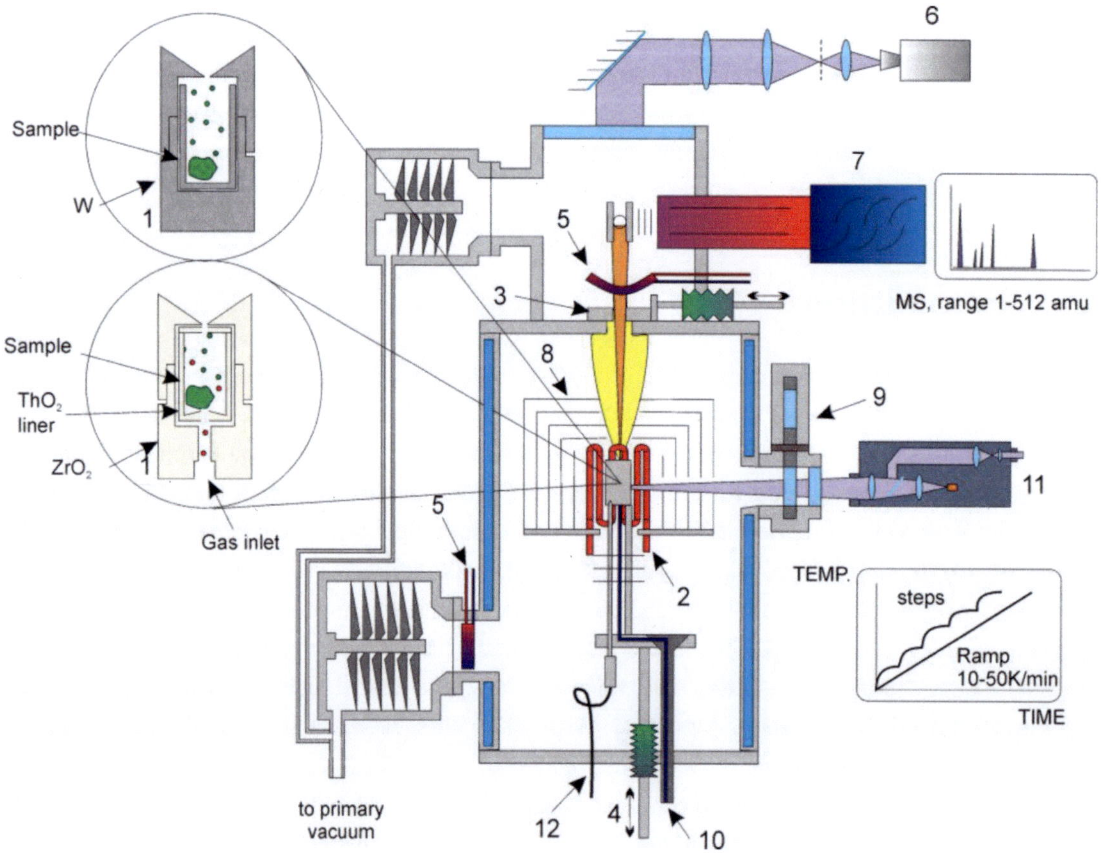

Figure 8.1: Schematic of the actual KCMS system at ITU. 1: Knudsen cell, 2: heating coil, 3: chopper, 4: lift, 5: LN2 cold trap, 6: camera, 7: mass spectrometer, 8: thermal shields, 9: revolving windows, 10: gas inlet, 11: pyrometer, 12: thermocouple.

8.4 Fission Product Realease from Irradiated UO_2

Within the First-Nuclide project we have examined a fine powder (diameter of ca. 50µm) of ~5mg quantity of the irradiated BWR UO_2 sample with average burn-up of 54 GWd/t_{HM} which was drilled from the core region of the fuel pin. The sample was put in the Knudsen cell and was heated under vacuum conditions at rate of 10 K/min until complete vaporization of the fuel at 2,460 K. During the measurement the atomic mass units corresponding to relevant isotopes of examined fission products have been recorded and a typical experimental output is given in Figure 8.2, showing the release of caesium-137. The figure indicates two release mechanisms, one with rather low intensity starting just below 1,000 K (marked with a red circle) and a significant release after 1,500 K. To semi-quantify the amount of fission products released at various steps, the so-called normalized fractional release (integrator of the total release normalized from 0 to 1) is plotted in Figure 8.3, showing all examined fission products (^{88}Sr, ^{87}Rb, ^{137}Cs, ^{136}Xe).

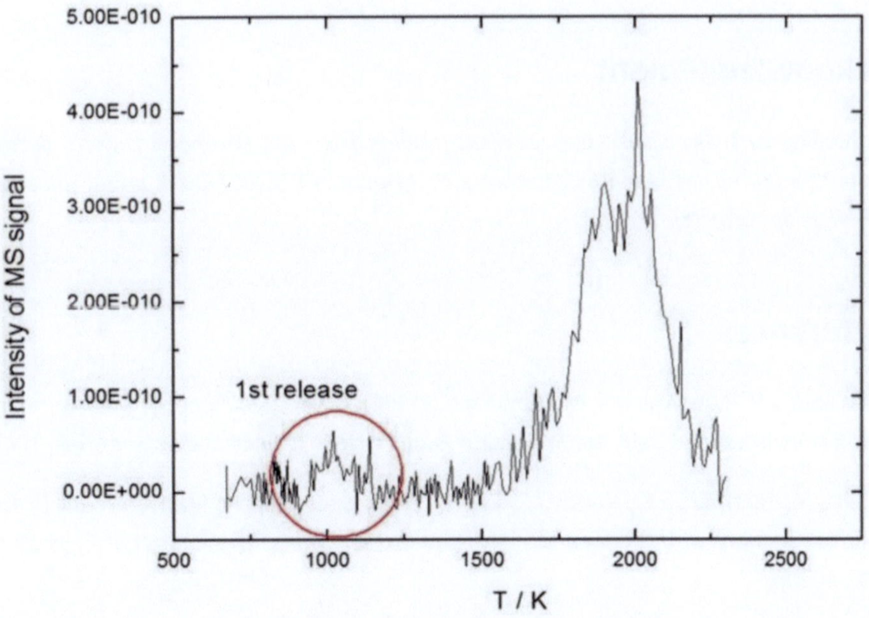

Figure 8.2: KEMS measurement of ^{137}Cs released from irradiated UO_2 fuel before leaching.

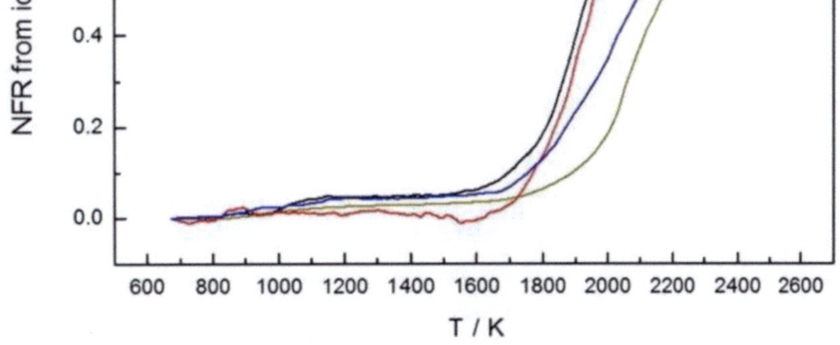

Figure 8.3: Normalized fractional release of selected fission products determined from the ion intensities measured by KEMS.

8.5 Conclusions and Future work

We have measured the fission product release of a BWR UO_2 irradiated fuel sample. In oncoming months a sample from the same region of this fuel that has been exposed to leaching conditions will be examined. From the comparison of these two sets of measurements the instant release fractions of relevant fission products will be identified (if applicable) and the results will be correlated with chemical analysis of the solvent used for the leaching experiments.

8.6 Acknowledgement

The research leading to these results has received funding from the European Union's European Atomic Energy Community's (Euratom) Seventh Framework Programme FP7/2007-2011 under grant agreement n° 295722 (FIRST-Nuclides project).

8.7 References

Capone, F., Hiernaut, J.P., Martellenghi, M., Ronchi, C. (1996). Mass Spectrometric Measurements of Fission Product Effusion from Irradiated Light Water Reactor Fuel. Nuclear Science and Engineering, 124(3), 436-454.

Capone, F., Colle, Y., Hiernaut, J.P., Ronchi, C. (1999). Mass Spetrometric Measurement of the Ionization Energies and Cross Sections of Uranium and Plutonium Oxide Vapors. The Journal of Physical Chemistry A, 103(50), 10899-10906.

Colle, J.Y., Hiernaut, J.P., Papaioannou, D., Ronchi, C., Sasahara, A. (2006). Fission Product Release in High-Burn-Up CO_2 Oxidized to U_3O_8. Journal of Nuclear Materials, 348(3), 229-242.

Drowart, J., Chatillon, C., Hastie, J., Bonnel, D. (2005). High-Temperature Mass Spectrometry: Instrumental Techinques, Ionization Cross-Sections, Pressure Meassurements, and Thermodynamic Data. Pure Applied Chemistry., 77(4), 683-737.

Gotcu-Freis, P., Colle, J.Y. Guèneau C., Dupin, N., Sundman, B., Konings, R.J.M. (2011a). A Thermodynamic Study of the Pu-Am-O System. Journal of Nuclear Materials, 414(3), 408-421.

Gotcu-Freis, P., Colle, J.Y., Hiernaut, J.P., Konings, R.J.M. (2011b). The Vaporization Behaviour of Americium Dioxide by Use of Mass Spetrometry. Journal of Nuclear Materials, 409(3), 194-198.

Gotcu-Freis, P., Colle, J.Y., Hiernaut, J.P., Konings, R.J.M. (2011c). (Solid + Gas) Equilibrium Studies for Neptunium Dioxide. The Journal of Chemical Thermodynamics, 43(3), 492-498.

Hiernaut, J.P., Ronchi, C. (2001). Fission Gas Release and Volume Diffusion Enthalpy in UO_2 Irradiated at Low and High Burnup. Journal of Nuclear Materials, 294(1-2), 39-44.

Hiernaut, J.P., Ronchi, C. (2004). Curium Vaporization from $(Cm,Pu)_2O_3$ and from Irradiated Oxide Fuel: Mass Spectrometric Measurement. Journal of Nuclear Materials, 334(2-3), 133-138.

Hiernaut, J.P., Colle, J.Y., Pflieger-Cuvellier, R., Jonnet, J., Somers, J., Ronchi, C. (2005). A Knudsen Cell-Mass Spectrometer Facility to Investigate Oxidation and Vaporisation Processes in Nuclear Fuel. Journal of Nuclear Materials, 344(1-3), 246-253.

Hiernaut, J.P., Gotcu, P., Colle, J.Y., Konings, R.J.M. (2008a). Thermodynamic Study of Actinides and Lanthanides during Total Vaporisation of a Very High Burn-Up UO_2 Fuel. Journal of Nuclear Materials, 378(3), 349-357.

Hiernaut, J.P., Wiss, T., Colle, J.Y., Thiele, H., Walker, C.T., Goll, W., Konings, R.J.M. (2008b). Fission Product Release and Microstructure Changes during Laboratory Annealing of a very High Burn-Up Fuel Specimen. Journal of Nuclear Materials, 377(2), 313-324.

Hiernaut, J.P., Wiss, T., Papaioannou, D., Konings, R.J.M., Rondinella, V.V. (2008c). Volatile Fission Product Behaviour during Thermal Annealing of Irradiated UO_2 Fuel Oxidised up to U_3O_8. Journal of Nuclear Materials, 372(2-3), 215-225.

Hiernaut, J.P., Wiss, T., Rondinella, V.V., Colle, J.Y., Sasahara, A., Sonoda, T., Konings, R.J.M. (2009). Specific Low Temperature Release of ^{131}Xe from Irradiated MOX Fuel. Journal of Nuclear Materials, 392(3), 434-438.

Hilpert, K. (1991). High Temperature Mass Spectometry in Materials Research. Rapid Communications in Mass Spectrometry, 5(4), 175-187.

Colle, J.Y., Hiernaut, J.P., Papaioannou, D., Ronchi, C., Sasahara, A. (2006). Fission Product Release in High-Burn-Up UO_2 Oxidized to U_3O_8. Journal of Nuclear Materials, 348(3), 229-242.

Ronchi, C., Hiernaut, J.P. (2004). Helium Diffusion in Uranium and Plunoium Oxides. Journal of Nuclear Materials, 325(1), 1-12.

Wiss, T.A.G., Damen, P.M.G., Hiernaut, J.P. Ronchi, C. (2004). Helium and Xenon Behaviour in Irradiated Am-Containing $MgAl_2O_4$ (Reactor Experiment EFTTRA_T4). Journal of Nuclear Materials, 334(1), 47-57.

Wiss, T.A.G., Hiernaut, J.P., Damen, P.M.G., Lutique, S., Fromknecht, R., Weber, W.J. (2006). Helium Behaviour in Waste Conditioning Matrices during Thermal Annealing. Journal of Nuclear Materials, 352(1-3), 202-208.

Wiss, T., Deschanels, X., Hiernaut, J.P., Roudil, D., Peuget, S., Rondinella, V.V. (2007). Helium Release from Plutonium and Cirium-Doped Zirconolite. Journal of Nuclear Materials, 362(2-3), 431-437.

9 WP2: Grain boundary diffusion in spent fuel and polycrystalline UO_2

Paul Carbol, Ilaria Marchetti, Detlef H. Wegen, Antonio Bulgheroni and Thierry Wiss*

European Commission, Joint Research Centre, Institute for Transuranium Elements (EC)
** Corresponding author: paul.carbol@ec.europa.eu*

9.1 Abstract

Emitted α-particles from a spent fuel create, due to radiolysis, a surficial slightly oxidative environment which leads to preferential oxidation, in comparison to UO_2 lattice, of the grain boundaries. Substantial oxidation of grain boundaries in combination with lattice change (U_4O_9 to U_3O_7) in the boundaries opens up the space between the grains for water penetration. The energy of α-particles, earlier absorbed in the UO_2 matrix, will now, partially, be absorbed by water molecules in the inter-granular space creating an elevated concentration of oxidative species, such as OH• and H_2O_2. The stepwise increased oxidative environment in the UO_2 or SNF (Spent Nuclear Fuel) matrix leads to disintegration of the matrix and release of elements present in the grain boundaries (IRF). It is therefore essential to determine the grain boundary diffusion coefficient of oxygen in UO_2 and SNF. The degree of grain boundary oxidation can be monitored using SIMS after exposure of UO_2 or SNF to $H_2{}^{18}O$ (10 mM NaCl + 2 mM HCO_3^-, pH 8). It could be concluded that under oxidising conditions and in presence of carbonate, the main part the UO_2 surface is oxidised and reacts with carbonate, which forms a carbonate layer on the surface. The modelling of the resulting ^{18}O penetration curve, under oxidative conditions, can only be explained by two parallel ^{18}O diffusion processes. The first process was assigned to the lattice diffusion of oxygen in UO_{2+x}, with $x<0.25$ (limiting phase is U_4O_9). The second process could not be singled out from the various possibilities but indicate a significant grain boundary diffusion contribution. The modelling results are supported by alternated SIMS depth profiling and ion mapping showing local crater areas with significantly varying ^{18}O concentrations. On the other hand, oxygen diffusion studies under reducing condition, in presence of carbonate, show that the measured ^{18}O depth profile curve follows Fick's 2nd law (lattice diffusion) and absence of grain boundary diffusion.

9.2 Introduction

In the event of exposure of spent nuclear fuel to groundwater in a final nuclear fuel repository, the preferential dissolution of grain boundaries rather than matrix is possible. The process driven by the corrosion of the grain boundaries with subsequent penetration by water - could cause a rapid increment of the fuel wet surface, and therefore increase the fraction of inventory becoming available for fast instant release. While this has a direct impact on the mobilization of radionuclides, the second effect of grain boundary corrosion is that it weakens the cohesion between UO_2 grains. This leads to degradation of the overall mechanical stability of the fuel pellet making it possible for water to reach each grain and through α-radiolysis considerably increase the amount of oxidants available for oxidative dissolution of the fuel grains.

Hence, the grain boundary diffusion of oxygen in polycrystalline UO_2 and spent fuel needs to be determined (Figure 9.1) together with the controlling mechanisms; lattice and grain boundary oxidation in combination with oxygen atoms reaction with ε-particles and the influence of hydrogen on these mechanisms.

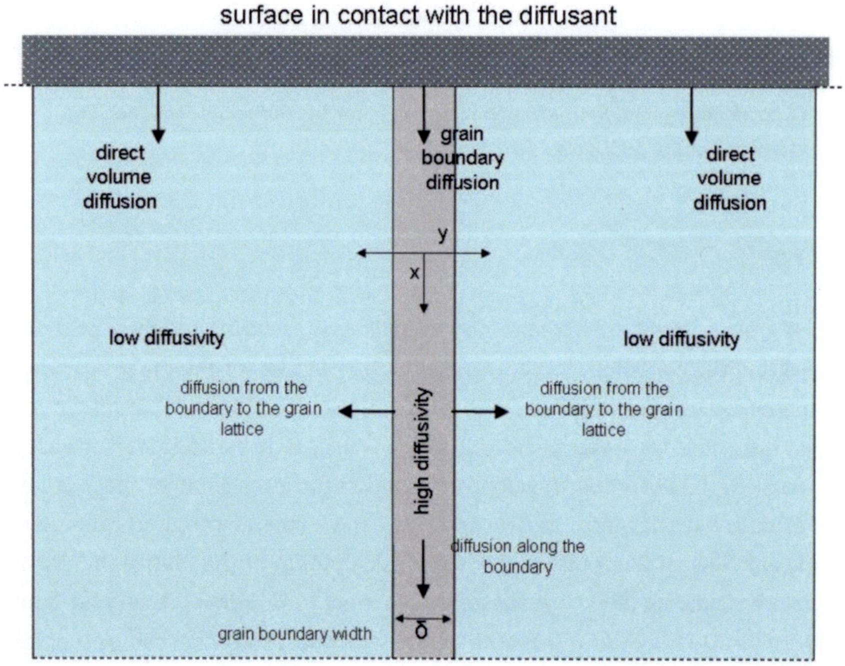

Figure 9.1: Model of lattice and grain boundary diffusion (Fisher, 1951).

9.3 Work progress

The literature review in open scientific publications related to experiments or predictions of oxygen lattice and grain boundary diffusion coefficients with or without coupling to the oxidation of the irradiated fuel has been reported (Carbol et al., 2013).

A comprehensive in-core fuel characterisation data of the selected fuel segment will be published in the FIRST-Nuclides WP1-report (Metz et al., 2013).

The preparation of SIMS tools, purchase of $H_2{}^{18}O$ tracer solution, ownership and preparation of spent nuclear fuel sample fuel as well as the foreseen autoclave construction has been reported (Carbol et al., 2013).

As reported in the last S+T report, the experimental part related to this task is delayed due to a broken hot cell autoclave. The FIRST-Nuclides consortium was informed about the delay and at the general assembly during the FIRST-Nuclides 2nd Annual Workshop it was decided that the delayed experiment will go on beyond the project time frame and therefore it will be made as a deliverable reported in a scientific journal.

In this report we summarise the ^{18}O diffusion studies in UO_2 matrix, performed parallel with the grain boundary experiment in spent nuclear fuel (SNF), and an important undertaking for the interpretation of the results awaited from the upcoming SNF diffusion experiment.

9.4 Selection, characterization and preparation of materials and set-up tools

Spent nuclear fuel sample

One standard UO_2 BWR fuel with an average burn-up of 42.2 GWd/t_{HM} will be used for this experiment. From a 160 mm long selected rod section a 50 mm long segment has been cut and is stored under oxygen free conditions. A 1 mm thick slice will be cut from this segment and de-cladded right before the start of the diffusion experiment. The experiment will be carried out using a fuel fragment from the de-cladded rod slice.

Physical and chemical characterization of the fuel sample

The fuel sample and the high burn-up structure will be characterized using SEM and image tool program. A detailed description of the characterization methodology and the results can be found in (Serrano et al., 2014).

Autoclave

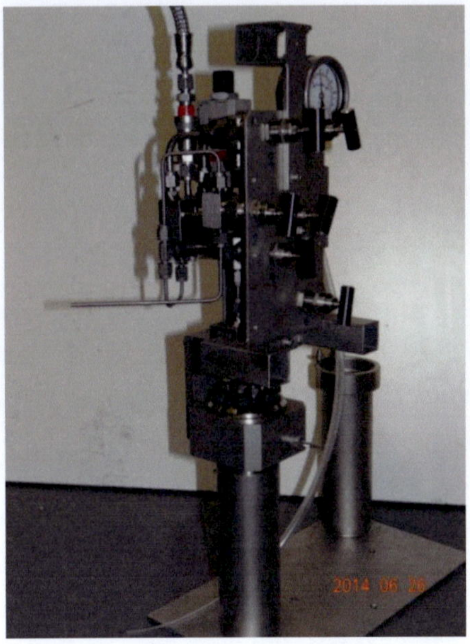

Figure 9.2: Manufactured autoclave undertaking 50 bar pressure test.

As discussed in the last S+T report a new autoclave needed to be designed, manufactured and tested. During the period July-August 2014 the new Ti-autoclave adapted for hot cell manipulations will undergo the necessary pressure and safety tests (Figure 9.2).

The inside of the Ti-autoclave is fitted with a PEEK (polyether ether ketone) insert (vessel, lid and dip-tube) to minimise the influence of metallic (foremost Fe) surfaces on the fuel corrosion process.

$H_2{}^{18}O$ solution

The ^{18}O-labelled water has been purchased and delivered to ITU.

SIMS analysis of SNF sample

After exposure of the fuel fragment to approximately 100 mL $H_2{}^{18}O$ in the autoclave at 40 bar H_2 the experiment will be terminated and the fuel fragment transferred to a SIMS holder. The sample in holder will be transferred a shielded SIMS 6F (Cameca, France) and the depth profiling made. A surface characterisation will be made of the SNF surface as outlined below.

9.5 Studies of ^{18}O diffusion in poly-crystalline UO_2

Parallel with the preparation work for the ^{18}O diffusion experiment with SNF, we have continued to study of oxygen diffusion in non-irradiated poly-crystalline UO_2. The outcome of these studies will help us to interpret the data from the ^{18}O diffusion experiments into spent fuel. Some outcome can be found in Marchetti et al. (2013) and Marchetti (2013), while the new results are related to recent measurements. SIMS $^{18}O/^{16}O$ depth profiling and ion mapping was employed to obtain new results. The ^{18}O tracer is used to study oxygen diffusion in the UO_2 lattice and grain boundaries. So far none of these diffusion studies has indicated any clear grain boundary diffusion.

Experimental

In general, two UO_2 disks exposed to oxidising (in air, sample 1B) and reducing (in presence of H_2, sample 2B) conditions, from previously performed ^{18}O diffusion experiments, were subject to an in-depth surface characterisation to understand the mechanism controlling/influencing the oxygen diffusion. The disks were in contact with a 10 mM NaCl + 2 mM HCO_3^- solution containing 98 wt.% $H_2{}^{18}O$ for 4 months at room temperature. The pH was around 8.2. After the termination of the experiment, the samples were dried without a preliminary distilled water ($H_2{}^{16}O$) washing in order not to change the experimentally established $^{18}O/^{16}O$ ratio. This procedure results in the fact that some sodium chloride will be present as dried salt on the UO_2 surface. More details concerning the experimental part can be found in Marchetti et al. (2013) and Marchetti (2013).

SIMS depth profiling of UO_2 corroded in air

One part of the SIMS ^{18}O depth profiling was made directly after the corrosion experiment while the new data were measured two years after the corrosion experiment (Figure 9.3). In general, the curve measured two years later (Figure 9.3 right, based on one sputtered crater) show a similar pattern as the one measured directly after the corrosion experiment (Figure 9.3 left, average of 6 craters). The difference between the two curves is mainly related to the diffusion process continuing between the two measurements. Observing the $^{18}O/^{16}O$ depth profile curve measured directly after terminating the corrosion experiment (Figure 9.3 left) one can conclude that the ^{18}O profile fits the Fick's 2nd law at a depth range >20 nm, while the superficial layer significantly deviates from Fick's 2nd law. From the shape of the ^{18}O diffusion curve one can deduce that the curve is composed of two independent curves, one with low and one with a higher diffusion coefficient. Comparison of a two paths diffusion model with the measured curve showed a good fit, but was not physically possible due to an unrealistic pre-exponential coefficient. For further discussions concerning modelling of the oxygen lattice diffusion, see Section 4.3.

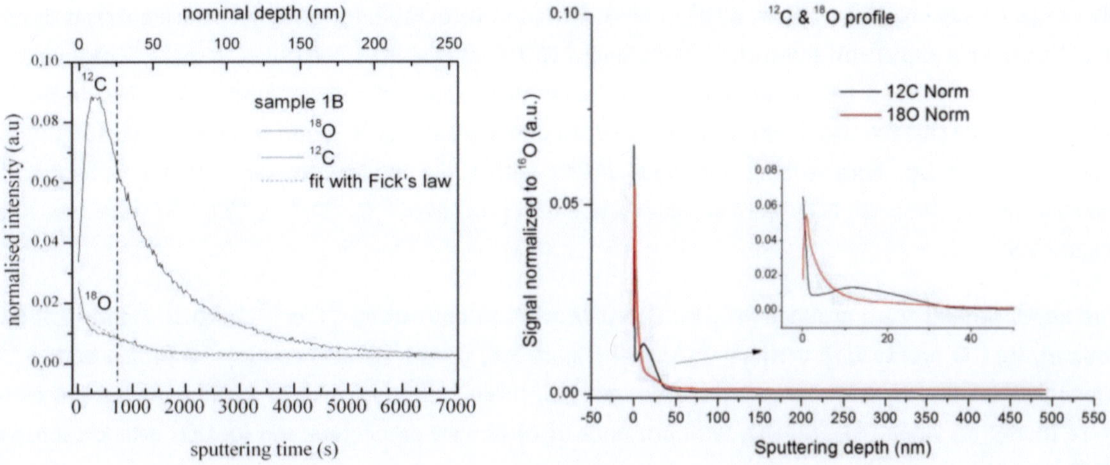

Figure 9.3: SIMS depth profiles of UO_2 sample 1B, measured 2010-03-20 (left) and 2014-03-04 (right), with ^{18}O and ^{12}C signals normalized to ^{16}O.

Concerning the ^{12}C signal the questions that need to be resolved is why the ^{18}O-signal reaches the background level within 50 nm while the ^{12}C-signal is present to a depth of 225 nm. A number of possible explanations are plausible: C contamination during preparation (polishing paste in pores present on the UO_2 surface), higher sputtering yield for ^{12}C in comparison to ^{18}O in combination with signals obtained from crater walls, or pores in the UO_2 surface where U carbonates have precipitated due to reached solubility limit. A further discussion is made in Section 0.

UO_2 surface characterisation by SIMS

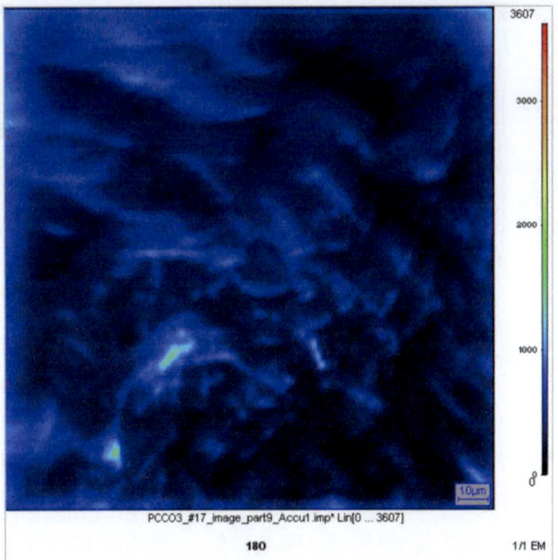

Figure 9.4: Ion filtered image at the nominal depth of 15 nm shows non-uniform distribution of ^{18}O.

To understand why the measured ^{18}O diffusion curves deviate from Fick's 2nd law the sample 1B (oxidising conditions) was re-analysed by SIMS. The difference was that this time the SIMS sputtering rate was orders of magnitude lower (3-10 pA instead of 2 nA) compared to the one used during the initial short-range measurements (Marchetti, 2013). The characterisation of the UO_2 surface by SIMS was made by cyclic imaging: this way one obtains a series of images (maximal 2,000 images) showing the spatial distribution of the studied masses (^{12}C, ^{16}O, ^{18}O, ^{35}Cl, ^{238}U). It is interesting to observe the evolution of the distribution of ^{18}O on the surface: the tracer is uniformly distributed on the very first images but soon after, areas with higher concentration are appearing as shown in Figure 9.4 at a nominal depth of 15 nm. Further below the overall concentration of ^{18}O is decreasing leaving a more uniform distribution.

There can be several explanations: a UO_2 grain orientation favourable for oxidation, a pore that is deeper than 15 nm or a particle on the surface containing ^{18}O. The only way to find out is to make SEM images of the same area or scan it with AFM. Both methodologies were applied and are further described in Section 4.4. The surface characterisation by SIMS was also used to explain the ^{12}C depth curve showed in Figure 9.3. This was made by characterising sub-areas (ROIs) within the shallow crater showing deviating ^{18}O concentrations. Ten small ROIs were selected and depth profiles of ^{12}C, ^{18}O, ^{16}O, ^{35}Cl, ^{238}U were measured (Figure 9.5).

The depth profiles were grouped as: (a) ^{18}O profile without any strong ^{12}C or ^{35}Cl signal (Figure 9.5, blue square), (b) ^{18}O profile with a strong ^{12}C signal (Figure 9.5, purple square) and (c) ^{18}O profile with a ^{35}Cl signal present (Figure 9.5, orange square). In general, three sub-areas with in a 150 µm x 150 µm crater were found: (a) virgin UO_2, (b) UO_2 with presence of carbonate-precipitate and (c) UO_2 with presence of NaCl crystals.

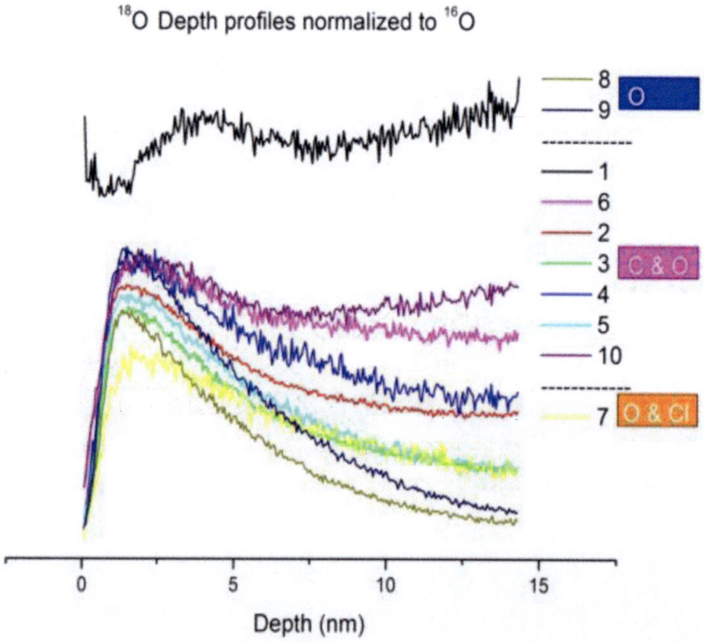

Figure 9.5: A posteriori depth profiles of selected local ROIs show significantly different diffusivity of ^{18}O. The ROIs where defined looking at the uneven distribution of ^{18}O at a nominal depth of 15 nm.

There is a correlation between the non-covered surface and surfaces covered with carbonate or sodium chloride and the ^{18}O profile. The non-covered UO_2 surface show a distinct ^{18}O peak that at shallow depth (<15 µm) approaches the background level while under carbonate covered surface the ^{18}O peaks are very broad and in some cases has a higher ^{18}O content at the bottom of the sub-area crater than at the outer disc surface. One can conclude that UO_2 surfaces that are highly oxidised have formed a surface complex with carbonate. Additionally, it is well known that the oxygen diffusion coefficient in oxidised uranium phases (UO_{2+x}) is by order of magnitudes higher than in $UO_{2.0}$, making it logical that the ^{18}O is found at a deeper disc location in the oxidised layer.

It can therefore be concluded that:

Characterisation of the outer (0-15 nm) corroded UO_2 surfaces gives additional information related to the surface reactions taking place at the interface between the solution and the UO_2 surface.

UO_2 surfaces subject to oxidation, mainly due to crystallographic orientation, show deeper ^{18}O penetration (higher D_L) and if oxidised to UO_{2+x} with x>0.25 (U_4O_9) form a complex with carbonate. Non-oxidised or slightly oxidised (UO_{2+x} with x<0.25) show no complex with carbonate and low ^{18}O penetration depth.

Based on a small number investigated sub-area craters the proportion between oxidised surfaces (UO_{2+x}, with x>0.25) and slightly or non-oxidised UO_2 surfaces is roughly ten to one.

SEM imaging of a shallow SIMS crater

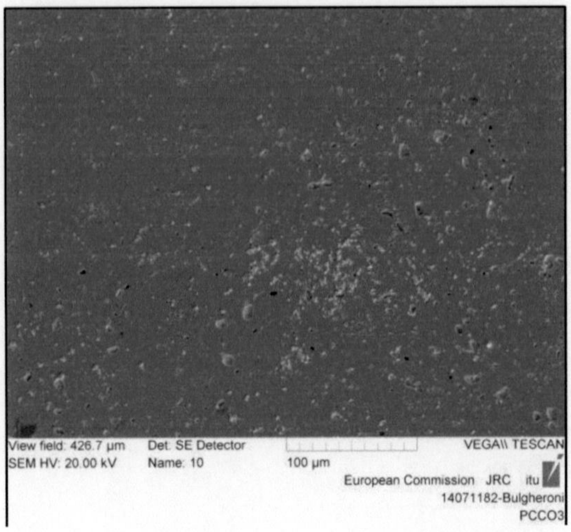

Figure 9.6: SEM image of a sample region containing a shallow crater.

To better understand the reason why the oxygen tracer is not uniformly distributed over the depth range between 1 to 50 nm, other imaging techniques (SEM and AFM) were used to analyse a shallow crater with an estimated depth of 15 nm. This depth is of the same order of magnitude as the root mean square roughness of the sample surface, thus making it difficult to visualise. Craters with a depth of hundreds of nm can easily be found with Scanning Electron Microscope, but craters with a depth of a few tens of nanometres are a challenge also for high-resolution SEM (Vega TESCAN, Czech Republic). Therefore, two reference points were used to align the frame of reference used by the SIMS with the reference frame of the SEM, so that high-resolution SEM images could be acquired of the region where the shallow crater is expected to be located (Figure 9.6).

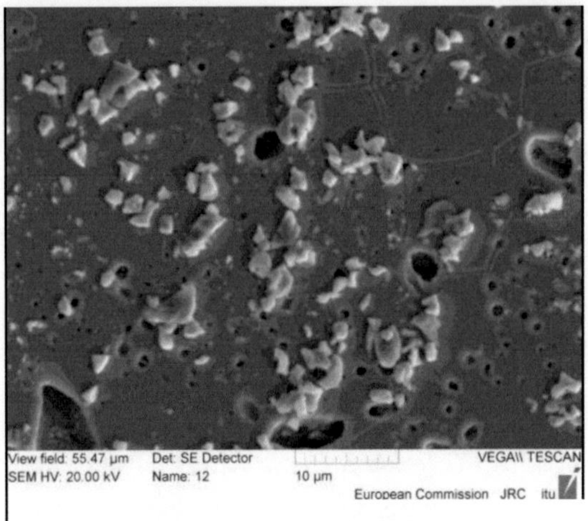

Figure 9.7: Magnification of the central area of the possible shallow crater bottom.

An increased level of roughness is visible in an area of 150 µm x 150 µm in this region, thus making it probable that this is the shallow crater. At a higher magnification of the central region of this area (Figure 9.7) revealed the presence of several deep pores, many crystal fragments and etched grain boundaries. A SEM-EDX spectrum of the crystal fragments show that they consist of UO_2 while spectra of the content in the pores show a higher fraction of carbon than the surrounding areas and would explain the deeper ^{12}C than ^{18}O profile, measured by SIMS depth profiling (Figure 9.3).

Anyhow, if the hypothesis that this area represents the shallow crater, then due to the significant roughness an independent estimate of the crater depth must be performed. A more precise measurement of the shallow crater effective depth will be made using an Atomic Force Microscope (AFM).

Modelling

The modelling of the ^{18}O diffusion was performed using the finite element method (FEM) using FlexPDE v6.35 as modelling tool. As discussed in Section 0 the ^{18}O depth curves in UO_2 (corroded in air) cannot be explained by simple Fick's diffusion. A number of approaches were modelled and compared with the measured curve. A logical approach was to assume an oxidation of the UO_2 from the surface and inwards during the corrosion experiment and, hence a sequence of UO_{2+x} phases (with increasing oxidation, x=0 to 1) starting from the UO_2 bulk and progressing outwards to the surface: $UO_{2.0}$, UO_{2+x}, U_4O_9, U_3O_7, U_3O_8 and UO_3. The corresponding O diffusion coefficient for each UO_{2+x} phase (insert in Figure 9.8) was taken from the scientific literature (Marchetti, 2013). Systematic changes of the thicknesses of each UO_{2+x} phase with its corresponding O diffusivity coefficient (Figure 9.8, green curve) and solving the partial differential equation with the variable diffusivities, using the FEM modelling and summing up the concentration at each UO_2 depth, a theoretical oxygen diffusion curve could be calculated (Figure 9.8, red curve). A comparison between the theoretical and experimental curve (Figure 9.8, black line) show significant difference.

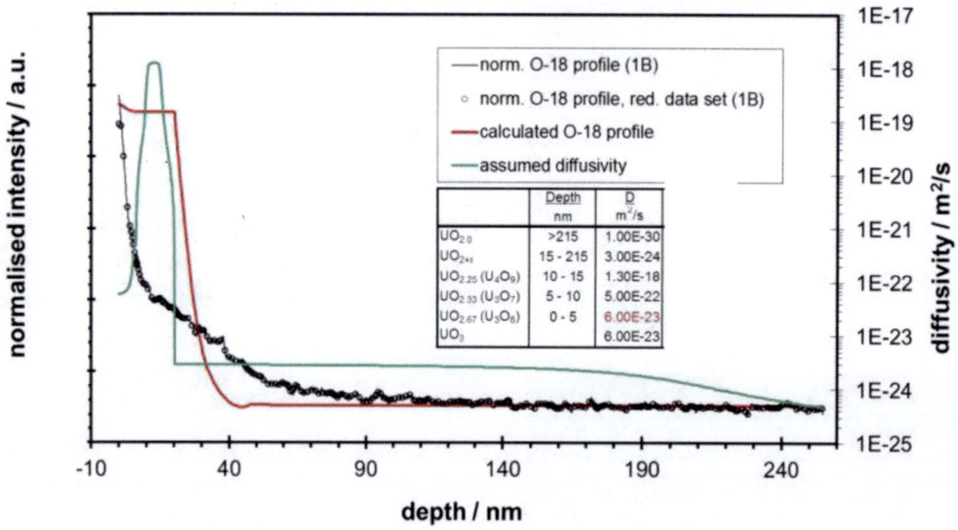

Figure 9.8: Modelling using variable diffusivity from published oxygen diffusivities of different uranium oxides.

The difference can be attributed to either error in the experiment and/or in modelling. The errors can originate from: a too short corrosion time, of four months (there is no time to develop the O profile in the UO_{2+x} layers, experimental error), that the superficial UO_3 layer was dissolved (experimental error), that the UO_2 walls inside pores on each individual grain has reacted with ^{18}O (related to the nature of fabricated UO_2), significant UO_2 surface roughness in comparison to the sputtering depth (experimental polishing error), UO_2 grain orientation (inherent UO_2 property that can affect both the diffusion coefficient and the SIMS sputtering rate), or that the diffusion coefficients in the literature are incorrect (model error).

Anyhow, the possible errors were checked and none could solely explain the large deviation between the modelled and measured curve. It was decided to revisit the original interpretation that the measured diffusion curve was composed of two curves, but this time we assumed that the processes were ongoing as two parallel processes.

Adapting this model by defining a UO_2 volume with a depth of 300 µm and with a fixed surface area, optimising the proportion between the areas subject to the two processes: the two volumes with different diffusivity: $D_L(U_3O_8)=1.15 \cdot 10^{-22}$ m^2/s (fast), $D_L(U_4O_9)=6 \cdot 10^{-25}$ m^2/s (slow) and the bottom volume consisting of UO_2 (bulk) with a $D_L(UO_{2.0})=1 \cdot 10^{-30}$ m^2/s. The oxygen diffusivity at the outermost surface of the volume was set to zero (maximal access to oxygen). The FEM program was run with the parallel diffusion model. The output of the FEM model, simulating a diffusion time of 4 months, gives concentrations at the different locations of the volume. Through integration of the concentrations in the sub-volumes at different depth a normalised concentration curve is obtained (Figure 9.9). A comparison of the modelled (Figure 9.9, blue line) and the measured curve (Figure 9.9, purple line) show a reasonable agreement.

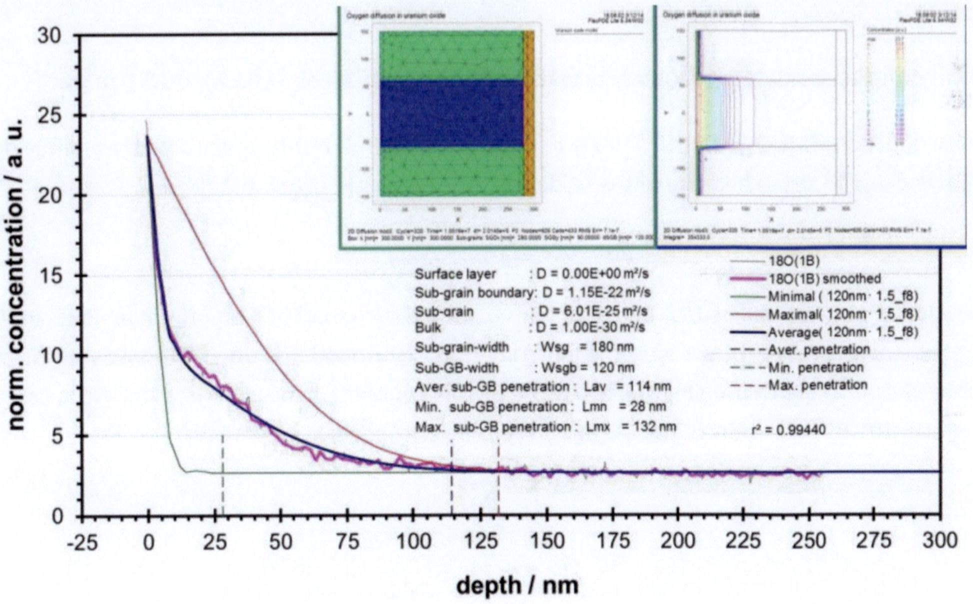

Figure 9.9: Two parallel ^{18}O diffusion processes in UO_2 where identified.

The fact that the model describes the measured results well and the initial intuitive interpretation of the composite diffusion curve gives an increased confidence of a correct interpretation. In Figure 9.9 the two curves, orange and green, represents two processes. It would be logical to assign the green curve to pure oxygen diffusion in the UO_2 lattice and the orange to interfering processes such as; pores, original surface roughness, orientated UO_2 grain in combination with increased grain oxidation and increased SIMS sputtering rate and grain boundaries diffusion.

The contribution of pores was ruled out since the pores are generally deeper than the sputtered depth and would therefore contribute to a raised $^{18}O/^{16}O$ background level, which we do not observe. Original surface roughness due to polishing was tested earlier (Marchetti, 2013) and was shown to have some contribution but would not explain the relatively large penetration depth of 132 nm.

Finally, the contribution from oriented grains needs to be discussed. Different scenarios are possible, but let us simplify and discuss two extremes cases of grain orientations present inside a crater. In the (100) UO_2 lattice orientation the oxidation is relatively fast (Rabone, 2014) and so is the oxygen diffusion. Although no literature data exists for the SIMS sputtering rate as a function of degree of UO_2 oxidation (UO_{2+x}) it can be assumed that a higher sputtering rate (and possible transmission yield) is expected for UO_{2+x} compared to UO_2. Taking in these parameters the (100) orientation will contribute to a deeper crater. The other extreme is the (111) orientation that makes the UO_2 grain difficult to get oxidised and sputter. This will contribute to a shallow crater. If the sputtered surface area is large enough all orientations should even out the two extremes.

Grain boundary diffusion will occur parallel with lattice diffusion, but if it is a slow diffusion, in relation to the 4 months diffusion experiment, it will be difficult to distinguish it from lattice diffusion.

The modelling results can be summarised as that:

The measured ^{18}O diffusion curves can be described by two parallel processes, one slow and one fast diffusion.

The slower diffusion is related to UO_2 lattice with a diffusion coefficient, $D_L(UO_2)=6\cdot10^{-25}$ m^2/s.

The faster diffusion ($D_L(UO_2)=1.2\cdot10^{-22}$ m^2/s) is most probably related to grain boundary diffusion and/or cracks formed due to stress because of the oxidation process (Desgranges, 2010).

SIMS depth profiling of disc corroded in presence of H_2

The corrosion of poly-crystalline UO_2 under 10 bar H_2, for a time period of 4 months in an PEEK vessel inside a Ti-autoclave, was studied. Direct after the experiment the corroded and dried UO_2 disc was introduced in the SIMS and the ^{18}O depth profiling started. More details concerning the experimental setup can be found in (Marchetti, 2013). The results of the ^{18}O and ^{12}C profile are shown in Figure 9.9.

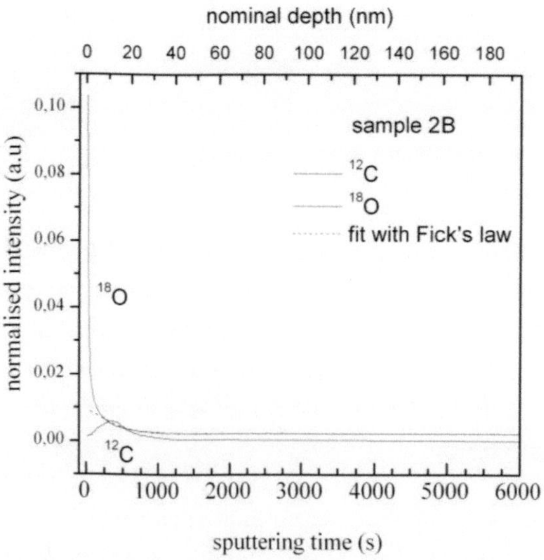

Figure 9.10: Depth profile for ^{18}O and ^{12}C (normalized to ^{16}O) in reducing condition (Marchetti, 2013).

In comparison to UO_2 corrosion in air (Figure 9.3) one directly observes the differences: such as the shallow penetration of ^{18}O into the matrix and the relatively small ^{12}C peak, both phenomena indicating a minimal or non-oxidised UO_2 surface.

The $^{18}O/^{16}O$ depth profile shows under reducing conditions a similar pattern of two parallel processes. The major part of the profile fits to the slow process while the faster one is much less pronounced.

9.6 Conclusions

Diffusion of oxygen and water penetration into UO_2 under both oxidising and reducing conditions has been studied. The two extremes cover the range expected in studies of oxygen diffusion in spent UO_2 fuel where the oxygen will be formed through α and β-radiolysis.

In oxidising conditions and in presence of carbonate, the main part the UO_2 surface is oxidised and reacts with carbonate which then forms a carbonate layer on the surface. Modelling of the resulting ^{18}O penetration curve, under oxidative conditions, can only be explained by two parallel ^{18}O diffusion processes. The first process was assigned to the lattice diffusion of oxygen in UO_{2+x}, with $x<0.25$ (limiting phase is U_4O_9). The second process could not be singled out from the various possibilities but indicate a significant grain boundary diffusion contribution. The modelling results are supported by alternated SIMS depth profiling and ion mapping showing local crater areas with significantly varying ^{18}O concentrations.

Oxygen diffusion studies under reducing condition show that the measured ^{18}O depth profile curve follows Fick's 2nd law (lattice diffusion).

This fundamental work and the obtained results performed within the FIRST-Nuclides project are crucial for the future interpretation of the planned but delayed experiments of grain boundary diffusion in spent nuclear UO_2 fuel.

9.7 Acknowledgement

The authors would like to especially acknowledge the hot cell colleague Edgar Ferreira-Teixeira and in the design office and workshop: Volkmar Ernst, Joachim Küst and Friedrich Blattmann.

The research leading to these results has received funding from the European Union's European Atomic Energy Community's (Euratom) Seventh Framework Programme FP7/2007-2011 under grant agreement n° 295722 (FIRST-Nuclides project).

9.8 References

Desgranges, L., Palancher, H., Gamaléri, M., Micha, J.S., Optasanu, V., Raceanu, L., Montesin, T., Creton, N. (2010). Influence of the U_3O_7 Domain Structure on Cracking during the Oxidation of UO_2. Journal of Nuclear Materials, 402 167-172.

Fisher, J. C. (1951). Calculations of Diffusion Penetration Curves for Surface and Grain Boundary Diffusion. Journal of Applied Physics, 22(1), 74–77.

Marchetti, I., Carbol, P., Himbert, J., Belloni, F., Fanghänel, T. (2013). Room-Temperature Diffusion Coefficients for Oxygen and Water in UO_2 Matrices: a SIMS Study. Surface and Interface Analysis, 45, 360-363.

Marchetti, I. (2013). Characterisation of Water Penetration into Polycrystalline UO_2. PhD- thesis defended at Ruprecht-Karls-Universität, Heidelberg, Germany, 19 April, 2013.

Marchetti, I., Belloni, F., Himbert, J., Carbol, P., Fanghänel, Th. (2011). Novel Insights in the Study of Water Penetration into Polycrystalline UO_2 by Secondary Ion Mass Spectrometry. Journal of Nuclear Materials, 408, 54-60.

Marin, J.F., Contamin, P. (1969). Uranium and Oxygen Self-Diffusion in UO_2. Journal of Nuclear Materials, 30, 16-25.

Matzke, H. (1996). Analysis of the Structure of Layers on UO_2 Leached in H_2O. Journal of Nuclear Materials, 238(1), 58-63.

McEachern, R.J., Taylor, P. (1998). A Review of the Oxidation of Uranium Dioxide at Temperatures below 400°C. Journal of Nuclear Materials, 254(2-3), 87-121.

McEachern, R.J. (1997). A Review of Kinetic Data on the Rate of U_3O_7 formation on UO_2. Journal of Nuclear Materials, 245(2-3), 238-247.

Metz, V., González-Robles, E., Müller, N., Bohnert, E., Herm, M., Lagos, M., Kienzler, B., Serrano-Purroy, D., Colle, J.Y., Beneš, O., Naisse, F., Wiss, T., Konings, R.J.M., Wegen, D.H., Papaioannou, D., Gretter, R., Nasyrow, R., Rondinella, V.V., Glatz, J.-P., Curtius, H., Müskes, H.W., Like, N., Bosbach, D., Günther-Leopold, I., Curti, E., Froideval Zumbiehl, A., Linder, H.P., Govers, K., Verwerft, M., Van Renterghem, W., Lemmens, K., Mennecart, T., Cachoir, C., Adriaensen, L., Dobney, A., Gysemans, M., Vandenborre, J., Traboulsi, A., Blain, G., Barbet, J., Fattahi, M., Sureda Pastor, R., Slonszki, E., Hózer, Z., Roth, O. (2013). Characterisation of spent nuclear fuel samples and description of methodologies and tools to be applied in FIRST-Nuclides. Deliverable D1.2 of the 7th EC FP CP FIRST-Nuclides project.

Rabone, J. (2014). DFT+U Calculation on UO_2. Private communication.

Sabionia, A.C.S., Ferrazb, W.B., Millotc, F. (2000). Effect of Grain-Boundaries on Uranium and Oxygen Diffusion in Polycrystalline UO_2. Journal of Nuclear Materials, 278(2), 364-369.

Serrano-Purroy, D., Clarens, F., González-Robles, E., Glatz, J.P., Wegen, D.H., de Pablo, J., Casas, I., Giménez, J., Martínez-Esparza, A. (2012). Instant Release Fraction and Matrix Release of High Burn-up UO_2 Spent Nuclear Fuel: Effect of High Burn-up Structure and Leaching Solution Composition. Journal of Nuclear Materials, 427, 249-258.

Tempest, P.A, Tucker, P.M., Tyler, J.W. (1988). Oxidation of UO_2 Fuel Pellets in Air at 503 and 543 K Studied Using X-ray Photoelectron Spectroscopy and X-ray Diffraction. Journal of Nuclear Materials, 151(3), 269-274.

10 Instant radionuclide release fraction from spent UO_2 triso coated particles and microstructure evolution

Hilde Curtius, Norman Lieck, Gabriele Kaiser, Murat Güngör, Martina Klinkenberg and Dirk Bosbach*

Institute of Energy and Climate Research, IEK-6: Nuclear Waste Management and Reactor Safety, Forschungszentrum Jülich (DE)
** Corresponding author: h.curtius@fz-juelich.de*

10.1 Abstract

The impact of burn up on the instant release fraction from spent fuel was studied using very high burn up UO_2 fuel (~100 GWd/t) from a prototype high temperature reactor (HTR). TRISO (TRi-structural-ISO-tropic) particles from the spherical fuel pebbles contain UO_2 fuel kernels (500 µm diameter) which are coated by four tight layers ensuring the encapsulation of fission products during reactor operation. Despite distinct differences between HTR and LWR spent fuel, these studies provide some unique information with respect to extremely high burn up and contribute to the European project FIRS-Nuclides.

After cracking of the tight coatings ^{85}Kr and ^{14}C as $^{14}CO_2$ were detected in the gas fraction. Xe was not detected in the gas fraction, although ESEM (Environmental Scanning Electron Micoscope) investigations revealed an accumulation in the buffer and inner pyrolytic coating.

For 276 days UO_2 fuel kernels were exposed to synthetic groundwater (19 mM NaCl and 1mM $NaHCO_3$) under oxic (air) and anoxic/reducing (Ar/H_2) conditions. Under both environmental conditions U concentration in the leachate was below the detection limit, indicating extremely low matrix dissolution. Also, the concentrations of the actinides (Pu, Am, Np, Cm) in solution were below their detection limits confirming that these radionuclides are not part of the IRF. The contribution of ^{90}Sr to the IRF was low (max. 0.2%) and comparable to LWR spent fuel. While Sr is highly soluble in the UO_2 matrix, this low value confirmed the extremely low matrix dissolution.

^{134}Cs and ^{137}Cs are not soluble in the UO_2 matrix and are excluded to grain boundaries during irradiation. In Ar/H_2 environment the maxima contributions to the IRF were reached within the first 5 days (up to 8% for ^{134}Cs and up to 8% for ^{137}Cs) and no further significant release was observed. In comparison, the labile Cs inventory (gap + cracks + grain boundaries) for LWR fuel (60 GWd/t) was determined to be about 1.4%. Depending on the environmental conditions, different release mechanisms were observed for Cs and for Sr. In air second relevant release steps occurred after ~ 120 days, while under Ar/H_2 the release was completed.

ESEM investigations were performed to study the impact of leaching on the microstructure. Before leaching the surface of the fuel matrix possessed a rough topography, few small pores (up to 2 µm) are visible and metallic precipitates appeared as hexagonal platelets. Under both environmental conditions leaching reduced the roughness of the surface of the fuel matrix and of the grain boundaries. Numerous small open pores acting as new accessible leaching sites were formed in oxic environment while under anoxic/reducing conditions these pores are not open and appear to be filled with metallic precipitates. In conclusion, leaching in different geochemical environments influenced the evolution of the microstructure and this directly had an impact on the IRF of radionuclides.

10.2 Introduction

Safety assessment of spent nuclear fuel disposal in a deep geological formation requires information about the release of radionuclides from the fuel after groundwater breaches the container and contacts the fuel. The fraction of the total inventory of a given radionuclide in the spent fuel available for instant release (instant release fraction, IRF) is used as source term in safety assessment models and contribute to the long-term safety analysis. Within the European project FIRST-Nuclides the impact of burn-up on the instant radionuclide release fraction from spent UO_2 fuel was investigated. Jülich concentrates on UO_2 based fuel elements developed for the prototype VHTR reactor. Compared to UO_2 LWR fuel, the VHTR fuel is very different in design and in applied irradiation conditions. A fuel pebble consist of up to 10,000 small UO_2 fuel kernels with diameters of about 500 µm, embedded in a moulded graphite sphere with a diameter of about 60 mm (Figure 10.1). The outer shell (5 mm in depth) of such a fuel pebble represents a fuel free zone. Each TRISO coated particle (Figure 10.1) possesses an UO_2 kernel which is coated with four layers (porous carbon buffer, inner dense pyrocarbon layer (IPyC), silicon carbide (SiC) and outer dense pyrocarbon layer (oPyC), represents a miniature fuel element of about 1 mm in diameter.

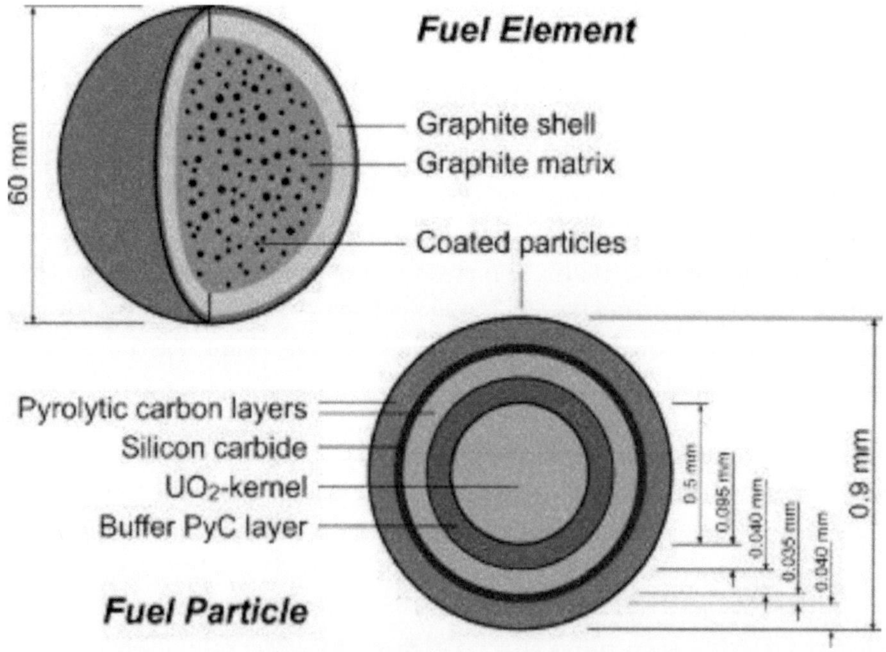

Figure 10.1: Design of a HTR fuel element and of a TRISO coated fuel particle.

In this paper the irradiation conditions, the radionuclide inventory, the radionuclides detected within the gas fraction and results of a polished specimen are summarized, while the contributions to IRF of radionuclides present as water soluble salts, as metallic inclusions and the effects of leaching on the microstructure are presented in more detail.

10.3 Irradiation data, radionuclide inventory, polished specimen and gas fraction

Five fuel pebbles from the German production line AVR GLE-4/2 were irradiated for 249 full effective power days at the High Flux Reactor in Petten (Fütterer et al., 2008). The burn-up of the pebble HFR-EU1bis/2 was determined to be 10.2% FIMA (Fissions per initial heavy metal atom). From this pebble Jülich obtained irradiated UO_2 TRISO coated particles. In Table 10.1 the irradiation characteristics (Fütterer et al., 2008) of the pebble HFR-EU1bis/2 are summarized.

Table 10.1: Irradiation data for the pebble HFR-EU1bis/2.

Enrichment	16.76 wt% ^{235}U
Irradiation at:	High Flux Reactor Petten
Reactor cycles:	10
Irradiation:	249.55 (efpds)
Thermal fluences:	2.23 x 10^{25} m^{-2}
Fast fluences:	3.98 x 10^{25} m^{-2}
Central temperature of pebbles:	1,250 °C
Power density:	30 W/cm^3
FIMA:	10.2%, (95.57 GWd/t)

The radionuclide inventory was calculated with the OCTOPUS code after a cool-down period of 1,749 days after end of irradiation on 18th October 2005. The main radioisotopes and their activities for a coated particle (CP) are summarized in Curtius et al. (2013). In order to compare these calculated values to measured values a gamma spectrometric measurement was performed. The determined activities of the elements Cs, Eu, Ce, Ru, Sb, Rh, Pr and Am of a coated particle agree with the calculated values.

A selective cracking, separation and dissolution/leaching step was performed to distinguish clearly between the elements in the fuel kernel and in the coatings. According to the obtained results, the elements Am, Pu, Cm, U, Eu, Ce, Sr, Tc, and Pr are located quantitatively within the kernel. However up to 95% of the activity of Cs was found in the coatings (Curtius et al., 2013). Under the irradiation conditions of the experiment HFR-EU1bis one can consider carbon oxidation leading to a strong decrease of fuel-kernel oxygen potential. Under this low oxygen potential (below -450 kJ/mol) ternary Cs compounds are not stable and Cs-atom diffusion by U vacancies and subsequent release is considered. This low oxygen potential is the key point in Cs behaviour in HTR TRISO particles, very different from UO_2 LWR fuel (Barrachin et al., 2011).

ESEM/EDX technique was used to study the microstructure of the fuel samples. A polished specimen revealed big pores, large grains and metallic precipitates. The elements Cs, O, U, Mo, Xe, Zr and Si were identified within the fuel kernel. However, the highest amounts of Cs were detected in the surrounding layers (buffer/iPyC) and confirm the results obtained by gamma spectrometry. Also, the highest amounts of Xe were found within the buffer and iPyC layer (Curtius et al., 2013). Under the irradiation conditions, Xe remained in atomic form and diffused to grain boundaries. Calculations performed with the MFRR code (Barrachin et al., 2011) assume that the Xe fraction released is in the range of 70%, indicating that the fission gas release in HTR fuel is much higher than that of irradiated UO_2 LWR fuels. It is thought that the release mechanism for Xe and Cs are very similar. At the periphery of the fuel kernel big grains and some small pores are visible (Curtius et al., 2014). The metallic phases appear as hexagonal platelets and the elements Mo, Zr and Tc were identified clearly. As expected for this fuel type under the performed irradiation conditions the low oxygen potential determines Mo as main contributor in the metallic precipitates. For UO_2 LWR spent fuel (around 10% FIMA) the oxygen potential is significant higher (-310 kJ/mol) and Mo is present as caesium molybdate and other parts are present in the matrix and in the metallic precipitates.

A crack device coupled with a gas sampling tool was developed and the gas components were measured by gas/radiogas chromatography. 35% of the ^{85}Kr inventory was released instantaneously. ^{14}C present as $^{14}CO_2$ was detected up to 15 Bq/fuel kernel. ^{14}C mainly is produced through fast neutron reactions with ^{14}N (present as impurity) and to a less amount by fission (calculated activity for a fuel kernel: 0.05 Bq) (Curtius et al., 2014).

10.4 Instant release fraction

Experimental procedure

Ten TRISO coated particles were cracked and the obtained UO_2 kernels were separated from the coatings. A low molar salt solution containing 19 mM NaCl and 1 mM $NaHCO_3$ (pH value of this solution was 7.4 ± 0.1) was used for the leaching procedure at room temperature. Five UO_2 kernels and their corresponding coatings were leached separately in 20 mL solution under oxic (air) conditions. Leaching also was performed under anoxic/reducing (Ar/H_2: 96% of Ar (type: 4.8 (4.8 indicate the pureness to 99.998%) and 4% of H_2 (type 3.0 (3.0 indicates the pureness to 99.9%)) conditions using equal amounts of UO_2 kernels and coatings respectively. At different time intervals 1.5 mL aliquots of the solutions were taken. In order to keep the solid/solution ratios constant 1.5 mL of the prepared salt solution was added to each leaching flask.

The aliquots were filtered (450 nm) and the pH values were determined to be in the pH range of 7.3 ± 0.1 to 7.5 ± 0.1. The stability of these pH values indicate a stable environment in which no dissolution process (consuming or releasing protons) occur so far. One milliliter of each filtrate was diluted with 9 mL of a 0.1M HNO_3 solution resulting in a 10 mL sample solution. Then 0.1 mL from each sample solution was vaporized directly on metal disks and an alpha-measurement (alpha-spectrometer: Octete, company Ortec, PIPS-detector (planar implanted passivated silicon)) was performed for 69.4 hours to analyze the activities of the radionuclides U, Pu, Am, Np and Cm.

The activity of ^{90}Sr was determined after the following selective separation steps; 1 mL of each sample solution was diluted with 0.67 mL of 14 M HNO_3 solution. Then 1 mL of this solution was used to quantify the ^{90}Sr activity. A column (6 mL in volume) was filled with a suspension of 1 g resin (Sr-Resin, Eichrom-Company) in 5 mL of a 2 M HNO_3 solution. The colum was washed two times with 5 mL of a 2 M HNO_3 solution and then 5 mL of a 8 M HNO_3 solution was added. Afterwards the sample solution was added to the column. The sample vial was rinsed with 1 mL of a 8 M HNO_3 solution and this solution was added to the colum as well. Then a washing step with 10 mL of a 8 M HNO_3solution was performed. The washing solution was collected. After the washing steps ^{90}Sr was eluated by using 10 mL of a 0.05M HNO_3 solution. Immediatly 1 mL of the eluate was used for the β-measurement (Liquid Scintillation Counter: Packard Canberra, TRICARB 2200 A). The measurement time was fixed to one hour and the detection limit for ^{90}Sr is 0.1 mBq/sample.

Then 5 mL from each sample solution was filled in a polyethylene vial and a γ-measurement was performed (HPGe detector type PGC 2,018, Bias: 2,500 V positiv, counting time: 24 hours, software: Gamma-W Version: 2.44 using ^{152}Eu as standard solution with the same geometry, distance 0.15 m). The uncertanties of the measured activities are in the range between 2% to 5% .

For all elements the measured values were calculated according to the following equations:

The Fraction of Inventory of an element *i* released in the Aqueous Phase (FIAP) is given by Equation 10.1:

$$FIAP_i = m_{i,aq}/m_{i,SNF} \tag{10.1}$$

where $m_{i,aq}$ is the mass of the element *i* in the aqueous phase (g) and $m_{i,SNF}$ is the mass of the element in the SNF sample (g). The Instant Release Fraction of an element *i* (IRF_i) is the Fraction of Inventory of the element *i* in the Aqueous Phase ($FIAP_i$) reduced by FIAP of uranium. For the reported time period no uranium was detected in solution, hence the $FIAP_i$ of the elements represent the IRF_i.

The Fractional Release Rate for an element *i* (FRR_i) in d^{-1} is given by the Equation (2):

$$FRR_i = \Delta FIAP_i/\Delta t \tag{10.2}$$

where Δt is the time range in days and $\Delta FIAP_i$ refers to the difference of the $FIAP_i$ values within this time range. Generally spoken, the FRR_i describes the elemental FIAP evolution with time.

Experimental results

Independent of the redox conditions, U concentrations in solution were below the detection limit ($4.3 \cdot 10^{-8}$ mol/L) indicating an extremely low matrix dissolution rate in the time frame of investigation (276 days). Also, Pu, Am, Np and Cm concentrations were below their detection limits in solution confirming that these elements do not contribute to the IRF.

For both environmental conditions the contributions (Figure 10.2) of ^{90}Sr to the IRF were low (max. 0.2%) and comparable to UO_2 LWR fuel (Roudil et al., 2007). ^{90}Sr is higly soluble within the UO_2 matrix and low release values are expected as long as matrix dissolution is low. Taking all FIAP data points into consideration three different regions are visible (Figure 10.2) under oxic and under anoxic/reducing conditions in the time frame of 276 days. The first region (0 to 5 days) correspond to the highest release rates at the beginning of the experiment, indicating the leaching of the grain boundaries which have an open access to the groundwater and including the contribution of fines leaching (fines located at the periphery of the fuel kernels). Under oxic conditions, the first FIAP value reached approximately 0.018% within this time period (5 days) while under anoxic/reducing conditions a FIAP value of 0.13% was determined. The second regions (5 to ~130 days under oxic conditions, 5 to 165 days under Ar/H_2 conditions) reveal lower release rates. It is assumed that the water needs to diffuse towards grain boundaries which have no direct open access. The third regions (oxic conditions: 130 - 276 days; Ar/H_2 conditions: 165 to 276 days) show a strong increase in release rates only under oxic conditions, while under Ar/H_2 the release was terminated.

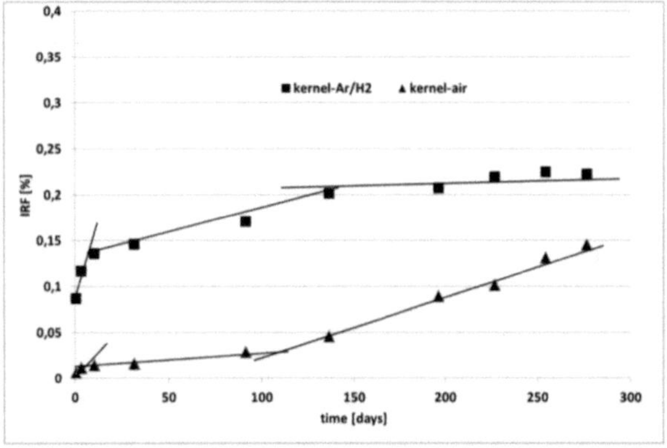

Figure 10.3: FIAP ^{90}Sr (%) vs. time under oxic and under anoxic/reducing conditions.

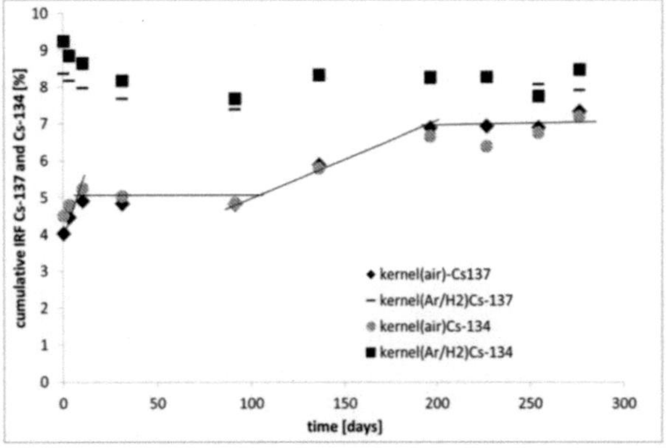

Figure 10.2: FIAP values for ^{134}Cs (%) and ^{137}Cs (%) under oxic and anoxic/reducing conditions.

Figure 10.3 reveal the calculated FIAP values representing the IRF for ^{134}Cs and ^{137}Cs vs. time under oxic and under anoxic/reducing conditions. In the time frame of investigations (276 days) the radiosisotopes ^{134}Cs and ^{137}Cs reacted similar (comparable curve progression). Within the first five days of experiment under air, $^{134/137}Cs$ contributed to the IRF up to 5%, indicating leaching of fines and of grain boundaries, which have an open access to the groundwater. No significant further contribution was observed in the next 90 days of leaching. Then a second relevant release step occurred for these second release steps FIAP contributions of around 2% for $^{134/137}Cs$ were obtained. After ~ 150 days the release was completed. Under anoxic/reducing conditions within the first five days of experiment the highest contributions for the IRF (~8% for $^{137/134}Cs$) were obtained. Taken all FIAP values into account over the leach peroid of 276 days no further significant contributions were obtained.

10.5 Microstructure investigations

Experimental procedure

UO_2 fuel kernels before and after leaching were investigated with an environmental scanning electron microscope (ESEM). The sample preparation was performed by sticking the kernels on a sample holder and then the analytic examination started.

A FEI Quanta 200 FEG instrument was used. The instrument is equipped with three detectors. The gaseous large field Secondary Electron-detector was used to characterize the morphology (grain boundaries, porosity). The back Scattered Electron-Detector was used to obtain chemical information due to the z-contrast (atomic number of the elements). With the Apollo X Drift detector the elemental mapping was performed (detection limit 0.1wt%).

Experimental results

In Figure 10.4a and 4b the periphery of the spent UO_2 kernel before leaching are presented. Figure 10.4a reveal the grains and the rough topography of the surface of the fuel matrix. Magnifying (Figure 10.4b) reveals the presence of numerous hexagonal platelets (blue squares in Figure 10.4b), hosting the metallic phases (the elements Mo, Tc and Zr were identified) and white spots (arrows) containing the elements Mo and Zr were detected. Figure 10.4c, 4d, 4e and 4f reveal SEM pictures of the leached kernels under oxic and under anoxic/reducing conditions. Independently of redox conditions, leaching reduced the roughness of the surface and the grain boundaries. However under oxic conditions, the white spots containing Mo and Zr exist, but the hexagonal platelets (metallic phases) could not be detected and numerous open pores are visible, accounting for new assessable leaching sites (Figure 10.4c and 4d). Under anoxic/reducing conditions the pores seemed to be filled with platelets hosting the metallic phases (the elements Mo, Tc, Ru and Zr were identified) (Figure 10.4e and 4f). In summary, leaching had an influence on the microstructure of the fuel and consequently on the IRF of the radionuclides.

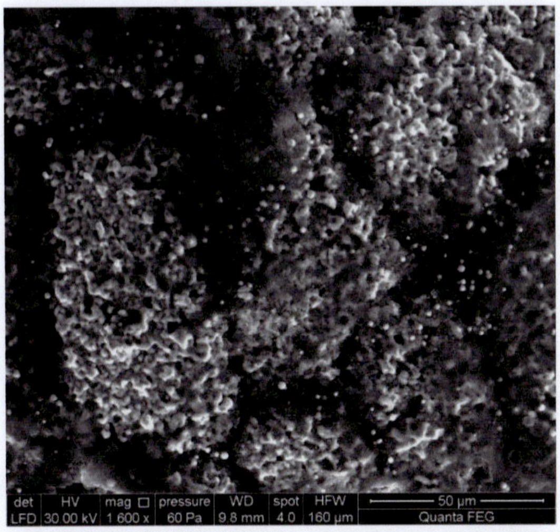

a

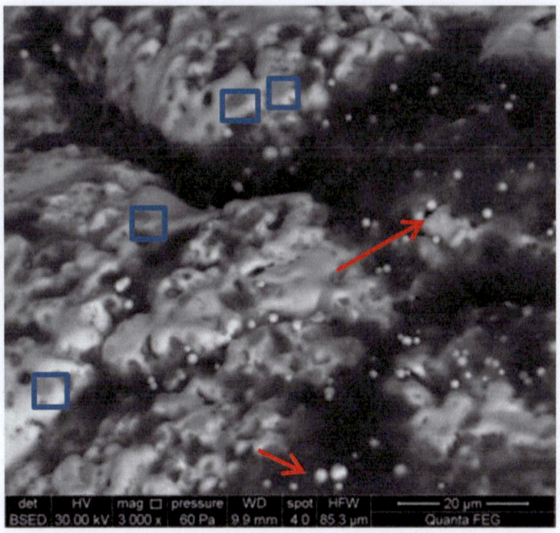

b

c

d

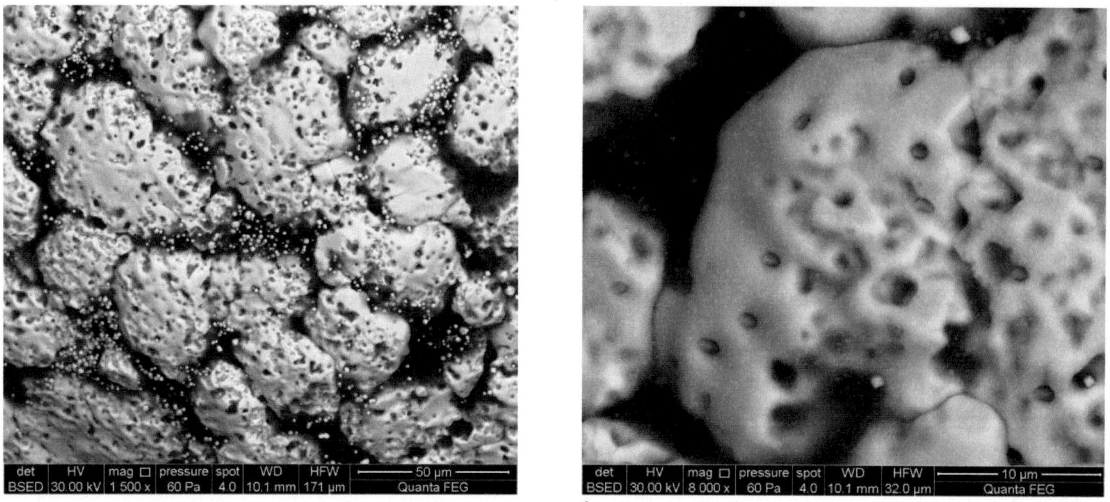

e f

Figure 10.4: SEM pictures of spent UO_2 kernels: a and b: before leaching; c and d: leaching under oxic conditions; and e and f: leaching under anoxic/reducing conditions.

10.6 Conclusions and Future work

UO_2 TRISO coated particles are used as very high-burn up (approximately 100 GWd/t) spent fuel samples within the joint project FIRST-Nuclides. These coated particles can be described as miniature fuel elements. Due to the tight coatings, no fission gas is released outside during the irradiation process, hence the complete activation/fission products are located in the TRISO coated particles.

By cracking of the tight coating an instant gas fraction was released. ^{14}C as $^{14}CO_2$ and ^{85}Kr (up to 35% of the inventory) were identified as gas components. Xe was not detected in the gas fraction but an accumulation within the buffer/iPyC layer was confirmed by ESEM investigations. The observed high fission gas release from the UO_2 matrix is different to UO_2 LWR spent fuel, because the used TRISO particles were irradiated under conditions which enabled a large part of Xe to be present in atomic form in the fuel lattice and to diffuse to the grain boundaries where it is released.

In a time frame of 276 days leaching experiments have been performed in order to determine the instant radionuclide release fractions under different geochemical environments. U was not detected in solution, indicating insignificant matrix dissolution effects. The contribution of ^{90}Sr to the IRF was max. 0.2% and comparable to LWR fuel (Roudil et al., 2007). The determined values of the IRF of ^{134}Cs and ^{137}Cs (8% reached within the first five days under anoxic/reducing conditions) represent the highest values, indicating leaching of fines and of open grain boundaries. In comparison, the labile Cs inventory (gap + cracks + grain boundaries) for LWR fuel (60 GWd/t) was determined to be about 1.4% (Roudil et al., 2007). Under the irradiation conditions carbon oxidation has to be considered, creating a low oxygen potential in which ternary Cs compounds are instable and Cs in atomic form is released similar to Xe.

In dependency of the environmental conditions different release mechanisms were observed for Cs and for Sr. Only in oxic environment a second relevant release step occurred after ~ 100 days, while under Ar/H_2 the release was completed.

ESEM investigations were performed in order to study the influence of leaching on the microstructure. Leaching under oxic and anoxic/reducing conditions reduced the rough topography of the fuel surface and of the grain boundaries. However under oxic conditions metallic precipitates were dissolved and numerous open pores were visible, while under anoxic/reducing the pores host the metallic precipitates. Especially the observed change of the microstructure under oxic conditions influenced the IRF of radionuclides. Open pores offer new accessible sites for solution attack and can explain the second relevant release steps observed for Cs and Sr.

Future investigations will focus on a detail process understanding of microstructure evolution (i.e. determination of intragranular pore and grain size, quantification of elemental composition of metallic precipitates before and after leaching under different environmental conditions and determination of their size).

10.7 Acknowledgement

Special thanks to Sander de Groot and Ralph Hania from Nuclear Research and Consultancy Group in Petten for the TRISO Coated Particles and for their kind support.

The research leading to these results has received funding from the European Union's European Atomic Energy Community's (Euratom) Seventh Framework Programme FP7/2007-2011 under grant agreement n° 295722 (FIRST-Nuclides project).

10.8 References

Barrachin, M., Dubourg, R., de Groot, S., Kissane, M. P., Bakker, K. (2011). Fission-Product Behaviour in Irradiated TRISO-Coated Particles: Results of the HFR-EU-1bis Experiment and Their Interpretation. Journal of Nuclear Materials, 415, 104-116.

Curtius, H., Müller, E., Müskes, H.W., Klinkenberg, M., Bosbach, D. (2013). Selection and Characterisation of HTR Fuel. 1st Annual Workshop Proceedings of the 7th EC FP CP FIRST-Nuclides project (eds. Kienzler et al.). KIT Scientific Reports 7639, 41-52.

Curtius, H., Müskes, H.W., Güngör, M., Lieck, N, Bosbach, D. (2014). First Results on Instant Radionuclide Release Fraction from Spent UO_2TRISO Coated Particles. 2nd Annual Workshop Proceedings of the 7th EC FP CP FIRST-Nuclides project (eds. Kienzler et al.). KIT Scientific Reports 7676, 97-103.

Fütterer, M.A., Berg, G., Marmier, A., Toscano, E., Freis, D., Bakker, K., de Groot, S. (2008). Results of AVR Fuel Pebble Irradiation at Increased Temperature and Burn-Up in the HFR Petten. Nuclear Engineering and Design, 238, 2877-2885.

Roudil, D., Jegou, C., Broudic, V., Muzeau, B., Peuget, S., Deschanels, X. (2007). Gap and Grain Boundary Inventories from Pressurized Water Reactor Spent Fuels. Journal of Nuclear Materials, 362, 411-415.

11 Radiolytic corrosion or uranium dioxide under He^{2+} irradiation

Johan Vandenborre[1], Ali Traboulsi[1], Guillaume Blain[1], Bernard Humbert[2], Ferid Haddad[3] and Massoud Fattahi[1]*

[1] *SUBATECH (FR)*

[2] *Institut de Matériaux Jean Rouxel (FR)*

[3] *Cyclotron ARRONAX (FR)*

* *Corresponding author: johan.vandenborre@subatech.in2p3.fr.*

11.1 Abstract

In this work, the oxidation of UO_2 by the radiolysis products of water at the solid/solution interface is investigated. Irradiation is realized by He^{2+} beam provided by the ARRONAX cyclotron (E = 64.7 MeV). The aim of this investigation is to determine the effect of the atmosphere (presence of gaseous H_2) and the dose on the UO_2 oxidation in order to couple for the first time (i) characterization of the secondary oxidized phases, (ii) quantification of H_2O_2 and H_2 produced by water radiolysis and (iii) determination of the quantity of uranium released into the solution. The temporal evolution of the solid surface oxidation is followed by Raman spectroscopy. H_2O_2 and H_2 are measured respectively by Ultra Violet-Visible (UV-VIS) spectrophotometry and micro Gas Chromatography (μ-GC). Inductively Coupled Plasma Mass Spectrometry (ICP-MS) is used to quantify the soluble uranium species released into the solution. Our results show that He^{2+} irradiation of water induced oxidation of the UO_2 surface which depended on: (1) the atmosphere (with or without gaseous H_2), (2) the dose rate (range between 0.8 – 4.4 kGy/min), (3) the accessibility of the radiolytic species (OH radical and/or H_2O_2) to the solid surface and (4) the surface morphology (with or without Grain Boundaries).

We have clearly shown that gaseous H_2 produced by water radiolysis completely inhibits the UO_2 corrosion. Moreover, when comparing to the yields obtained for ultra-pure water, $G(H_2)$ obtained in presence of UO_2 decreases by a factor of 2/3 whereas $G(H_2O_2)$ is quite similar. From this, we can conclude that H_2 has an inhibition effect on the UO_2 oxidation and this inhibition does not take place by direct effect of H_2O_2 but by interaction between H_2 and the UO_2 surface.

In order to determine the dose rate effect onto the UO_2 radiolytic corrosion, we have proposed a new parameter close to the classical dissolution rate but expressed as a function of the absorbed dose (Gy) and not the contact time with water (day). For the lower dose rate (0.8 kGy/min), we obtained a dissolution rate ten-fold higher than that found for the higher one (4.4 kGy/min). We should note that the lower dose rate is of the order of magnitude of the spent fuel dose rate (about 100 kGy/min).

We developed a particular experimental setup which permits to vary the thickness of the irradiated water and then the distance between the He^{2+} beam and the UO_2 surface (from 0.4 to 2 cm). At low distance (0.4 cm), all the radiolytic species (radical (OH) and molecular (H_2 and H_2O_2) ones) can access to the UO_2 surface and interact with it whereas at high distance (2 cm), only the molecular species are concerned. We found that in presence of OH radicals, not only the UO_2 corrosion is more important but also that no H_2O_2 have been produced during water radiolysis. Then, when formed, OH radicals interact directly with the UO_2 surface. They do not have time to interact with each other to form H_2O_2.

The corrosion of the UO_2 particles was investigated with and without Grain Boundaries (GB). For the same dose rate and absorbed dose, solid analysis and soluble U quantification showed that in absence of GB, the UO_2 radiolytic corrosion decreases drastically. To clarify this observation, we have performed a Raman map which permits to analyze the Raman spectra on a large space scale of the UO_2 corroded surface. The results show that the UO_2 radiolytic corrosion under He^{2+} radiation occurs essentially on the GB and not on the crystallized UO_2 grains.

11.2 Introduction

This paper deals with the radiolytic corrosion at the UO_2 surface. We study the impact of water radiolysis on the corrosion of the grain boundaries (GB) present on the TRI-structural ISO-tropic (TRISO) particle surface. The influence of gaseous H_2 produced by water radiolysis on the UO_2 corrosion is also studied. More details on the context of this study is published elsewhere (Kienzler et al., 2012; Traboulsi et al., 2014a). The aim of this work, performed in the context of WP2 of the FIRST-Nuclides project, was to investigate experimentally the effect of the radiolytic species produced by localized $^4He^{2+}$ radiolysis of water on UO_2 corrosion as a function of different parameters:

Water in contact with UO_2 particles has been irradiated by $^4He^{2+}$ beam under different atmospheres in order to evaluate the effect of H_2.

The irradiation experiments have been performed with different dose rates in order to understand their effect onto the UO_2 corrosion process.

Different radiolytic oxidative species, either molecular (H_2O_2) or radical (OH), have been studied in this work. Hence, we have developed in the context of WP1 of the FIRST-Nuclides project different irradiation cells where the localization of $^4He^{2+}$ radiolysis can be initiated more or less far from the UO_2 surface.

The IRF is clearly linked to the GB of the UO_2 surface. That is the reason why in the context of WP1 we have prepared and characterized the surface of two sets of UO_2 TRISO samples described in Kienzler et al. (2012) and Metz (2013) which present GB or not. The UO_2 radiolytic corrosion experiments have been performed with these samples with and without GB in order to localize the corrosion process at the surface.

To fulfill this work, we chose to investigate the corrosion of the UO_2 surface by Raman Spectroscopy which is known to be very efficient in surface characterization of solid state materials, see the literature in Traboulsi et al. (2014a). Soluble uranium species leached into the solution were analyzed by Inducted Coupled Plasma Mass Spectrometry (ICP-MS) technique. H_2O_2 and H_2 produced by water radiolysis were respectively quantified by UV-VIS Spectrophotometry and Gas Chromatography (µ-GC) in order to calculate their radiolytic yields, compare them to those of pure water and clarify the role of these two species in the UO_2 corrosion.

11.3 Material and Methods

WP1 Work

The properties of the TRISO samples, the preparation of the two sets of samples (with or without GB) and the characterization of the UO_2 surface by SEM are described in (Kienzler et al., 2012; Metz, 2013). Moreover, the average length of the $^4He^{2+}$ particle, produced by the ARRONAX cyclotron facility (E = 64.7 MeV), is about 4 mm in ultra-pure water. It was determined by the SRIM 2008 simulation code (Ziegler et al., 1985; Ziegler et al., 2010). Development of the irradiation cells where the radiolytic corrosion experiments were performed with a controlled range distance (0.4 – 2 cm) between localized water radiolysis and the UO_2 surface is also detailed in Kienzler et al. (2012) and Traboulsi et al. (2014b). These distances are chosen to study the effect of the molecular species only, such as H_2O_2, (for the largest one) and the cumulative effect of molecular and radical (OH radical) for the lowest one.

The experimental setup developed in the context of this WP1 task, for the surface analysis by *in situ* Raman spectroscopy, the solution analysis by *in situ* UV-VIS spectrophotometry and the gas measurement by *ex situ* μ-GC is described in Kienzler et al. (2012), Traboulsi et al. (2014a) and Traboulsi et al. (2014b).

WP2 Work

Table 11.1 presents the irradiation and experimental conditions of the UO_2 radiolytic corrosion experiments studied in this work. The He^{2+} beam energy is determined at 64.7 MeV for a LET average of 22.7 $keV.\mu m^{-1}$ and a dose range of 4.3 – 87 kGy deposited in the solution. The dose rate determination by Fricke dosimetry is described elsewhere (Traboulsi et al., 2014a; Traboulsi et al., 2014b) and the dose rate is measured at a range between 0.8 and 4.4 $kGy.min^{-1}$.

Table 11.1: Experimental and irradiation conditions used in this study: dose rate, distance between UO_2 surface and localized water radiolysis, atmosphere, surface state (+GB: with grain boundaries, -GB: without grain boundaries).

Exp.	Parameter Studied	Dose Rate kGy/min	Atm.	Distance UO_2/α (cm)	Surface ± GB
(1)	H_2 effect	4.4	Ox	2.2	+ GB
		4.4	Red	2.2	+ GB
(2)	Dose Rate	4.4	Ox	2.2	+ GB
		0.8	Ox	2.2	+ GB
(3)	OH Radical effect	0.8	Ox	2.2	+ GB
		0.8	Ox	0.5	+ GB
(4)	GB effect	0.8	Ox	2.2	+ GB
		0.8	Ox	2.2	- GB

11.4 Results

Table 11.2 summarizes the results of each UO_2 radiolytic corrosion experiment performed in the FIRST-Nuclides project *vs.* the studied parameters (H_2, dose rate, OH radical, GB).

Table 11.2: Experimental Results (H_2 and H_2O_2 Radiolytic yields, U concentration in solution, Dissolution rate r') measured during the UO_2 radiolytic corrosion experiments.

Exp.	Parameter Studied		$G(H_2)$ µmol/J	$G(H_2O_2)$ µmol/J	[U] 10^{-7} Mol/L	Diss. rate r' µg/m²/Gy
(1)	H_2 effect	Ox	x	0.06	11.5	2.01
		Red	0.02	0.10	5.5	1.11
(2)	Dose Rate	4.4 kGy.min^{-1}	x	0.06	11.5	2.01
		0.8 kGy.min^{-1}	x	0	99.1	20.75
(3)	OH Radical effect	2.2 cm	x	0.06	11.5	2.01
		0.4 cm	x	0	99.1	20.75
(4)	GB effect	+ GB	x	0.06	11.5	2.01
		- GB	x	x	5.6	x

It has been shown that the UO_2 radiolytic corrosion starts by an oxidation process which leads to the formation of metastudtite ($UO_2(O_2)$ $2H_2O$) detected by Raman spectroscopy on the particles surface as seen in Figure 11.1 (Kienzler et al., 2012; Traboulsi et al., 2014a) .

In experiment (1), irradiation was performed either with open to air atmosphere, or with closed to air atmosphere. In this last case, it is known that the high H_2 production (measured at 3.5%) implies a reductive atmosphere which may affect the UO_2 oxidation. That is why, in this study, we decided to determine the effect of H_2 on the UO_2 corrosion by comparing the two sets of data. When the UO_2 particles were irradiated in closed to air atmosphere, H_2 produced by water radiolysis was quantified by calculating its radiolytic yield (G) in order to compare it to that of pure water and explain any gas consumption due to the inhibition of the UO_2 corrosion process. In our work, the solution used during irradiation is ultra-pure water and the mean $G(H_2)$ obtained when irradiated in presence of UO_2 is 0.02 ± 0.002 µmol/J. In Crumière et al. (2013), ultra-pure water irradiated in same conditions gives $G(H_2)$ = 0.061 ± 0.006 µmol/J which is three-fold higher than that found in this investigation. Hence, for the system UO_2/ultra-pure water, a non negligible part (around 2/3) of the H_2 produced by water radiolysis was consumed by inhibition of the UO_2 oxidation process. Moreover, we have observed the variation of the soluble uranium species concentration leached into the solution after irradiation as a function of the dose and the irradiation time in both open and closed atmospheres. Regardless the type of irradiation atmosphere, the quantity of soluble uranium species increased with the dose until 43.7 kGy and then remained constant.

At this latter dose, concentration of soluble uranium species found in closed atmosphere ($5.5\ 10^{-7}$ mol/L) is two-fold lower than that observed in open atmosphere ($11.5\cdot10^{-7}$ mol/L) due to inhibition of UO_2 oxidation by H_2 produced by water radiolysis. In both, open and closed atmospheres, concentrations of H_2O_2 produced by water radiolysis were determined in order to analyze the effect of the irradiation atmosphere on its consumption by the UO_2 oxidation. It showed that concentration of H_2O_2 produced by water radiolysis increased with the dose and then reached a steady state at 21.8 and 65.5 kGy in open and close atmospheres respectively. At these latter doses, $[H_2O_2]$ values were respectively $1.4\cdot10^{-4}$ and $5.5\cdot10^{-4}$ mol/L which means that H_2O_2 concentration was four-fold higher when the UO_2 particles were irradiated in presence of H_2. It seems that in the latter conditions consumption of H_2O_2 and thus oxidation of the UO_2 surface was inhibited by H_2 produced by water radiolysis. Radiolytic yields of H_2O_2 were also calculated and the values obtained were 0.06 ± 0.006 and 0.10 ± 0.01 µmol/J in open and closed atmospheres respectively. $G(H_2O_2)$ obtained in open atmosphere is two-fold lower than that obtained in closed one. In this last condition, H_2O_2 yield is equal to that of ultra-pure water (LaVerne, 2004; Pastina and LaVerne, 1999; Yamashita et al.,

2008a; Yamashita et al., 2008b) which means that this molecule was not consumed whereas a non-negligible part of H_2 was consumed. From this, we can conclude that in open atmosphere, half of the H_2O_2 produced by water radiolysis was consumed by oxidation of the UO_2 surface leading to the formation of metastudtite secondary phase. We can conclude too that H_2 has an inhibition effect on the UO_2 oxidation and this inhibition does not take place by direct effect on H_2O_2 but by interaction between H_2 and the UO_2 surface. A complete data discussion has been published elsewhere (Traboulsi et al., 2014a), but the main result is that we have experimentally observed that H_2, produced by water radiolysis, can be considered as an efficient UO_2 corrosion inhibitor.

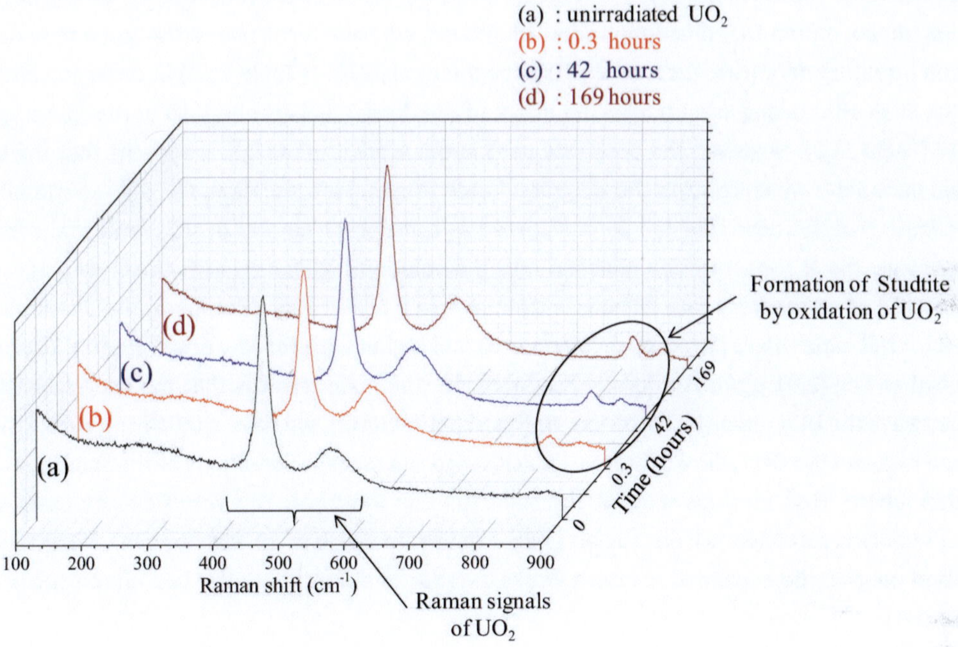

Figure 11.1: Raman spectra of unirradiated UO_2 ((a)) and a sample irradiated at 8.73 kGy under open atmosphere ((b), (c) and (d)). The spectra (b), (c) and (d) are obtained 0.3, 42 and 169 hours after contact with the irradiated solution.

In order to determine the dose rate effect, we have chosen to perform experiments (1) and (2) with two different dose rates of 4.4 and 0.8 kGy/min and compare the results. For the experiment (1), values of [U] measured in our system at the steady state are presented in Table 11.3 with the normalized leaching rate (r). Table 11.3 presents also UO_2 dissolution rate values obtained from the literature in order to compare them to our data. When UO_2 particles were irradiated in closed atmosphere, concentration of Uranium species remained practically constant at $(5.5 \pm 0.5)\cdot10^{-7}$ mol/L. These values are comparable to most solubility data reported in the literature (Casas et al., 2009; Grambow et al., 2004; Jégou et al., 2003; Shoesmith and Sunder, 1992). However, we believe that in our work the thermodynamic equilibrium of the UO_2 particle dissolution has not been achieved due to their composition of grains and grain boundaries. In these conditions, the dissolution rate was equal to 13.5 $g/m^2/d$. A comparison of our dissolution rates with literature data is given in Table 11.3. Despite the similarity with [U] reported in the literature, r values obtained in this investigation are at least 10^3 fold higher. It seems that with the irradiation conditions used in our work, the UO_2 oxidation/dissolution mechanism occurred with a faster kinetic. This is very probably due to the impact of the high doses deposited into the irradiated solution for this work. In order to apprehend the dose impact on the UO_2 corrosion, we propose here a new methodology to express the dissolution rate. This methodology consists in calculating the dissolution rate as a function of the dose deposited into the solution and not of the irradiation time. Thus, the new dissolution rate r' is defined as follow:

$$r' = \frac{V}{S}\frac{d[U]}{dd}M_{UO_2}$$

where V is the volume of the corroded particles, S their surface (*V/S* = 0,18 m) and *d[U]/dd* (mol/m^3) the variation of is the variation of soluble uranium species as a function of the dose deposited into the solution. In our case the r' values are 2.01 µg/m^2/Gy for the open atmosphere experiment and 1.11 µg/m^{2}Gy for the closed atmosphere. These results confirm that the reducing atmosphere (in presence of H_2) has a strong diminishing impact on the UO_2 dissolution rate. Moreover, we have recalculated the same new dissolution rate r' from literature data (Grambow et al., 2004; Jégou et al., 2003) (Table 11.3) in order to compare it to our results. Even after taking into account the effect of the dose, r' values obtained in this investigation are still higher (Table 11.3). However, the dose rate used in our work is very high. It appears that the dose rate is the main parameter which controls the UO_2 dissolution rate under irradiation. For the experiment (2), our results validate that the dose rate can be considered as a major parameter in UO_2 radiolytic corrosion. In fact, surprisingly the U concentration released into the solution highly increases (from $11.5 \cdot 10^{-7}$ mol/L to $99.1 \cdot 10^{-7}$ mol/L) when the dose rate decreases (from 4.4 to 0.8 kGy/min). Moreover, the dissolution rate r' calculated for this experiment (20.75 µg m^{-2} Gy^{-1}) is 10 fold higher with the low dose rate (0.8 kGy/min) than with the higher one (2.01 µg/m^2/Gy for 4.4 kGy/min). So, the main result is that the dose rate decreasing did not decrease the UO_2 radiolytic corrosion. In this experiment (2), we have studied the cumulative effect of the dose rate and the OH radical in order to be closer to the realistic conditions of the spent fuel radiolytic corrosion where $^4He^{2+}$ particles react at the solid/solution interface. This result can be used when the spent fuel radiolytic corrosion will be studied (with a dose rate range of 25-100 Gy/min). Nevertheless, this investigation needs to be expanded in future works in order to differentiate the two effects (dose rate and radical species).

Table 11.3: Steady state U concentrations, Dissolution rates (r and r') of UO_2 and experimental conditions used in this study and others found in the literature.

Ref.	Physical aspect UO_2	$^4He^{2+}$ beam	Dose rate Gy min^{-1}	Atm.	[U] 10^{-7} mol l^{-1}	Diss. rate r Mg/m^{2}d	Diss. rate r' µg/m^{2}Gy
This work	Particle	External source*	4400	Ox / Red	11.5 5.5	13500 7000	2.01 1.11
(Shoesmith and Sunder, 1992)	Pellet	Internal source	No	Ox.		0.5	X
(Casas et al., 2009)	Particle	No	0	Red.		23.3	X
(Grambow et al., 2004)	Colloid	External source*	52.7	Red.		15.7	0.246
(Jégou et al., 2003)	Pellet	Internal source**	18 1.8	Ox. Ox.		2.5 0.2	0.095 0.076

* $^4He^{2+}$ Beam produced by cyclotron facility

** Doped UO2 with Pu

In experiment (3), the irradiation cell was enhanced, in the context of WP1, in order to decrease the distance (0.4 cm instead of 2.2 cm) between the He^{2+} beam and the UO_2 surface (see Figure 11.2). As already shown, the UO_2 dissolution highly increases in presence of OH radical at the solid/solution interface (higher U concentration and dissolution rate). We should mention that in these conditions, no H_2O_2 was found in the solution whereas it has been measured in the other experiments. The absence of H_2O_2 indicates a total consumption of the OH radical (main precursor of H_2O_2) by the UO_2 surface.

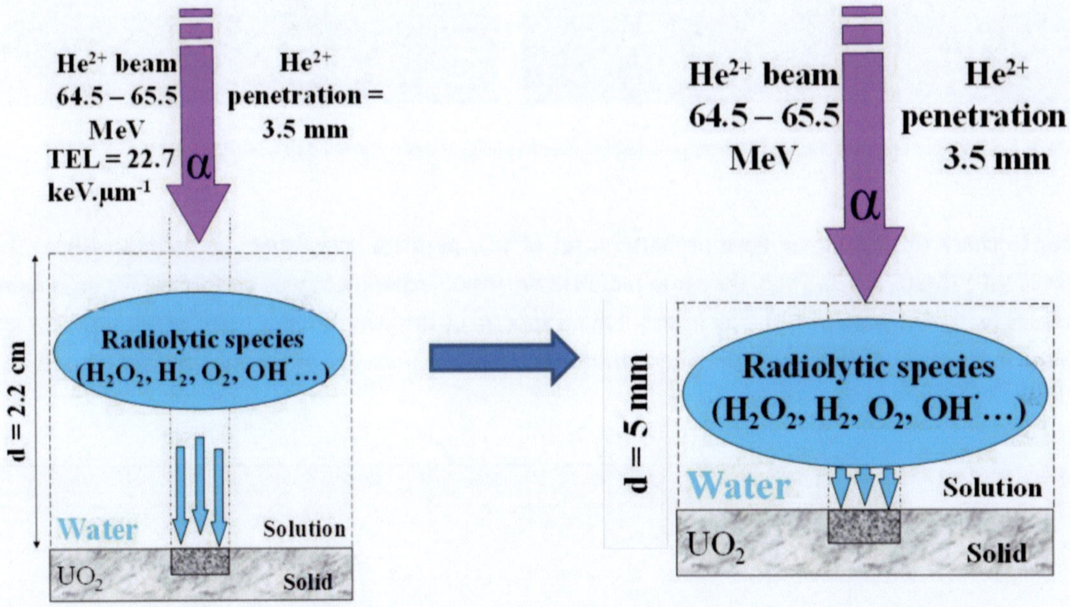

Figure 11.2: Schematic picture of the variation in the distance between the localized radiolysis of water induced by the He^{2+} beam and the UO_2 surface.

The experiment (4) was performed in order to concretely localize the UO_2 oxidation. During the FIRST-Nuclides project, our sample was characterized by SEM before and after irradiation experiments (see Figure 11.3). An UO_2 TRISO particle is constituted of an UO_2 based kernel (500 µm of diameter) composed by grains and grain boundaries. The grain size has been measured between 10 and 20 µm. After irradiation, SEM pictures of the particles surface were realized in order to determine the zones most affected by the corrosion process. Figure 11.3 clearly shows that regardless the irradiation atmosphere, the grain boundaries are a lot more degraded than the UO_2 grains. It appears too that the corrosion did not occur homogeneously on the entire surface. Some regions of the sample surface are indeed more corroded. The corrosion can be considered as localized to the GB.

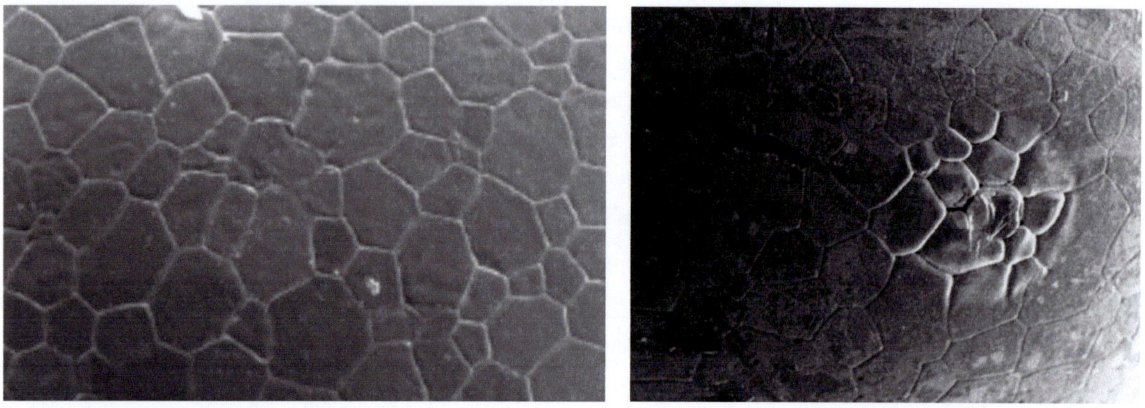

Figure 11.3: SEM pictures of the surface of a nonirradiated UO_2 particle (left) and of a corroded UO_2 particle (right).

In order to check this result, we have prepared a set of UO_2 particles prewashed, i.e. without GB, in the context of WP1 (Metz, 2013). Then, the same radiolytic corrosion experiment was performed for both types of particles (with and without GB). The *in situ* Raman spectra of the two samples (with or without GB) are presented in Figure 11.4. They show that metastudtite is formed in presence of GB only (red spectrum).

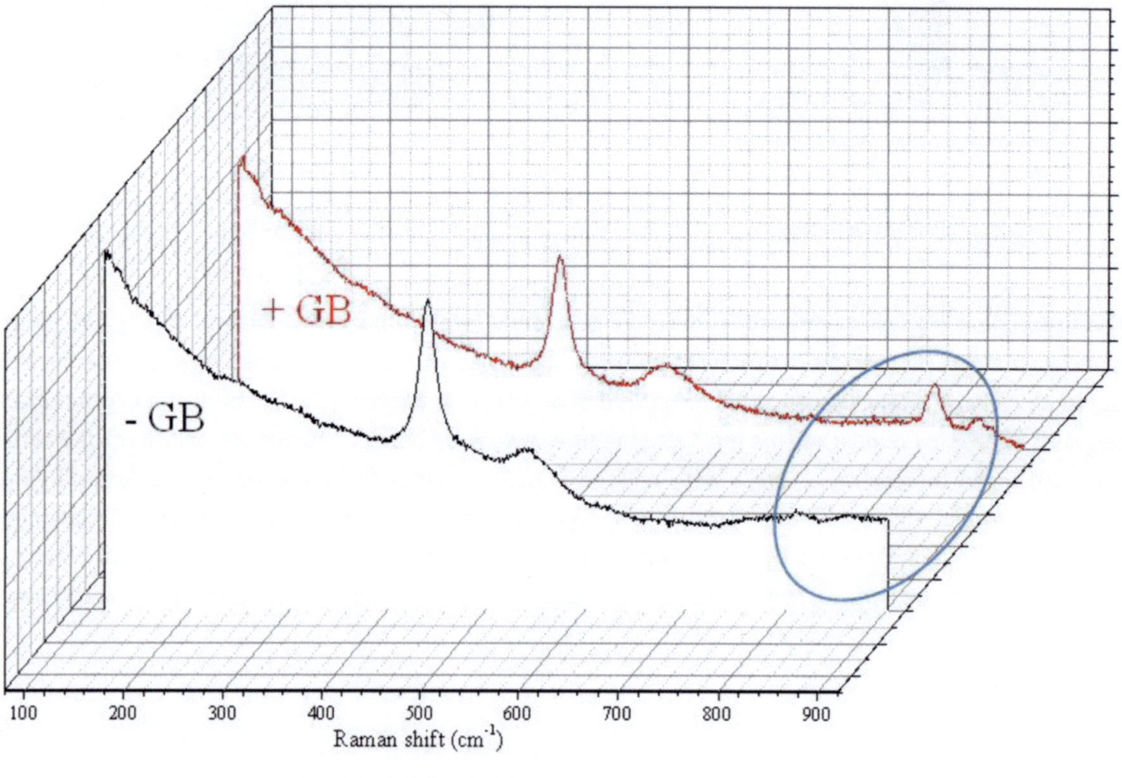

Figure 11.4: Raman spectra of corroded UO_2 with GB (red) and without GB (black).

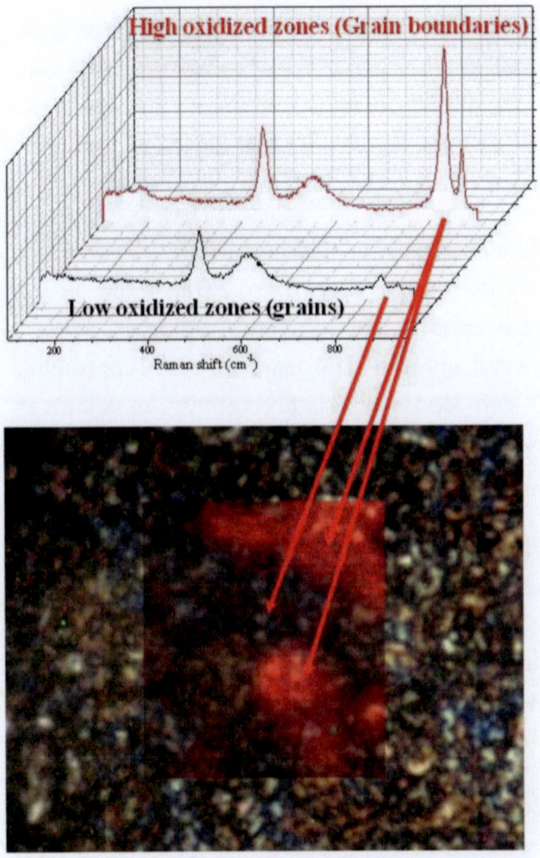

Figure 11.5: Space-resolved Raman spectra of corroded UO_2 with GB (red) and Grains (black).

In the last part of this experiment, we have analyzed the corroded UO_2 surface by space-resolved Raman spectroscopy in order to detect and localize the UO_2 oxidation at the solid surface. The analysis was done on a surface of 50·50 μm^2 with a step of 1 µm of the UO_2 particle. Results are presented in Figure 11.5, where are shown in black color the poor oxidized part of the surface (low ratio between UO_2 and metastudtite peaks) and in red color the strong oxidized one (high ratio). We have to notice that the black area corresponds to the UO_2 grains on the particle surface and the red one to the GB. This result confirms that the UO_2 radiolytic oxidation mainly occurs on the GB.

11.5 Conclusions and Future work

In conclusion, this work sheds some light on the radiolytic corrosion of UO_2 by identification of (1) the oxidized secondary phase formed (metastudtite), (2) the H_2 role as an inhibitor agent, (3) the oxidative role of H_2O_2 and (4) the localization of the oxidation on the GB of the UO_2 surface. Detailed mechanisms of UO_2 corrosion/oxidation were proposed taking into account the phenomena of water radiolysis.

Our results show that He^{2+} irradiation of water induced oxidation of the UO_2 surface depends on: (1) the atmosphere (with or without gaseous H_2), (2) the dose rate (range between 0.8 – 4.4 kGy/min), (3) the accessibility of the radiolytic species to the solid surface (OH radical and/or H_2O_2) and (4) the surface morphology (with or without grain boundaries). In details:

(1) We have clearly shown that gaseous H_2 produced by water radiolysis completely inhibits the UO_2 corrosion. Moreover, when comparing to the yields obtained for ultra-pure water, $G(H_2)$ obtained in presence of UO_2 decreases by a factor of 2/3 whereas $G(H_2O_2)$ is quite similar. From this, we can conclude that H_2 has an inhibition effect on the UO_2 oxidation and this inhibition does not take place by direct effect of H_2O_2 but by interaction between H_2 and the UO_2 surface.

(2) In order to determine the dose rate effect on the UO_2 radiolytic corrosion, we have proposed a new parameter close to the classical dissolution rate but expressed as a function of the absorbed dose (Gy) and not the contact time with water (day). For the lower dose rate (0.8 kGy/min), we obtained a dissolution rate ten-fold higher than that found for the higher one (4.4 kGy/min). We should note that the lower dose rate is of the order of magnitude of the spent fuel dose rate (about 100 Gy/min).

(3) We developed a particular experimental setup which permits to vary the thickness of the irradiated water and then the distance between the He^{2+} beam and the UO_2 surface (from 0.4 to 2 cm). At low distance (0.4 cm), all the radiolytic species (radical (OH) and molecular (H_2 and H_2O_2) ones) can access to the UO_2 surface and interact with it whereas at high distance (2 cm), only the molecular species are concerned. We found that in presence of OH radicals, not only the UO_2 corrosion is more important but also that no H_2O_2 has been produced during water radiolysis. Then, when formed, OH radicals interact directly with the UO_2 surface. They do not have time to interact with each other to form H_2O_2.

(4) The corrosion of the UO_2 particles was investigated with and without grain boundaries (GB). For the same dose rate and absorbed dose, solid analysis and soluble U quantification showed that in absence of GB, the UO_2 radiolytic corrosion has drastically decreased. To clarify this observation, we have performed a Raman map which permits to analyze the Raman spectra on a large space scale of the UO_2 corroded surface. The results show that the UO_2 radiolytic corrosion under He^{2+} radiation occurs essentially on the GB and not on the crystallized UO_2 grains.

Therefore, we can conclude that the UO_2 radiolytic corrosion occurs onto the Grain Boundaries area and can be considered as the main mechanism which induces the behaviour of the IRF at the spent fuel surface.

11.6 Acknowledgement

We acknowledge J.Y. Mevellec from the IMN laboratory for support in Raman measurements and N. Stephant for SEM measurements. The authors acknowledge ARRONAX staff for the efficient performing of irradiation runs onto the cyclotron facility.

The research leading to these results has received funding from the European Atomic Energy Community's (Euratom) Seventh Framework Programme FP7/2007-2011 under grant agreement n° 295722 (FIRST-Nuclides project).

11.7 References

Casas, I., de Pablo, J., Clarens, F., Giménez, J., Merino, J., Bruno, J., Martinez-Esparza, A. (2009). Combined Effect of H_2O_2 and HCO_3^- on UO_2(s) Dissolution Rates under Anoxic Conditions. Radiochim. Acta, 97, 485-490.

Crumière, F., Vandenborre, J., Essehli, R., Blain, G., Barbet, J., Fattahi, M. (2013). LET Effects on the Hydrogen Production Induced by the Radiolysis of Pure Water. Radiation Physics and Chemistry, 82, 74-79.

Grambow, B., Mennecart, T., Fattahi, M., Blondiaux, G. (2004). Electrochemical Aspects of Radiolytically Enhanced UO_2 Dissolution. Radiochimica Acta, 92, 603-609.

Jégou, C., Broudic, V., Poulesquen, A., Bart, J.M. (2003). Effects of a and g Radiolysis of Water on Alteration of the Spent UO_2 Nuclear Fuel Matrix. MRS Online Proceedings Library, 807.

Kienzler, B., Metz, V., Duro, L., Valls, A. (2012). Collaborative Project "Fast / Instant Release of Safety Relevant Radionuclides from Spent Nuclear Fuel" , Budapest 09 - 11 October 2012 (KIT Scientific Reports ; 7639). Proceedings 1st Annual Workshop (7th EC FP CP FIRST-Nuclides).

Metz, V. (2013). Characterisation of Spent Nuclear Fuel Samples and Description of Methodologies and Tools to Be Applied in FIRST-Nuclides. FIRST-Nuclides Deliverable 1.2.

Shoesmith, D.W., Sunder, S. (1992). The Prediction of Nuclear Fuel (UO_2) Dissolution Rates under Waste Disposal Conditions. Journal of Nuclear Material, 190, 20-35.

Traboulsi, A., Vandenborre, J., Blain, G., Humbert, B., Barbet, J., Fattahi, M. (2014a). Radiolytic Corrosion of Uranium Dioxide: Role of Molecular Species. Journal of Physical Chemistry C, 118, 1071-1080.

Traboulsi, A., Vandenborre, J., Blain, G., Humbert, B., Fattahi, M. (2014b). In Situ Method for Exhaustive Investigation of Radiolytic Corrosion at Solid/Liquid Interface: Example of application on UO_2. Analytical Chemistry (submitted).

Ziegler, J.F., Biersack, J.P., Littmark, U. (1985). The Stopping and Range of Ions in Matter. Pergamon Press, New York, USA.

Ziegler, J.F., Ziegler, M.D., Biersack, J.P. (2010). SRIM the Stopping and Range of Ions in Matter. Nuclear Instruments and Methods in Physics Research, Section B, 268, 1818-1823.

12 Laser ablation study of irradiated standard UO_2 fuel and AL/CR doped UO_2 fuel

Anders Puranen, Michael Granfors and Olivia Roth*

Studsvik Nuclear AB, Hot Cell Laboratory (SE)
** Corresponding author: anders.puranen@studsvik.se*

12.1 Abstract

Laser Ablation Inductively Coupled Plasma Mass Spectroscopy analysis was applied on fuel cross-sections from a standard UO_2 fuel (sample 5A2) and from a Al/Cr additive fuel (sample C1). For both fuels the samples were cut out from neighbouring pellets to samples used for leaching studies. The findings of the laser ablation study on both pellets indicate cesium and iodine profiles that are very similar and generally appear to follow the radial burnup profile (as indicated by ^{140}Ce). Cesium and iodine appear to collect in some fuel cracks. Selenium was identified by the reasonable fit of the isotopic ratios of mass 77, 79 and 82 with the calculated selenium inventory. For the additive pellet chromium and especially aluminum are heterogeneously distributed in the pellet.

12.2 Introduction

This paper presents results from experiments performed at Studsvik Nuclear AB within the EURATOM FP7 Collaborative Project "Fast / Instant Release of Safety Relevant Radionuclides from Spent Nuclear Fuel (CP FIRST-Nuclides)". The overall aim of the project is to study the fraction of fission and activation products that are fast/instantly released from spent nuclear fuel upon contact with aqueous media.

In order to complement the fuel leaching experiments undertaken at Studsvik (Roth et al., 2014) laser ablation was applied to study the radial distribution of potentially mobile fission products (mainly I, Cs and Se) in two different samples of irradiated nuclear fuel. One sample was taken from a fuel with Al/Cr doped UO_2 pellets and one from a fuel with standard UO_2 pellets. The samples were taken adjacently to samples used for leaching studies within this project.

12.3 Experimental details

Fuel samples

Laser Ablation Inductively Coupled Plasma Mass Spectroscopy (LA-ICP-MS) analysis was applied on fuel cross-sections from the standard UO_2 fuel 5A2 and from the Al/Cr additive fuel C1. Basic sample data are given in Table 12.1, further details are found in (Roth et al., 2014).

For both fuels the samples were cut out from neighbouring pellets to the samples used for the leaching studies, as reported in (Roth et al., 2014). The samples were prepared by cutting approximately 20 mm long samples from each fuel rod. The samples were thereafter embedded in epoxy and slightly grinded.

Table 12.1: Fuels selected for laser ablation investigation at Studsvik.

Sample name	Reactor type	Fuel type	FGR [%]	Calculated BU (rod average) [MWd/kg$_U$]
5A2	BWR	Std UO_2	~2.4	57.1
C1	BWR	Al/Cr doped UO_2	~1.4	59.1

Laser ablation equipment

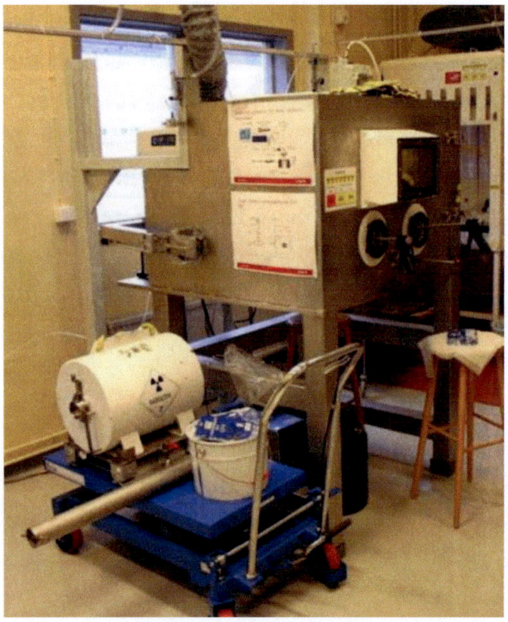

Figure 12.1: The laser ablation hot cell and the sample transfer cask.

The LA technique consists of a pulsed laser that ablates the material to be studied. The ablation equipment consists of a New-Wave UP-213 Nd:YAG laser mounted on a motorized stage in connection to an ablation chamber that is housed in a hot cell (Figure 12.1). A carrier gas (Ar, He or <1% H_2 in He) transports the created aerosol for analysis to an Inductively Coupled Plasma Mass Spectrometer (ICP-MS) installed in a glove box.

The ablating laser operates at a wavelength of 213 nm with a pulse length of < 4 ns. The spot size can be varied between 5-160 µm with an ablation frequency between 1-20 Hz. For scans across a fuel pellet cross section line speeds of 10-30 µm/s are commonly employed. In this study a 20 Hz ablation frequency, spot size between 10 to 60 µm and line speed of 10 to 30 µm/s was used.

Optimisation and calibration of the system is performed by ablation on SRM NIST glass standards (Pearce et al., 1997; Jochum et al., 2011). For investigations of spent fuel the data is normalized as fractions of ^{238}U. The results of this method have been shown to be in good agreement with determinations by dissolution analysis and the results of theoretical calculations (Granfors et al., 2012).

12.4 Results

Laser ablation was performed on cross section samples from the fuel rods O3C1 and 5A2 respectively. The results from this study have partly been reported in (Roth et al., 2013).

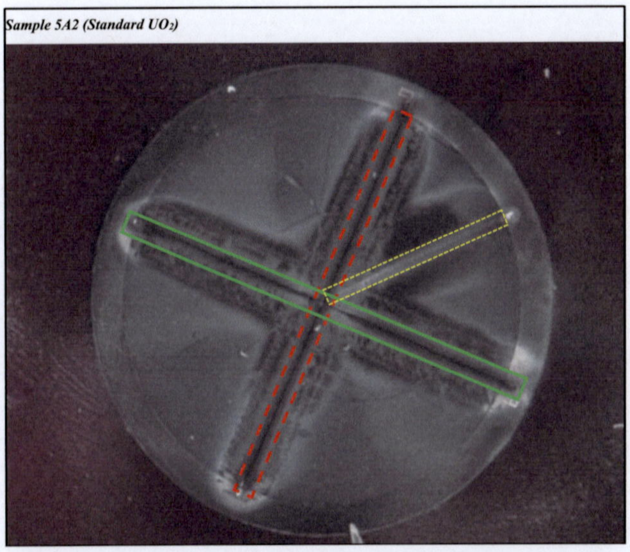

Figure 12.2: Overview of sample 5A2 after laser ablation.

Figure 12.2 and Figure 12.7 show overview images of the fuel samples after laser ablation. Figure 12.3-Figure 12.6 show ablation profiles obtained for sample 5A2 and Figure 12.8-13 show ablation profiles obtained for sample C1.

The red dashed line in Figure 12.2 corresponds to the radial laser ablation profiles in Figure 12.3, the green solid line corresponds to the profiles in Figure 12.4. The dotted yellow line corresponds to the profiles in Figure 12.5.

As can be seen in Figure 12.3 and Figure 12.44 it appears that the cesium and iodine profiles follow each other. The cesium and iodine peaks, such as the one at ~1,000 µm in Figure 12.3 are typically associated with visible cracks in the fuel. Overall the cesium, iodine and xenon profiles are rather flat in the interior of the pellet (no excessive linear heat rates during operation).

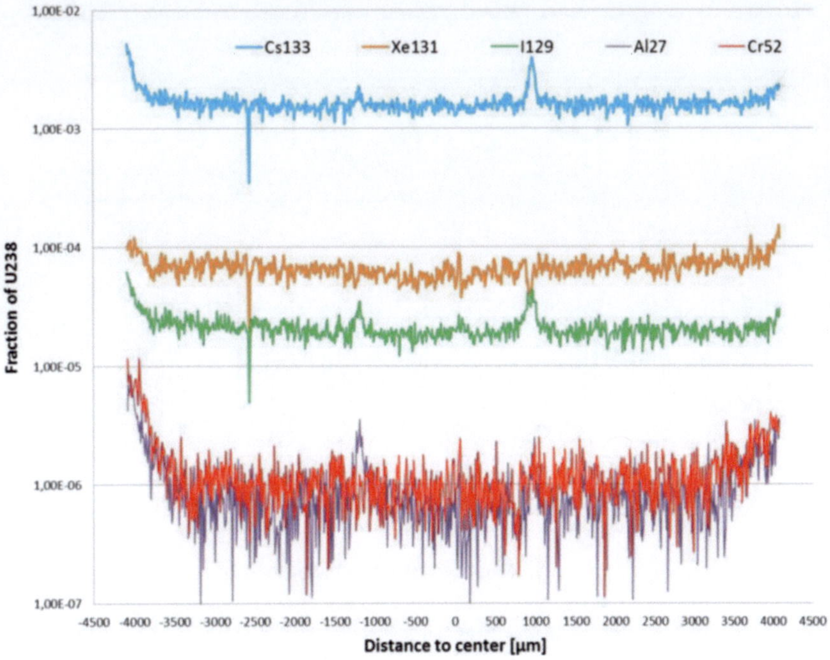

Figure 12.3: Ablation profiles on 5A2 along the red dashed track in Figure 12.2 (60 µm beam).

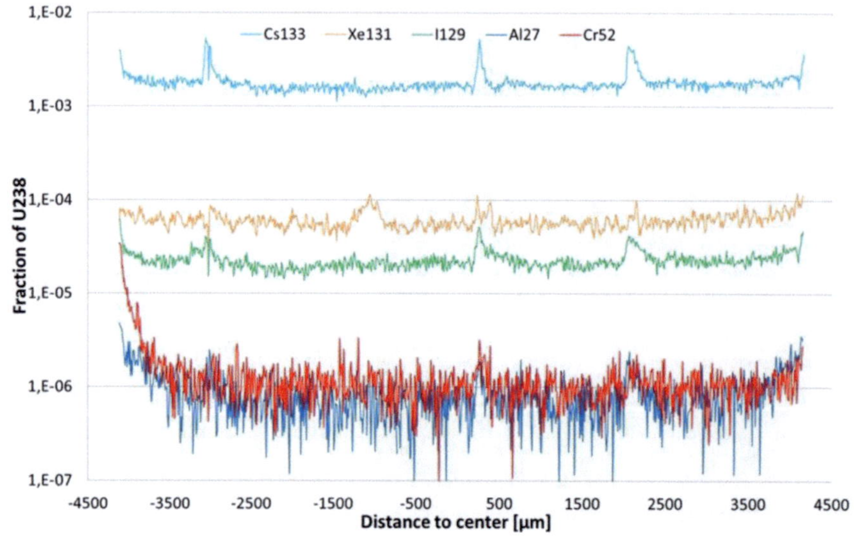

Figure 12.4: Ablation profiles on 5A2 along the green track in Figure 12.2 (60 µm beam).

Figure 12.5 shows the profiles of ^{133}Cs, ^{129}I, ^{235}U and the rather immobile lanthanide ^{140}Ce, which approximately follows the radial burnup profile of the fuel. The profiles are normalised to the pellet center intensity of the respective nuclides. Since the radial ^{133}Cs, ^{129}I & ^{140}Ce profiles are nearly identical (10 µm beamsize) it suggests very limited cesium and iodine redistribution during the power operation.

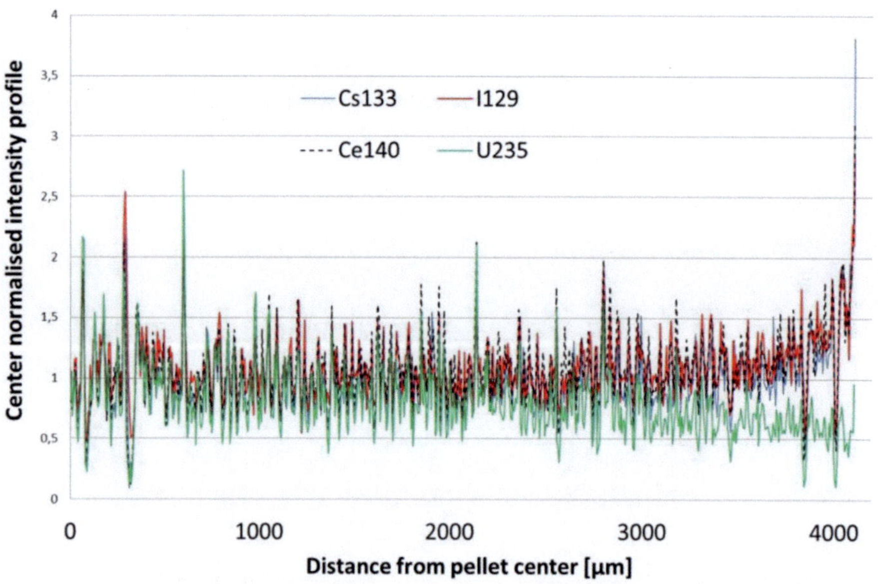

Figure 12.5: Ablation profiles on 5A2 along the yellow dotted track in Figure 12.2. (10 µm beam).

As can be seen in Figure 12.6 ^{133}Cs appears to be more strongly associated with fuel cracks (at ca -2,600 and 300 µm) than ^{99}Tc and ^{82}Se. The higher concentration of Cs in the rim area might be due to the lower ^{239}Pu fission yield of ^{83}Se compared to ^{235}U.

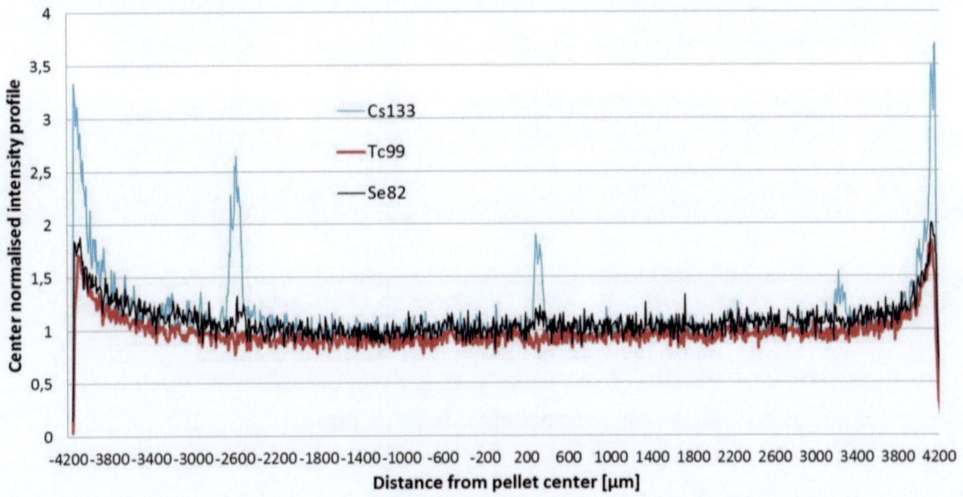

Figure 12.6: Normalised intensity profiles on 5A2 (60 µm beam) of ^{133}Cs, ^{99}Tc and ^{82}Se.

Sample C1 (Al/Cr doped UO_2)

Fi gure 12.7: Overview of additive sample C1 after laser ablation.

Figure 12.7 shows an overview image of the fuel cross section from the additive sample C1 after laser ablation.

As can be seen in Figure 12.8-10, the cesium, iodine and xenon profiles are rather flat. The aluminium additive on the other hand shows a very heterogeneous distribution, which is expected given the low solubility of Al_2O_3 in the UO_2 matrix. The chromium additive appears to be rather homogenously distributed in the fuel, although there appears to be some correlation with the aluminium rich areas. It should be noted that a significant fraction of the Cr_2O_3 could be expected to be dissolved in the UO_2 matrix. Figure 12.9 shows the result of repeated reanalysis (4th passage) of the same laser ablation track as in Figure 12.8. The passages have ablated away a few µm of material. The iodine and cesium profiles appear to be rather consistent during the multiple passages. The xenon intensities does however decrease, this is probably due to beam induced xenon depletion of the material (fission gas release due to the vibration from the impacting beam). The peaks in the chromium and aluminium profiles also change with multiple passages, supporting that the peaks for these elements are highly localized, such as in grain boundaries.

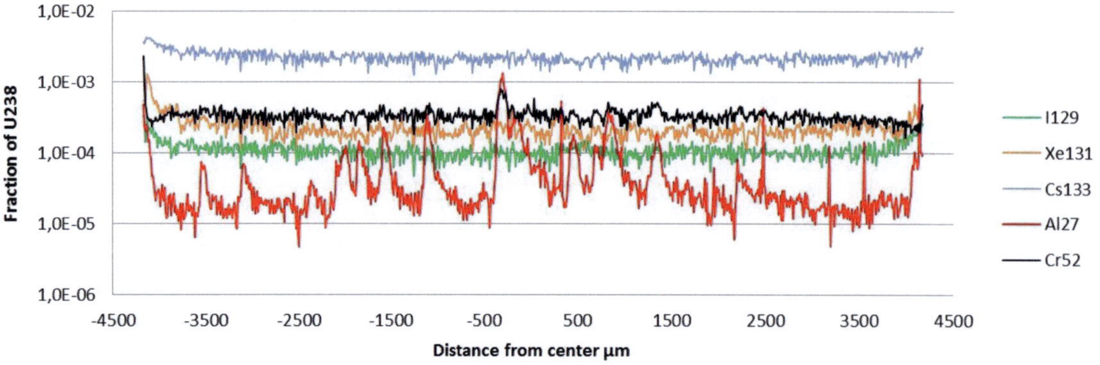

Figure 12.8: Ablation profiles on sample C1 along the track marked 1 in (60 µm beam).

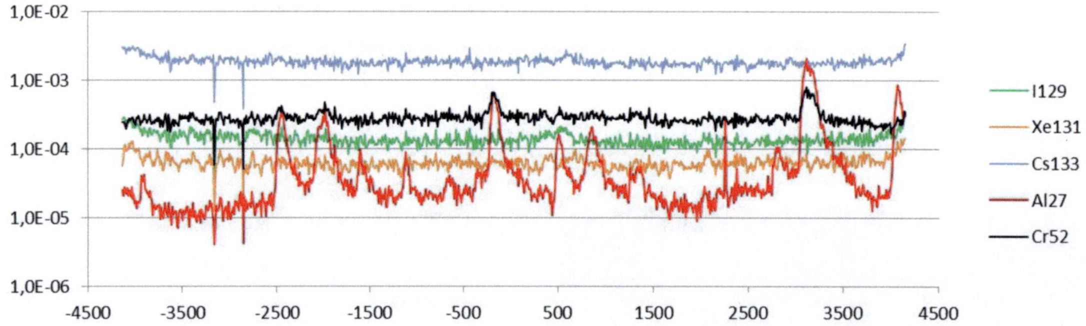

Figure 12.9: Ablation profiles on sample C1 at the 4th passage in the track marked 1 in (60 µm beam).

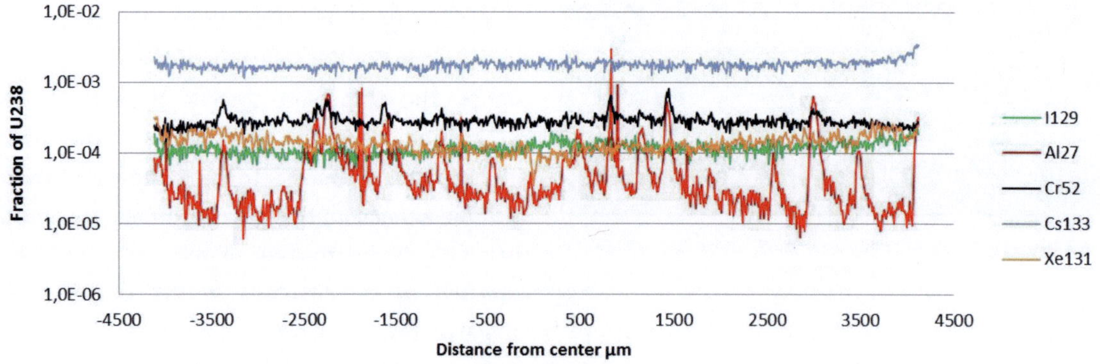

Figure 12.10: Ablation profiles on sample C1 along the track marked 2 in (60 µm beam).

Figure 12.11 (left) shows that the intensity profiles of cesium, cerium and iodine are very similar also for the additive fuel. Figure 12.11 (right) shows the very highly heterogeneous Cr and Al profiles for the additive fuel.

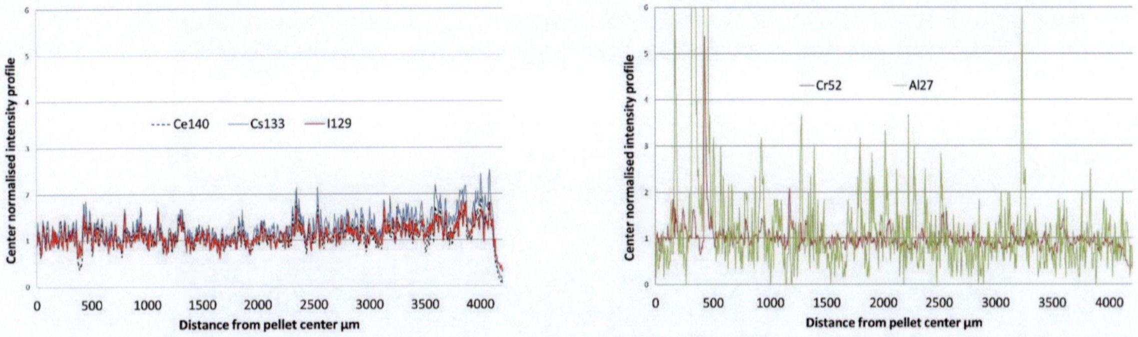

Figure 12.11: Ablation profiles on C1 in the fine track between 1 & 3 in (10 µm beam).

Figure 12.12 shows another radial profile from sample C1 that crosses a pellet crack at ~550 µm. As can be seen from the ^{129}I and ^{133}Cs profiles there is a tendency for accumulation at the crack. The insert shows the overlaid profiles at the location of the crack. Figure 12 also displays a potential effect of azimuthal linear power/heat variations due to the peripheral position of the rod in the fuel bundle given the asymmetrical profiles at -4,200 and 4,200 µm.

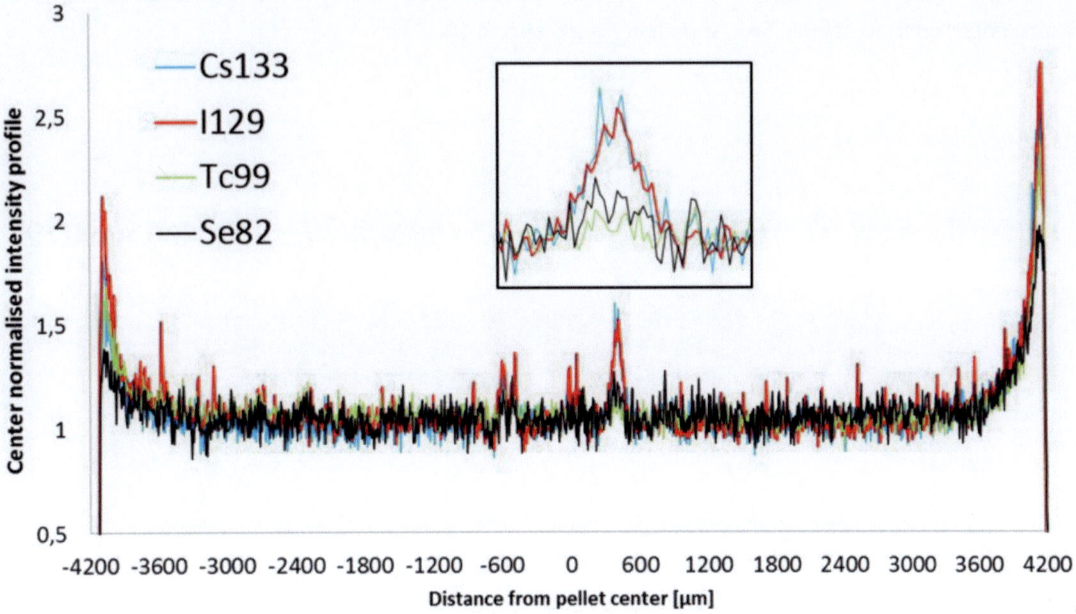

Figure 12.12: *Ablation profiles on sample C1 (60 µm beam).*

Figure 13 shows the radial profiles of mass 77, 79 and 82 from sample C1, which should all be dominated by isotopes of selenium.

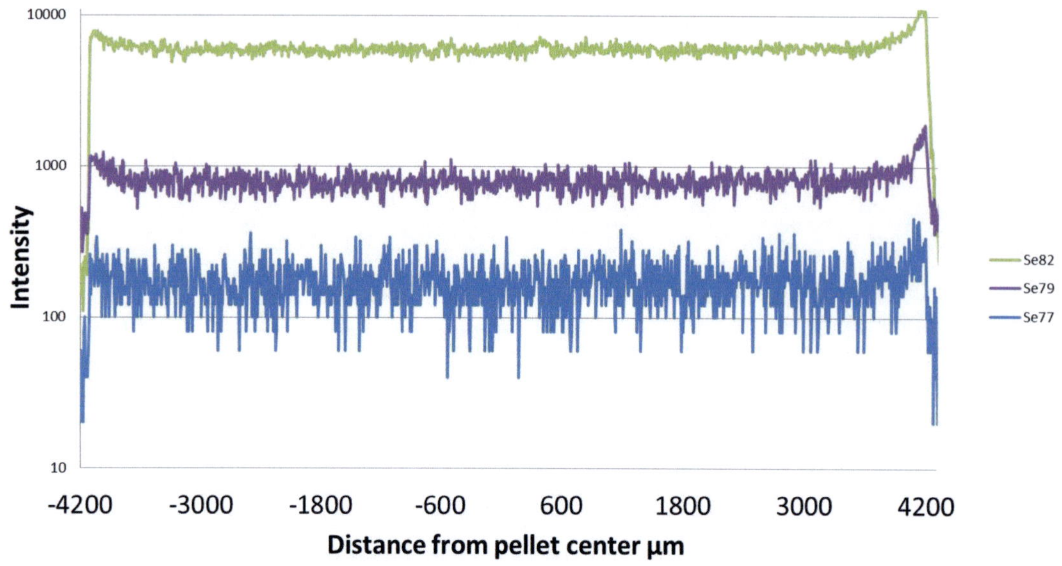

Figure 12.13: ^{82}Se, ^{79}Se and ^{77}Se ablation profiles on sample C1 (60 µm beam).

Figure 14 shows the radial isotopic ratios of mass 82/79 and by 82/77 obtained by LA and the good agreement with the calculated average $^{82}Se/^{79}Se$ and $^{82}Se/^{77}Se$ ratios obtained by the Origen code. Similar results were also obtained from sample 5A2, shown in Figure 15 and 16.

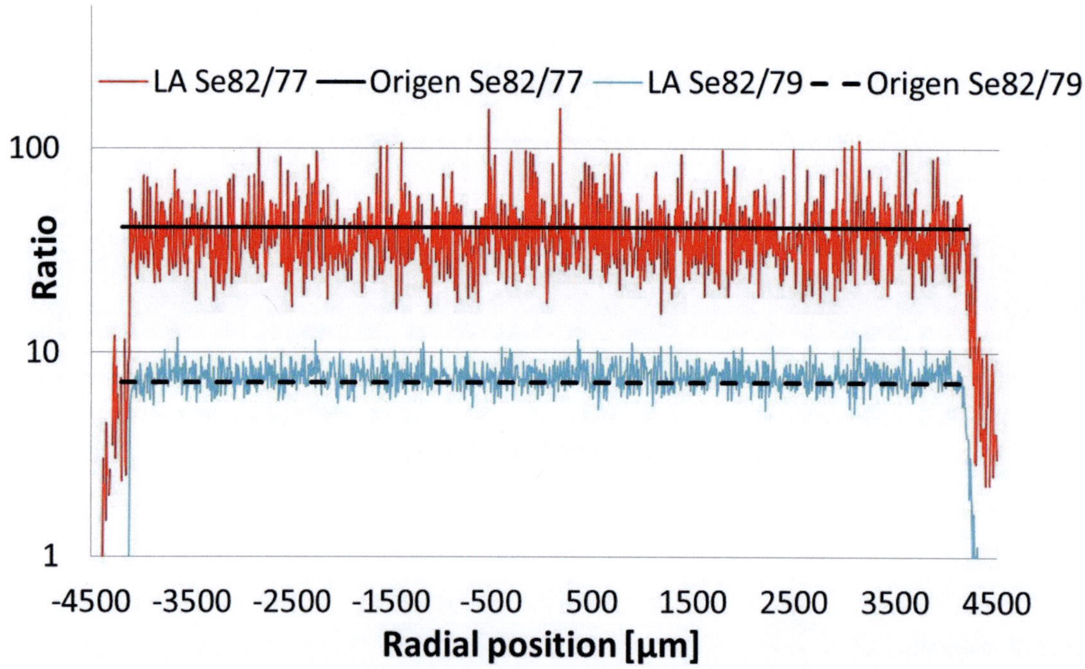

Figure 12.14: Comparison of Se isotope ratios from LA and pellet average values calculated by the Origen code, sample C1 (60 µm beam).

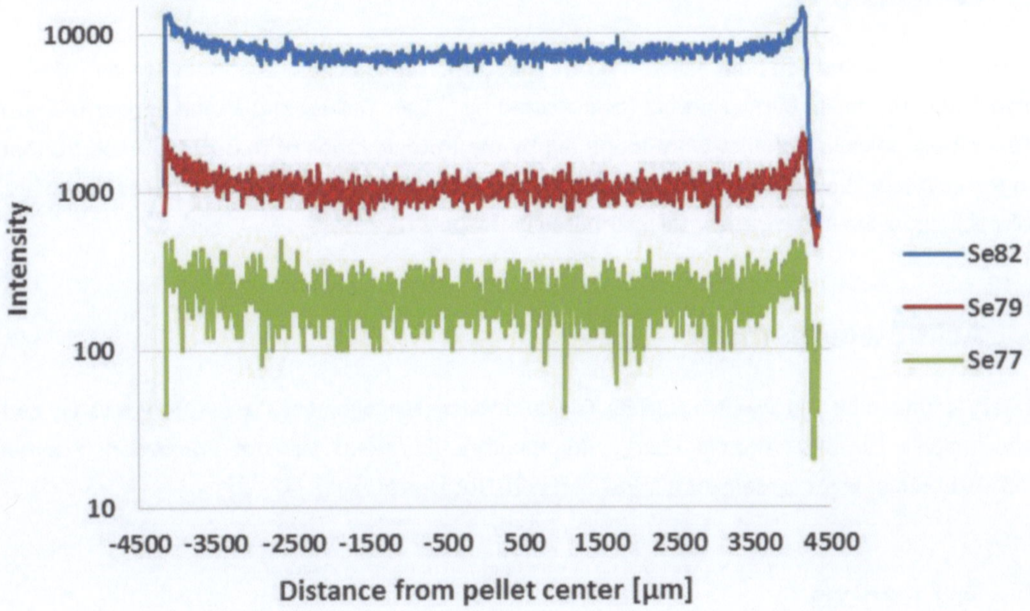

Figure 12.15: ^{82}Se, ^{79}Se and ^{77}Se ablation profiles on sample 5A2 (60 µm beam).

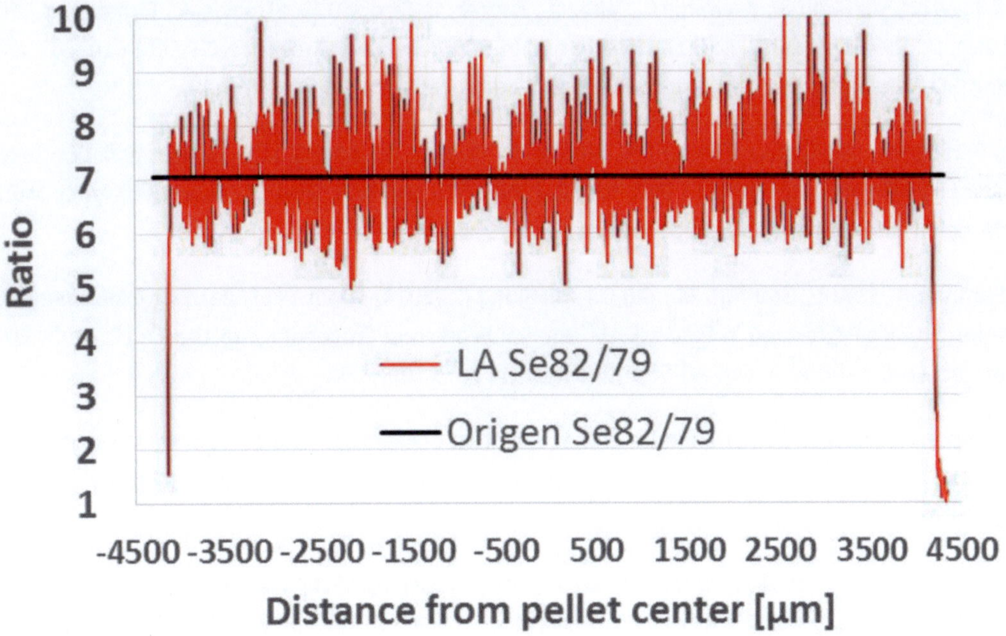

Figure 12.16: Comparison of $^{82/79}$Se isotope ratios from LA and pellet average values calculated by the Origen code for sample 5A2 (60 µm beam).

The Se isotopic ratios also agrees with those obtained in an analysis of selenium in the leaching experiments using adjacent fuel samples (Puranen et al., 2014). The laser ablation study was however performed without any suppression of isobaric interferences preventing evaluation of ^{78}Se and ^{80}Se.

12.5 Conclusions

The findings of the LA study on both pellets indicate that the cesium and iodine profiles are very similar and appear to follow the radial burnup profile (as indicated by ^{140}Ce). Cesium and iodine appear to collect in some fuel cracks. Selenium profiles were identified by the isotopic ratios of mass 77, 79 and 82 that are close to the calculated inventory and the Se leaching analysis results. For the additive pellet chromium and especially aluminum are heterogeneously distributed in the pellet.

12.6 Acknowledgement

This project is funded by The Swedish Nuclear Fuel and Waste Management Co. (SKB) , Posiva Oy and the European Union's European Atomic Energy Community's (Euratom) Seventh Framework Programme FP7/2007-2011 under grant agreement n° 295722 (FIRST-Nuclides project).

12.7 References

Granfors, M., Puranen, A., Zwicky, H-U. (2012). Radial Profiling and Isotopic Inventory Analysis of Irradiated Nuclear Fuel Using Laser Ablation ICP-MS. Proceedings of Top Fuel 2012 Transactions.

Jochum, K.P., Weis, U., Stoll, B., Kuzmin, D., Yang, Q., Raczek, I., Jacob, D.E., Stracke, A., Birbaum, K., Frick, D.A., Günther, D. and Enzweiler, J. (2011). Determination of Reference Values for NIST SRM 610–617 Glasses Following ISO Guidelines. Geostandards and Geoanalytical Research, 35(4), 397–429.

Pearce N., Perkins W., Westgate J., Gorton M., Jackson S., Neal C., Chenery S. (1997). A Compilation of New and Published Major and Trace Element Data for NIST SRM 610 and NIST SRM 612 Glass Reference Materials. The Journal of Geostandards and Geoanalyses, 21(06/97), 115-144.

Roth O., Puranen A., Low J., Granfors M., Cui D., Askeljung C. (2014) Spent Fuel Leaching Experiments and Laser Ablation Studies Performed in Studsvik. 2nd Annual Workshop Proceedings of the 7th EC FP CP FIRST-Nuclides project (eds. Kienzler et al.). KIT Scientific Reports 7676, 163-174.

Roth O., Puranen A., Askeljung C., and Cui D., (2014) Selection of Materials and Characterization of Samples Used in Spent Fuel Leaching and Laser Ablation Studies. Final Workshop Proceedings of the 7th EC FP CP FIRST-Nuclides Project (eds. Kienzler et al.).

Roth O., Askeljung C., Puranen A., Granfors M., Cui D., Low J. Leaching of High Burn up Spent Fuel With and Without Matrix Dopants. Final Workshop Proceedings of the 7th EC FP CP FIRST-Nuclides Project (eds. Kienzler et al.).

Puranen A., Granfors M., Roth O. (2014). Aqueous Leaching of 79Se from Spent Nuclear Fuel. Final Workshop Proceedings of the 7th EC FP CP FIRST-Nuclides Project (eds. Kienzler et al.).

13 Leaching experiments with cladded pellet and fragment of high burn-up nuclear fuel rod segment under argon/ H_2 atmosphere

Ernesto González-Robles, Markus Lagos, Elke Bohnert, Nikolaus Müller, Michel Herm, Volker Metz and Bernhard Kienzler*

Karlsruhe Institute of Technology (KIT) (DE)
** Corresponding author: ernesto.gonzalez-robles@kit.edu*

13.1 Abstract

This paper presents the current status of leaching experiments performed at KIT-INE within work-package 3 of the EURATOM FP7 Collaborative Project, "Fast / Instant Release of Safety Relevant Radionuclides from Spent Nuclear Fuel (CP FIRST-Nuclides)". The studied spent nuclear fuel material was sampled from a fuel rod segment, which experienced a burn-up of 50.4 GWd/t_{HM} in the Gösgen pressurized water reactor. Two static leaching experiments were performed with a cladded pellet and fragments of spent nuclear fuel in bicarbonate water - under Ar / H_2 atmosphere with a pH_2 of 3 bar. The experiment with the cladded pellet was finished after 333 days of leaching, whereas the experiment performed with fragments is still on-going.

13.2 Introduction

The disposal in deep bedrock repositories is considered as the preferred option for the management of spent nuclear fuel (SNF) in many countries (Johnson and Shoesmith, 1988; Shoesmith, 2000; Bruno and Ewing, 2007). The aim is to permanently and safely dispose of the highly radioactive material so that it is isolated from the biosphere for an appropriate length of time. A multi-barrier system is interposed between the SNF and the environment considering the SNF itself as the first technical barrier. In safety assessments for disposal of SNF in deep underground repository, failure of canisters and loss of the integrity of fuel rods is considered in the long term. Assessing the performance of SNF in a potential future geological disposal system requires the understanding and quantification of the radionuclide release in case of water access.

The release of the radionuclides into water occurs by two main processes (Johnsson et al., 2005; Poinssot et al., 2005): i) short term release of the so-called instant release fraction (IRF): ii) long term release dominated by the dissolution of the UO_2 grains, which is referred as matrix contribution.

Within the EURATOM FP7 Collaborative Project, "Fast / Instant Release of Safety Relevant Radionuclides from Spent Nuclear Fuel (CP FIRST-Nuclides)" the objective is to have a better comprehension of the IRF. The IRF is due to the segregation of a part of the radionuclide inventory to the gap interface between the cladding and the pellet, the pellet fractures as well as to fuel grain boundaries. The radionuclides that will be segregated are: fission gases (Kr and Xe), volatiles elements with the isotopes ^{129}I, ^{137}Cs, ^{135}Cs, ^{36}Cl and ^{79}Se and other elements or isotopes such as ^{99}Tc, ^{107}Pd and ^{126}Sn (Johnson et al., 2005). The degree of segregation of the various radionuclide is highly dependent on in-reactor fuel operating parameters such as linear power rating, fuel temperature, burn-up and ramping processes. In the case of the fission gases, the gas release occurs by diffusion to grain boundaries, grain growth accompanied by grain boundary sweeping, gas bubble interlinkage and intersection of gas bubbles by cracks in the fuel (Johnson and Shoesmith, 1988).

The present paper focuses on the determination of the concentration of the Xe and Kr in the gas phase as well as the concentration of ^{90}Sr, ^{99}Tc, ^{129}I, ^{137}Cs and ^{238}U in solution in two experiments with a cladded pellet and SNF fragments.

13.3 Spent nuclear fuel

The studied SNF samples were taken from the SBS1108 N0204 fuel rod segment, which was irradiated in the pressurized water reactor placed at the Gösgen nuclear power plant in Switzerland. The irradiation was carried out in four cycles for a period of time of 1,226 days with an average linear power of 260 W/cm and achieving an average burn-up of 50.4 GWd/t_{HM}. The fuel rod segment was discharged the 27th May 1989 that implies a cooling time of 24 years before characterisation and cutting of the segment. Characteristic data of the studied SBS1108 N0204 fuel rod segment are given in Metz et al. (2014). Based on results of a puncturing test of the fuel rod segment, the fission gas release was previously calculated to be 8.35% (Gonzalez-Robles et al., 2014).

13.4 Spent nuclear fuel sample preparation

From the fuel rod two cladded fuel pellets were cut. The complete process is described elsewhere (Wegen et al., 2012). The first leaching experiment was performed with one cladded pellet. The second experiment was performed with SNF fragments. Therefore, the second pellet was decladded in order to get fragments and to separate the fuel from the Zircaloy (the cladding was not used in this leaching experiment). Prior to the start of the experiments, the samples were weighed; in the case of the cladded pellet, the length and diameter of the sample were measured, too.

13.5 Experimental set-up of leaching experiments and leachant composition

Both static leaching experiments were carried out in 250 mL stainless steel Ti-lined VA autoclaves (Berghof Company, Eningen, Germany) with two valves in the lid to allow solution and gas sampling. The composition of the bicarbonate leaching solution used in each of the experiments is 19 mM NaCl + 1 mM $NaHCO_3$ with a pH of (8.9 ± 0.2) and Eh of (-116.3 ± 50) mV. The leachants were prepared in a glove-box under Ar atmosphere. The experiments were carried out in an Ar / H_2 atmosphere with a total pressure of (40 ± 1) bar (pH_2: 3 bar) at room temperature.

In the case of the cladded pellet, the sample was mounted in a Ti sample holder to ensure the contact of both pellet surfaces with the solution, whereas the fragments were kept in a glass basket, see Figure 13.1.

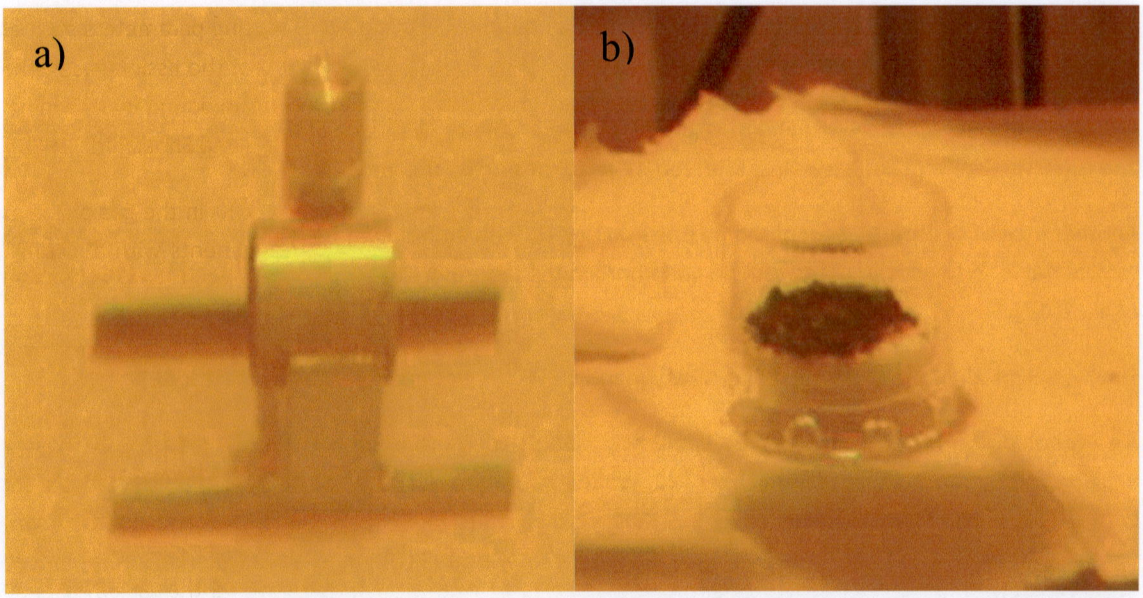

Figure 13.1: a) Ti sample holder with cladded pellet; b) glass sample holder containing SNF fragments.

Once the samples were placed into the autoclaves, the autoclaves were closed and fluxed with Ar to avoid the possible presence of air. Afterwards, a pre-leaching test with duration of 1 day was performed by filling each autoclave with 220 mL of bicarbonate water. The solution was transferred into the autoclaves using syringes of (50 ± 1) mL under constant Ar flux to avoid air contact. This pre-leaching step was performed in order to reduce the amount of $^{135,137}Cs$ in the solution and to remove potential pre-oxidation layers present at the surface of the samples.

After the pre-leaching, the solution was completely replenished and a gas sample was collected. Then each autoclave was again refilled with 220 mL of bicarbonate solution as described previously. Gaseous (50 ± 1 mL) and liquid (15 ± 1 mL) samples were taken at certain time steps. After the sampling, the gas volume of the autoclaves was purged with Ar, and the initial conditions (40 bar of Ar + H_2 mixture) were again established. The solution was not renewed after sampling, thus the remaining leachant volume was reduced at each sampling step.

In the case of the of the cladded pellet experiment, samples were taken after 1, 7, 21, 56, 84, 176, 245 and 332 days. The leaching experiment performed with fragments is still on-going; samples have been taken at: 1, 7, 27, 71 and 114 days.

13.6 Analyses of released radionuclides in liquid and gaseous samples

Gas samples were collected in stainless steel single-ended miniature sampling cylinders (SS-4CS-TW-50, Swagelok Company, USA) to determine the amount of Kr and Xe released during the leaching experiments and to control the gas atmosphere. These gas sampling cylinders are characterised by a length of 159 mm, an outer diameter for tube fitting of 9.5 mm and an inner diameter for tube socket weld connection of 6.4 mm. Since the volume of the cylinders was known, the moles of released gases could be calculated.

From the liquid samples obtained during each campaign different aliquots were prepared to determine the amount of ^{90}Sr, ^{99}Tc, ^{129}I, ^{137}Cs and ^{238}U:

An aliquot of 1 mL was acidified and measured by γ-spectrometry by means of Ge-detectors (EGC-15-185-R and GX3018, Canberra Industries Inc, Meriden, USA) to determine the amount of ^{137}Cs.

Another aliquot of 1 mL was prepared by precipitating Cs with ammonium molybdate phosphate (AMP) to reduce the ^{135,137}Cs activity of the sample and improve the determination of ^{129}I by γ-spectrometry (Johnson et al., 2012).

Finally, a 5 mL aliquot was acidified and, subsequently, ^{135,137}Cs was again precipitated with AMP:

An aliquot of 2.5 mL was analysed by liquid scintillation counting using a Packard Tri-Carb 3110TR low activity scintillation analyser (Perkin Elmer INC, Waltham, USA) to quantify ^{90}Sr. Prior to the LSC analysis ^{90}Sr was extracted by chromatography using a Sr-Resin crown ether (4,4'(5') -di-t-butylcyclohexano-18-crown-6), and the solution aliquots were homogenized with a LSC-Cocktail (Ultima Gold LLT, Perkin Elmer) before measurement.

Another aliquot of 2.5 mL was analysed by means of Inductively Coupled Plasma Mass Spectrometry (ELAN 6100 Perkin Elmer Inc, Waltham, USA) to quantify ^{99}Tc ^{238}U dissolved in solution.

13.7 Results and Discussion

Gas phase in leaching experiments

The concentration of the fission gases (Xe+Kr) in the gas phase as well as the concentration in solution of ^{90}Sr, ^{99}Tc, ^{129}I, ^{137}Cs and ^{238}U of the experiments are shown Figure 13.2.

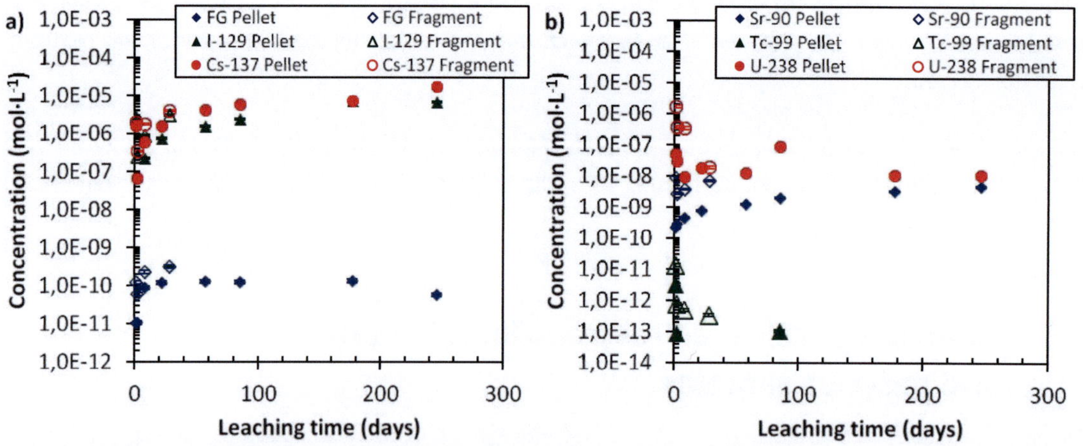

Figure 13.2: Concentration as a function of leaching time in cladded pellet and fragment experiments: a) ^{129}I, ^{137}Cs and normalized content of FG (Xe+Kr); b) ^{90}Sr, ^{99}Tc and ^{238}U.

Both in the cladded pellet experiment and in the fragments experiment an increase in concentration is observed in the case of ^{129}I, ^{137}Cs and ^{90}Sr as well as in the release of fission gases. On the other hand the concentrations of ^{99}Tc and ^{238} U decrease after a higher initial release reaching almost a constant value in the case of cladded pellet experiment (there are not enough values to draw such a conclusion in the case of the fragments experiment). Since Tc and U are redox sensitive elements, it is expected to find them in solution as Tc(IV) and U(IV) with low solubility values in the long run. However the relatively high initial release may be associated to the presence of oxidised layers on surfaces of the samples, that release Tc(VII) and U(VI) to the solution. Furthermore it is hypothesized that these elements will be reduced as a consequence of the strongly reducing conditions present in the experiments.

Comparing the values obtained in both experiments, there is no significant difference in the initial release of ^{129}I and ^{137}Cs despite the difference of the SNF samples accessible surfaces. In the case of ^{90}Sr, ^{99}Tc, ^{238}U as well as FG considerably higher initial concentration values are reached during the pre-leaching of the fragments experiment than in the pre-leaching of the cladded pellet. This fact may be explained for the different particle size degree of oxidation on the surface of the samples, higher in the case of fragments, which will cause a higher initial release.

In order to evaluate how much of the inventory has been released through the experiment, the fraction of inventory released for an element *i* is calculated following Equation 13.1:

$$\text{Fraction release} = \frac{m_i}{m_{UO_2} \text{ x } H_i} \tag{13.1}$$

where $m_{i,}$ is the mass of element (g)*i* in the gas or liquid phase, m_{UO2} is the initial oxide mass (g) in the fuel sample and H_i corresponds to the fraction of inventory for the element *i* (g_i/g_{UO2}) based on inventories calculated at the sample with program the *webKorigen* software package (Nucleonica GmbH 2011).

Afterwards, the cumulative fraction of the inventory released is calculated as the summary of the release for each contact period as described in Equation 13.2:

$$\text{Cumulative release fraction} = \sum \text{Fraction release}_i \tag{13.2}$$

In Figure 13.3, the cumulative release fraction as function of the leaching time for both experiments is plotted. From Figure 13.3 the cumulative fraction released from the inventory after 246 days of leaching time of a cladded pellet was: $1.5 \cdot 10^{-1}$ for FG, $7.6 \cdot 10^{-2}$ for ^{129}I, $2.2 \cdot 10^{-5}$ for ^{90}Sr, $5.2 \cdot 10^{-7}$ for ^{238}U, $1.1 \cdot 10^{-8}$ for ^{99}Tc (data obtained at 85 days, the other data were under the detection limit). In the case of the experiment carried out with the SNF fragments, the cumulative fraction released from the inventory after 27 days of leaching time was: $1.1 \cdot 10^{-1}$ for FG, $9.7 \cdot 10^{-2}$ for ^{129}I, $1.1 \cdot 10^{-5}$ ^{90}Sr, $1.6 \cdot 10^{-5}$ ^{238}U, $4.5 \cdot 10^{-8}$ for ^{99}Tc.

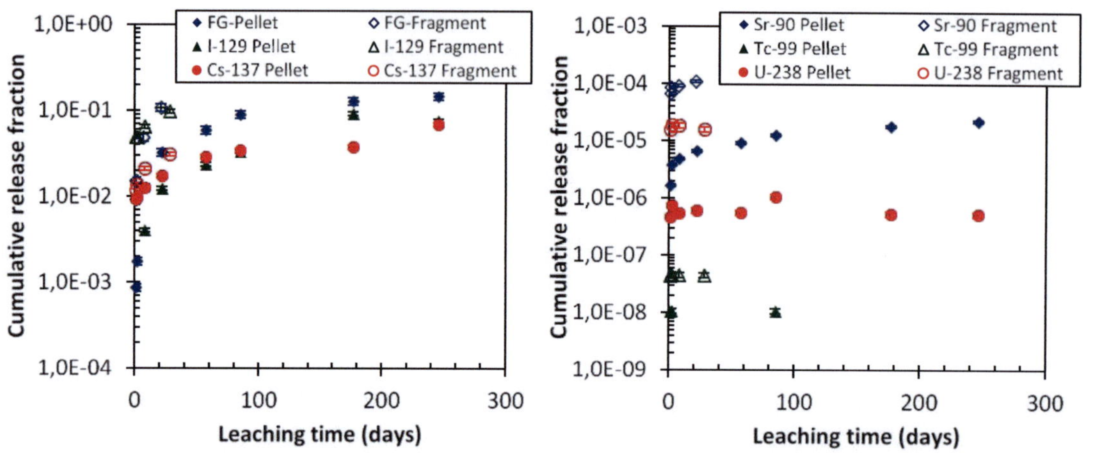

Figure 13.3: Cumulative release fraction as a function of leaching time in cladded pellet and fragment experiments: (left) FG (Xe+Kr), ^{129}I and ^{137}Cs; (right) ^{90}Sr, ^{99}Tc and ^{238}U.

13.8 Summary

Two static leaching experiments were carried out in 250 mL stainless steel Ti-lined VA autoclaves in bicarbonate water (19mM $NaHCO_3$ + 1mm NaCl) under strongly reducing conditions (pH_2: 3 bar). The experiments were performed with a cladded pellet and SNF fragments (ongoing experiment) of the same fuel rod segment. The concentration of FG, ^{90}Sr, ^{99}Tc, ^{129}I, ^{137}Cs and ^{238}U were measured. A higher initial release of FG, ^{90}Sr, ^{129}I, ^{99}Tc and ^{238}U is observed in the experiment with the SNF fragments than in the experiment with the cladded pellet. This fact may be attributed to the difference in accessible surfaces of the samples as well as the existence of pre-oxidized layer on the surface of the fragments. Finally the cumulative release was calculated after 246 days of leaching time in the case of a cladded pellet and after 27 days of leaching time in the case of the fragments.

13.9 Acknowledgement

The research leading to these results has received funding from the European Union's European Atomic Energy Community's (Euratom) Seventh Framework Programme FP7/2007-2011 under grant agreement n° 295722 (FIRST-Nuclides project).

13.10 References

Bruno, J., Ewing, R.C. (2006). Spent Nuclear Fuel. Elements, 2(6), 343-349.

González-Robles, E., Bohnert, E., Müller, N., Herm, M., Metz, V., Kienzler, B. (2014). Determination of the Fission Gas Release in the Segment N0204 and Gas Phase Result of Anoxic Leaching Experiment. 2nd Annual Workshop Proceedings of the 7th EC FP CP FIRST-Nuclides project (eds. Kienzler et al.). KIT Scientific Reports 7676, 37-40.

Johnson, L., Shoesmith, D.W. (1988). Spent Fuel in Radioactive Waste Forms for the Future. (Eds.W. Lutze and R.C. Ewing), North-Holland, Amsterdam, 635-698.

Johnson, L., Ferry, C., Poinssot, C., Lovera, P. (2005). Spent Fuel Source-Term Model for Assessing Spent Fuel Performance in Geological Disposal. Part I: Assesment of the Instant Release Fraction. Journal of Nuclear Materials, 346, 56-65.

Metz, V., González-Robles, E., Kienzler, B. (2014). Characterization of PWR UOX Fuel Segments Irradiated in the PWR Gösgen. 2nd Annual Workshop Proceedings of the 7th EC FP CP FIRST-Nuclides project (eds. Kienzler et al.). KIT Scientific Reports 7676, 55-60.

Nucleonica GmbH (2011) Nucleonica Nuclear Science Portal (www.nucleonica.com), Version 3.0.11.

Poinssot, C., Ferry, C., Lovera, P., Jegou, C., Gras, J.-M. (2005). Spent Fuel Radionuclide Source Term Model for Assessing Spent Fuel Performance in Geological Disposal. Part II: Matrix Alteration Model and Global Performance. Journal of Nuclear Materials, 346, 66-77.

Shoesmith, D.W. (2000). Fuel Corrosion Process under Waste Disposals. Journal of Nuclear Materials, 282, 1-31.

Wegen, D.H., Papaioannou, D., de Weerd, W., Rondinella, V.V., Glatz, J.-P. (2012a). Fission Gas Release Measurement on 50.4 GWd/t_{HM} PWR Fuel. 1st Annual Workshop Proceedings of the 7th EC FP CP FIRST-Nuclides project (eds. Kienzler et al.). KIT Scientific Reports 7639, 201-205.

Wegen, D.H., Papaioannou, D., Gretter, R., Nasyrow, R., Rondinella, V.V., Glatz, J.-P. (2012b). Preparation of Samples for IRF Investigation and Post Irradiation Examinations from 50.4 GWd/t_{HM} PWR Fuel. 1st Annual Workshop Proceedings of the 7th EC FP CP FIRST-Nuclides project (eds. Kienzler et al.). KIT Scientific Reports 7639, 193-199.

14 Development of methods for the determination of ß emitters (^{79}Se, ^{126}Sn, ^{135}Cs, ^{90}Sr) based on sequential injection flow analysis coupled to ICP-MS

Laura Aldave de las Heras[1], Daniel Serrano-Purroy[1], Stefaan Van Winckel[1], Miguel Sandow[1], Sylvain Millet[1], Rosa Sureda Pastor[2], Albert Martinez Torrents[2] and Joan de Pablo[2,3]*

[1] *European Commission, Joint Research Centre, Institute for Transuranium Elements (EC)*
[2] *Fundació CTM Centre Tecnològic (ES)*
[3] *Universitat Politècnica de Catalunya (ES)*
* *Corresponding author: laura.aldave-de-las-heras@ec.europa.eu*

14.1 Abstract

Analytical methods for the determination of some key instant release fraction (IRF) pure ß emitters (^{79}Se, ^{126}Sn, ^{135}Cs, ^{90}Sr) at trace levels by high resolution ICP-MS coupled to an automated flow injection analysis (FIA) system have been developed. Flow injection techniques for pre-concentration and/or separation of radionuclides present in spent nuclear fuels leachates analogues without sophisticated and time consuming procedures have been established. Analytical columns packed with a specific stationary phase that selectively retains key radionuclides, for separation from the matrix elements, followed by a concentration step have been optimised in order to enhance sensitivity and improve detection limits of the methods developed. The analytical methods developed using natural isotopes in analogue spent fuel leachates show high potential to be applied in actual spent fuel samples with detection limits below the expected concentration levels and a reduced sample preparation time.

14.2 Introduction

In the long-term safety assessment of spent fuel disposal, the rapid release is often referred to as the instant release fraction or IRF and is considered to include (1) the release from the fuel-to-sheath gap which occurs in the first weeks to months of contact with groundwater and (2) release of material segregated at grain boundaries. The IRF is of particular interest in safety assessments, because some of the preferentially released radionuclides (e.g. ^{129}I, ^{36}Cl ^{135}Cs, ^{79}Se, ^{14}C, ^{99}Tc and ^{126}Sn) are both long-lived and geochemically mobile (Nagra, 2002).

ß-emitting nuclides pose a challenging task for reliable, quantitative determination because both radiometric and mass spectrometric methodologies require prior chemical purification for the removal of interfering activity and isobars, respectively. Inductively coupled plasma mass spectrometry (ICP-MS) has been used increasingly for the determination of long-lived radionuclides. This technique combines several advantages over classical radiochemical techniques such as high sensitivity and selectivity, low detection limits, and the possibility to couple ICP-MS on-line to different separation techniques. Flow injection and sequential injection techniques for pre-concentration and/or separation of radionuclides represent a versatile handling methodology that can be used for automating radiochemical separations. Automated flow injection analysis coupled to ICP-MS performs on-line pre-concentration on specific radionuclides to reduce matrix-related interferences and enhance sensitivity eliminating lengthy off-line sample preparation, providing timeliness analytical results and improving detection limits for radioisotopes by ICP-MS (Aldave de las Heras et al., 2013).

The aim of the present study is the development of analytical methods for the determination of key pure ß emitters at trace levels by high resolution ICP-MS coupled to an automated FIA system. More reliable analytical results and speciation analysis will provide a more realistic and true status of those nuclides in waste matrices and their possible transport in the environment using more accurate estimation and modelling scenarios.

14.3 Experimental

Table 14.1 summarises the key radionuclides and the developed method.

Table 14.1: *Key ß emitters radionuclides.*

Radionuclide	Interferences	Method feature	FIA method: *Column*/Eluents
^{79}Se	$^{38}Ar^{40}ArH^+$, $^{79}Br^+$, $^{39}K^{40}Ar^+$, $^{158}Gd^{2+}$, $^{63}Cu^{16}O^+$	Pre-concentration/ Separation	Nobias ^{a}Acetate 0.5 mol/L pH 4 $^{b}HNO_3$ 8%
^{79}Se		Speciation	Anion Exchange Phosphate 0.2 mol/L, pH 8
^{126}Sn	$^{126}Te^+$, $^{126}Xe^+$, $^{110}Pd^{16}O^+$, $^{110}Cd^{16}O^+$	Pre-concentration/ Separation	Nobias ^{a}Acetate 0.5 mol/L pH 6 $^{b}HNO_3$ 8%
^{135}Cs	$^{135}Ba^+$	Separation	AMP-PAN $^{a}HNO_3$ 2 mol/L
^{90}Sr	$^{180}W^{2+}$, $^{180}Hf^{2+}$, $^{58}Ni^{16}O_2^+$, $^{74}Ge^{16}O^+$, $^{52}Cr^{38}Ar^+$, $^{50}V^{40}Ar^+$, $^{54}Fe^{36}Ar^+$, $^{50}Ti^{40}Ar^+$, $^{90}Zr^+$	Pre-concentration/ Separation	$^{b}NH_4NO_3$ 5 mol/L *CG5A* $^{a}HNO_3$ 2 mol/L *CG12A* $^{a}HNO_3$ 50 mmol/L Sr-resin $^{a}HNO_3$ 4 mol/L ^{b}MQW

^{a} loading solution, ^{b} eluent

14.4 FAST system and ICP-MS instrumentation

Flow injection analyses were performed using the FAST system (Elemental Scientific, Omaha, Nebraska), a fully-automated online system that improves elemental detection limits in spent nuclear leachates by simultaneously preconcentrating analyte and eliminating matrix components. The targeted elements from an aliquot of sample are trapped on a mixed resin pre-concentration column while the matrix ions pass through to the waste. Selective eluents then elute the concentrated analyte from the column into the ICP-MS nebuliser.

Figure 14.1 represents the two FAST system configurations used for the pre-concentration and matrix removal (A) and for the speciation analysis (B).

The FAST system in pre-concentration uses two high purity valves to take up an aliquot of sample, retain and concentrate the radionuclide of interest removing the matrix on a PFA column packed with a radionuclide selective column. The preconcentrated radionuclide is eluted by backpressure into a PFA nebulizer attached to the ICP-MS spray chamber.

In the Se speciation analysis configuration, the FAST system uses one high purity valve to take up an aliquot of sample and separate Se species on a PFA column packed with an ion exchange resin. Se species are eluted into a PFA nebulizer attached to the ICP-MS spray chamber.

All ICP-MS measurements were carried out in a double focusing sector field ICP-MS (Element 2, Thermo Finnigan MAT GmbH, Bremen, Germany). Element 2 is equipped with a PC3 Peltier cooled spray chamber, a Fassel torch and a 27 MHz generator. The instrument is placed in a clean room facility class 1000. Instrumental settings and measurement parameters were optimised for each radionuclide determined. Typical sensitivity is $2 \cdot 10^6$ cps per µg/L of uranium in low resolution. Se measurements were performed in medium resolution.

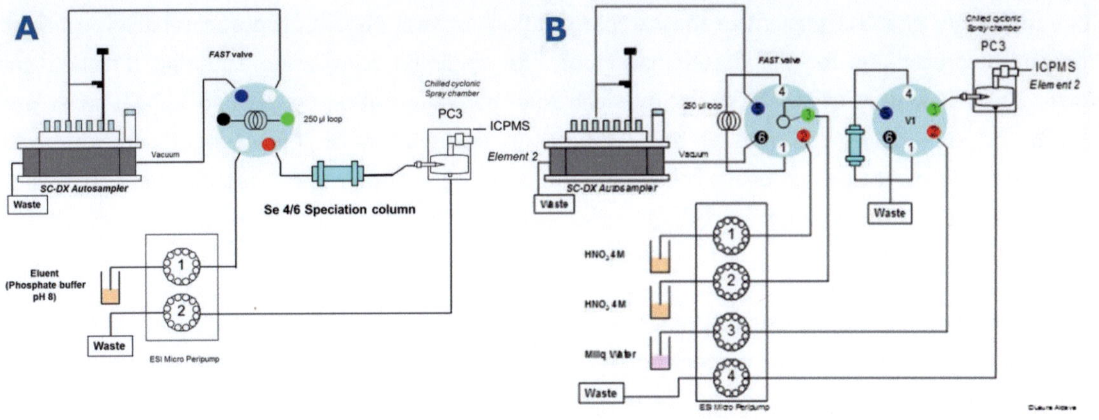

Figure 14.1: (A) FAST system configurations for speciation analysis (A) and for pre-concentration and/or separation (B) purposes.

14.5 Standards, Reagents and Samples

For the preparation of all solutions, high-purity water (18.2 MΩ/cm) from a Milli-Q Element system designed for ultratrace analysis (Millipore, Milford, MA) was used. Nitric acid, suprapur grade from Merck (Darmstadt, Germany), was purified using a quartz sub-boiling distillation unit. Both the water purification system and the sub-boiling distillation unit were operated in a clean room. Suprapur grade reagents (Acetic acid, NH_3, H_3PO_4, NH_4NO_3, NaCl and $NaHCO_3$) were used to clean materials, prepare solutions and preserve and analyse samples.

Natural element standards were obtained from CPI international (Amsterdam, The Netherlands) as 1,000 µg/mL stock standard solutions and diluted as necessary with 1% sub-boiled nitric acid. Se(IV) and Se(VI) standards were provided by Inorganic Ventures (Christiansburg, VA, USA) as 1,000 µg/mL stock standard solutions.

Nuclear spent fuel leachates analogues were prepared in a buffer of NaCl 19 mol/L and $NaHCO_3$ 1 mmol/L pH 8.06 with a uranium concentration from $1{\cdot}10^{-7}$ to 10^{-5} mol/L.

14.6 Results

Se pre-concentration in leachates

Due to isobaric interferences on several of its most abundant isotopes and its high ionisation potential (9.75 eV), Se is one of the most difficult elements to quantify by ICP-MS. Selenium minor isotopes are usually used for its determination since the use of the major isotopes is limited by the presence of argon dimers. In the case of ^{79}Se, many elements or molecular species arising from the sample matrix (Br, Gd, Dy, Cu, K) or from the plasma (^{38}Ar^{40}ArH) can induce interferences that are very difficult to correct for. Before the quantification of Se, preliminary separations are therefore necessary to remove potentially interfering species as well as major β/γ emitters present in the sample solution. Several papers proposed relatively complex radiochemical procedures for the determination of ^{79}Se by liquid scintillation counting (Frechou and Aguerre, 2002, Dewberry et al., 2000), or by electrothermal vaporisation coupled to ICP-MS after pre-treatment by ion exchange (Comte et al., 2003). The introduction of ICP-MS equipped with collision/reaction cells overcomes problems linked with the occurrence of polyatomic interferences. The elimination of the spectral background due to argon species allows for selenium determinations at concentrations down to 5 pg/g.

A method has been developed using a chelating resin in an online pre-concentration manifold with high resolution inductively coupled plasma spectrometry (ICP-MS) detection for the analysis of selenium in spent nuclear fuel leachates. Nobias Chelate-PA1 resin is a hydrophilic methacrylate with immobilised ethylenediaminetriacetic and iminodiacetic acid functional groups. This resin has an excellent affinity to trace metals ions (Sohrin et al., 2008) and can concentrate Se ions in a pH range from 4 to 6 in acetate medium.

Figure 14.2 shows the Se profile (Se^{IV}) after pre-concentration and separation from the matrix components in the Nobias column. The detection limits based on three blanks measurements and according to Currie (Currie, 1968) is around 140 fg/g at m/z 77. All measurements were performed in medium resolution to resolve some interferences at m/z 77 (typically ^{40}Ar^{36}ArH^{+}).

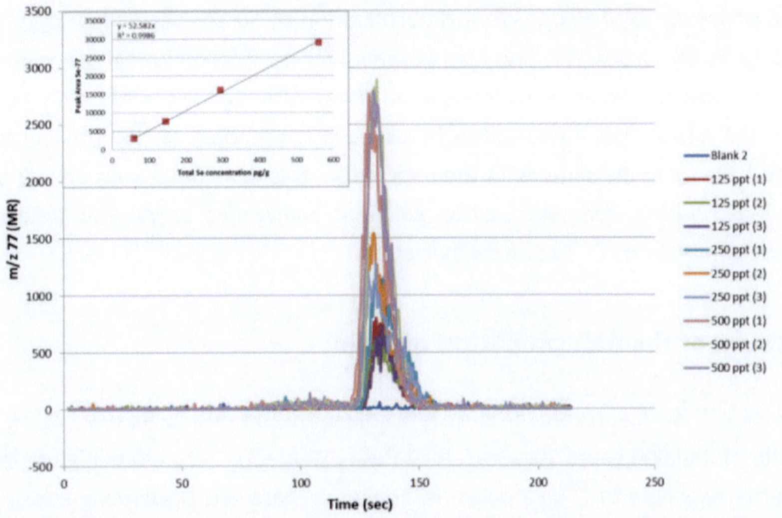

Figure 14.2: Pre-concentration of Se in nuclear spent fuel analogue samples in a Nobias column at different Se concentrations. Se(IV) species.

Se speciation in leachates

A method for the determination of Se species at trace levels by high resolution ICP-MS coupled to an automated chromatographic system has been developed. The FAST system uses one high purity valve to take up an aliquot of sample and separate Se species on a PFA column packed with an ion exchange resin in phosphate media at pH 8 (Table 14.1). Se species are eluted into a PFA nebulizer attached to the ICP-MS spray chamber. Figure 14.3 shows the Se species elution profiles at different concentration in bicarbonate water (19 mmol/L NaCl +1 mmol/L $NaHCO_3$) at pH 7.6 in spent nuclear leachates analogues. It also shows the analytical figures of merit of the method developed. A fit for purpose curve is obtained which does not introduce an extra uncertainty component showing that the method has potential for the determination of Se species in nuclear spent fuel leachates. The detection limit is 25 pg/g, representing an absolute amount of 6 pg. The repeatability, on the basis of three repetitions, is between 0.2 and 1% in this concentration interval.

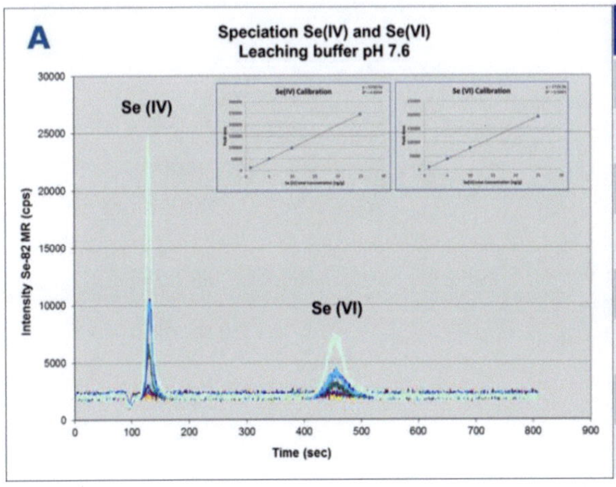

B Analytical figures of merit:

	Se (IV)	Se(VI)
Repeatability (%)	0.2 -1	
Total Se		
Detection limits (pg/g)	23.5	25.3
Absolute amount (pg)	5.9	6.3
^{82}Se		
Detection limits (pg/g)	2.2	2.3
Absolute amount (pg)	0.5	0.6

Figure 14.3: A: Se species elution profiles at different concentrations. B: Analytical figures of merit.

The new method offers an easy and efficient determination of Se species at low pg/g concentrations in spent nuclear leachates analogues. It shows high potential for application in real spent fuel leachates as the detection limits are below the expected concentration levels. Although the method only detects ^{82}Se, which is not relevant for the safety case, it does allow to calculate the amount of ^{79}Se present using the K-Origen code and the local burn up in the sample. In the near future this developed method will be applied using a new ICP-MS equipped with a collision/reaction cell that overcomes polyatomic interferences and that allows the direct determination of ^{79}Se and its speciation.

Preliminary results of the determination of ^{126}Sn

The major difficulty in ^{126}Sn measurement is its low concentration due to its low fission yield of 0.065 %. ^{126}Sn has a half-life of 230,000 years. Its short-lived decay product, ^{126}Sb, emits high-energy gamma radiation, making external exposure to ^{126}Sn a potential concern. There is considerable lack of literature regarding the analysis of ^{126}Sn. Previous studies on ^{126}Sn measurements were mainly intended to confirm its half-life (Oberli et al., 1999; Catlow et al., 2005; Bienvenue et al., 2009). Asai et al. (2013) have recently reported the content of ^{126}Sn in spent nuclear fuel samples. The determination of ^{126}Sn by mass spectrometry suffers from isobaric interference of ^{126}Te which cannot be resolved even with modern high resolution mass spectrometers. Therefore, if Te is not chemically separated, it can yield incorrect ^{126}Sn concentration in nuclear waste analysis. Nobias chelate-PA1 chelating resin has been employed in an online pre-concentration manifold with high resolution inductively coupled plasma spectrometry (ICP-MS) detection for the analysis of Sn in spent nuclear fuel leachates. The different behaviour of Sn and Te in the chelating column allows determining ^{126}Sn because Te is not retained on the column. Figure 14.4 shows the preliminary results of the determination of ^{126}Sn in diluted spent nuclear fuel leachates.

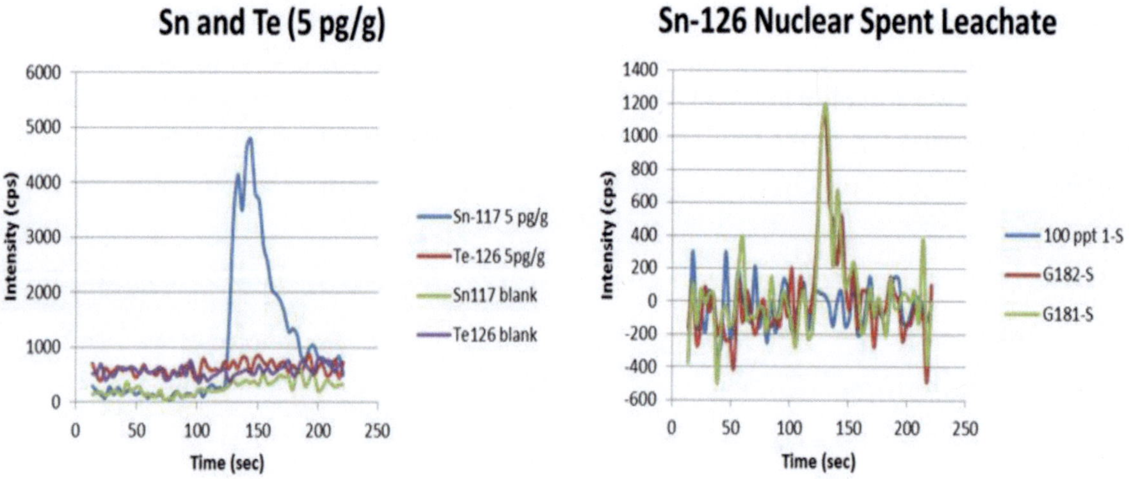

Figure 14.4: (Left) Behaviour of Sn and Te in the chelating column. (Right) Detection of ^{126}Sn in nuclear spent fuel leachates (dilution 1/10).

Preliminary results of the determination of ^{135}Cs

While ^{135}Cs is difficult to measure by radiochemical methods because of its long half-life and low energy ß-decay, ICP-MS instruments have reported detection limits of pg/ml of Cs. Moreover a separation procedure to remove isobaric interferences from ^{135}Ba is required for the identification and the quantification of ^{135}Cs.

AMP-PAN composite sorbent has been successfully used to concentrate and/or remove caesium radioisotopes from large volumes of environmental or nuclear waste samples (Brewer et al., 1999; Herbst et al., 2002; Pike et al., 2012). Exploiting the different behaviour of Cs and Ba in the AMP-PAN resin, a home-made column has been used to determine Cs isotopes in the presence of fission Ba and/or natural Ba isotopes.

Other separations schemes for the separation of Cs and Ba have been performed: the first is based on the use of mixed bed columns (cationic and anionic exchangers) of CG5A and CS5 (Dionex) and 1 mol/L nitric acid as eluent. The second separation scheme consisted of the use of CG12A and 20 mmol/L methanesulfonic acid as eluent. The CG12A is a hydrophilic, carboxylate /phosphonate functionalised cation exchanger that permits the elution of mono- and divalent cations using dilute acid solutions as eluent. Figure 14.5 shows the preliminary results of the potential determination of ^{135}Cs in nuclear spent fuel leachates.

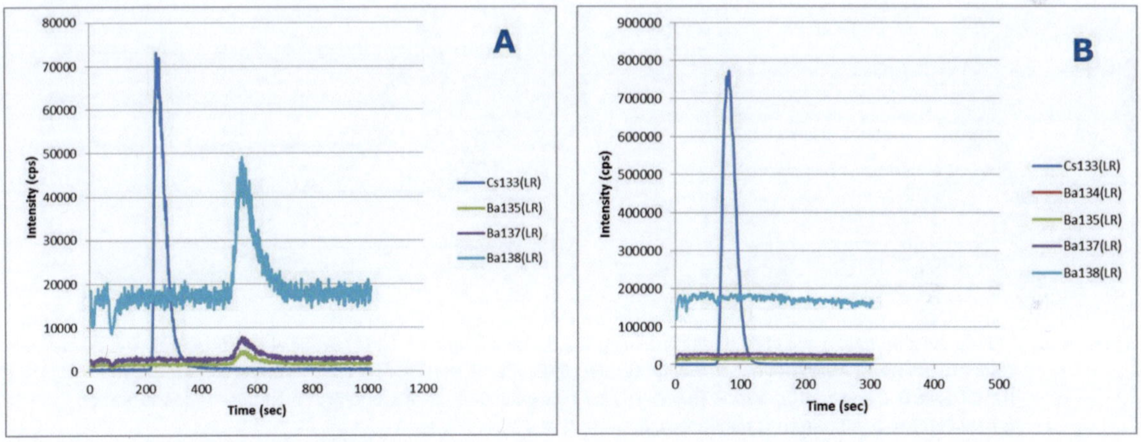

Figure 14.5: Cs and Ba separation using CG12A, injection volume 100 µL (A) and CG5A, injection volume 250 µL (B) coupled to FAST-ICP-MS. Natural Cs and Ba concentration is 1 ng/g.

Other ß radionuclides: ^{90}Sr

For the determination of ^{90}Sr very complicated chemical separations from ^{90}Y are necessary before ß counting. The determination of ^{90}Sr by ICP-MS suffers from a different set of analytical challenges from those associated with beta counting. Depending on the matrix composition of the sample, isobaric interferences by molecular or atomic ions can be expected at m/z 90 affecting the detection limit, accuracy and precision of determination of ^{90}Sr by ICP-MS. Complete removal of the isobar ^{90}Zr to below background intensity is necessary to quantify ^{90}Sr. Low instrumental detection limits (< pg/g) are required because the high natural abundance of stable Sr, present in environmental samples, limits pre-concentration. High concentrations of stable Sr in the final sample solution require sufficient abundance sensitivity to resolve the peak tail of ^{88}Sr. Finally, an efficient sample preparation protocol is needed in order to remove other elements present in the sample attaining a final solution with low total dissolved solids.

The method proposed exploits the coupling of FIA to ICP-MS using two high purity valves to take up an aliquot of sample, retain and concentrate Sr removing the matrix on a PFA column packed with Sr®-Spec resin. The pre-concentrated Sr is eluted by backpressure into a PFA nebulizer attached to the ICP-MS spray chamber.

In Figure 14.6A the elution profiles obtained for solutions of 4 mol/L HNO_3 containing 10, 25, 50 and 100 pg/g of ^{90}Sr are reported. As can be seen, ^{90}Sr is eluted in a total time of 120s. A calibration curve was obtained by using the peak area of ^{90}Sr versus the total ^{90}Sr concentration (Figure 14.6). A fit for purpose curve is obtained which does not introduce an extra uncertainty component. The detection limit of ^{90}Sr was calculated by means of repeated measurements of the blank and according to Currie. The detection limit is 1.6 pg/g, taking into account that the injection volume is 250 µL that represents an absolute amount of 0.25 pg (5 Bq) of ^{90}Sr. The repetability of the method, based on the relative standard deviation of the peak area calculated on the basis of three repetitions is always less than 2% in this concentration range.

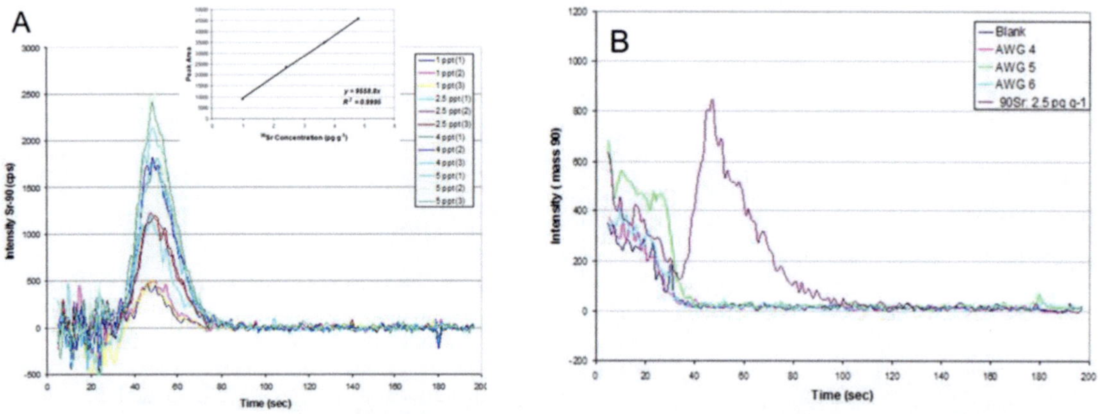

Figure 14.6: ^{90}Sr elution chromatograms of solution containing 1, 2.5, 4 and 5 pg/g of ^{90}Sr (A) Artificial ground waters (AGW) elution chromatograms (B) AGW1: Ti: 0.2, V: 0.2, Cr: 0.1, Ge: 0.05, Zr: 0.2, and Fe: 6; AGW2: Ti: 0.2,V: 0.7, Cr: 0.2, Ge: 0.05, Zr: 0.2, and Fe: 10;. AGW3: Ti: 0.2, V: 0.3, Cr: 0.5, Ge: 0.05, Zr: 2.5, and Fe: 17. Natural Sr concentration is between 2 to 6. All concentrations are in µg/L.

Artificial ground waters samples were prepared and analysed in order to determine the impact of possible interferences in the m/z 90 due to the presence in the water of some elements (Ti, V, Cr, Fe, Ge, Zr, Sr) and the peak tailing of ^{88}Sr on m/z 90. The concentration of natural strontium in the artificial ground water samples used was between 2 to 6 ng/g. Figure 14.6B shows that the peak tailing of the ^{88}Sr isotope is not disturbing the peak area of ^{90}Sr when the concentration of Sr is around 5 ng/g. It is clear that the method developed can be applied for the determination of Sr in leachates of spent fuel in ground water. In spite of the presence of some elements interfering at m/z 90, those elements are not retained in the column and have no impact on the quantitative determination of ^{90}Sr. The influence of the abundance sensitivity of ^{88}Sr will only depend of the total concentration of natural Sr in the ground water used for the leaching experiments. Indeed, the concentration of fission ^{88}Sr in the spent fuel is of the same order of magnitude as ^{90}Sr.

14.7 Conclusions and Future work

Flow injection analysis coupled to ICP-MS offers easy and efficient determination of low pg/g levels of the key ß radionuclides studied in natural samples. Moreover, with its automated and versatile sample uptake and introduction capabilities, reduces sample preparation time while achieving good detection limits. The analytical methods developed using natural isotopes in analogue spent fuel leachates show high potential to be applied in actual spent fuel samples. In the near future, the method will be tested and implemented in a new nuclearised ICP-MS equipped with a collision reaction cell (CRC technology is a tool to minimise the effect of many potential interferences) and a FAST system.

14.8 Acknowledgement

The research leading to these results has received funding from the European Union's European Atomic Energy Community's (Euratom) Seventh Framework Programme FP7/2007-2011 under grant agreement n° 295722 (FIRST-Nuclides project).

14.9 References

Aldave de las Heras, L., Sandow, M., Olszewski, G., Serrano-Purroy, D., van Winckel, S., Glatz, J.P. (2013). Determination of Traces of Radionuclides by Hyphenated Techniques Coupled to Inductively Coupled Plasma Mass Spectrometry. Revista de la Societat Catalana de Química, 2, 61-67.

Asai, S., Toshimitsu, M., Hanzawa, Y., Suzuki, H., Shinohara, N., Inagawa, J., Okumura, K., Hotoku, S., Kimura, T., Suzuki, K., Kaneko, S. (2013). Isotope Dilution Inductively Coupled Plasma Mass Spectrometry for Determination of ^{126}Sn Content in Spent Nuclear Fuel Sample. Journal of Nuclear Science and Technology, 50, 556-562.

Bienvenu, P., Ferreux, L., Andreoletti, G., Arnal, N., L´Epy, M.-C., Comte, J., Be, M.-M. (2009). Determination of ^{126}Sn Half-Life from ICP-MS and Gamma Spectrometry Measurements. Radiochimica Acta, 7, 687-694.

Brewer, K.N., Todd, T.A., Wood, D.J., Tullock, P.A., Sebesta, F., John, J., Motl, A. (1999). AMP-PAN Column Tests for the Removal of ^{137}Cs from Actual and Simulated Ineel High-Activity Wastes. Czechoslovak Journal of Physics, 49, 959-964.

Catlow, S.A., Troyer, G.L., Hansen, D.R., Jones, R.A. (2005). Half-life Measurement of ^{126}Sn Isolated from Hanford Nuclear Defense Waste. Journal of Radioanalytical Nuclear Chemistry, 263, 599-603.

Comte, J., Bienvenu, Ph., Brochard, E., Fernandez, J.M., Andreoletti, G. (2003). Determination of Selenium-79 in Solutions of Fission Products after Pre-Treatment by Ion Exchange Chromatography and ETV-ICP-MS. Journal of Analytical and Atomic Spectrometry, 18, 702-707.

Currie, L.A. (1968). Limits for Qualitative Detection and Quantitative Determination. Application to radiochemistry. Analytical Chemistry, 40, 586-593.

Dewberry, R.A., Leyba, J.D., Boce, W.T. (2000). Observation and Measurement of ^{79}Se in Savannah River Site High Level Waste Tank Fission Product Waste. Journal of Radioanalytical Nuclear Chemistry, 245, 491-500.

Frechou, C., Aguerre, S. (2007). Improvement of a Radiochemical Separation for Selenium 79: Applications to Effluents and Nuclear Wastes. Talanta, 72, 1166-1171.

Herbst, R.S., Law, J.D., Tood, T.A. (2002). Integrated AMP-PAN, TRUEX, and SREX Testing. I. Extended Flowsheet Testing for Separation of Surrogate Radioradionuclides from Simulated Acidic Tank Waste. Separation Science and Technology, 37, 1321-1351.

Nagra (2002). Project Opalinus Clay. Nagra NTB 02-05, Nagra, Wettingen, Switzerland.

Oberli, F., Gartenmann, P., Meier, M., Kutschera, W., Suter, M., Winkler, G. (1999). The Half-Life of ^{126}Sn Refined by Thermal Ionization Mass Spectrometry Measurement. International Journal Mass Spectrometry, 184, 145-152.

Pite, S.M., Buesseler, K.O., Breier, C.F., Dulaiova, H., Stastna, K., Sebesta, F. (2012). Japan Using AMP-PAN Resin and Quantification Via Gamma Spectroscopy and Inductively Coupled Mass Spectrometry. Journal of Radioanalytical Nuclear Chemistry, 296(1) , 369-374, DOI 10.1007/s10967-012-2014-5.

Sohrin, Y., Urushihara, S., Nakatsuka, S., Kono, T., Higo, E., Minami, T., Norisuye, K., Umetani, S. (2008). Multielemental Determination of GEOTRACES Key Trace Metals in Seawater by ICP-MS after Pre-Concentration Using an Ethylenediaminetriacetic Acid Chelating Resin. Analytical Chemistry, 80, 6267-6273.

15 WP3: Dissolution based release IRF-corrosion-test of commercial UO_2 BWR spent nuclear fuel

Daniel Serrano-Purroy[1], Laura Aldave de las Heras[1], Stefaan Van Winckel[1], Albert Martínez Torrents[2], Rosa Sureda[2], Jean Paul Glatz[1] and Vincenzo V. Rondinella[1]*

[1] *European Commission, Joint Research Centre, Institute for Transuranium Elements (EC)*
[2] *Fundació CTM Centre Tecnològic (ES)*
* *Corresponding author: daniel.serrano-purroy@ec.europa.eu*

15.1 Abstract

This report summarizes the work carried out at ITU in the frame of WP3 during the third year of the FIRST Nuclides project. It includes IRF results with a SNF cladded piece and with SNF powders selected from the centre and from the periphery of the fuel. All experiments were carried out under oxidising conditions.

15.2 Introduction

In the frame of the FIRST Nuclides project, ITU's main objective in WP3 is the quantification of the instant/fast release fraction of fission products into bicarbonated aqueous phase (19 mmol/L NaCl + 1 mmol/L $NaHCO_3$) during corrosion leaching test using a commercial BWR spent nuclear fuel. The work was carried out using samples from the same SNF, but from different pellet regions and with different morphologies: a cladded segment of about 2 mm length and powder samples with a mean particle size of 50 µm selected from the centre and from the periphery of the fuel pellet.

During the reporting period the experiment which started in 2013 using a SNF segment was completed; a second pellet was used for inventory determination; in addition, two powder samples were prepared, characterised and corrosion experiments were carried out. The results of these tests are presented in this report.

15.3 Experimental

Samples

Cladded segment samples

According to a prepared cutting plan, two neighbouring cladded segments of approximately 2 mm length selected from the middle of a pellet were cut: one to carry out the corrosion experiment and one to determine the inventory by acid dissolution. Sample characteristics as well as macroscopic views of the selected pellet can be found in (Serrano-Purroy et al., 2013).

Powder samples

Two different Spent Nuclear Fuel (SNF) powder samples corresponding to the central radial position (labelled CORE) and the periphery of the SNF pellet (labelled OUT) were prepared.

First, from the selected pin, several segments from a neighbouring position to the previously selected cladded segments were cut and drilled in order to obtain the CORE SNF fraction. The drilling was carried out starting from the centre in a number of different steps to minimise the friction between the tip of the drilling tool and the SNF and, consequently, minimise the alteration of the core surface. The segment radius without cladding was of 4.8 mm. Samples from different radial positions were collected separately (0–1.5, 1.5–3 and 3–4 mm). To avoid any risk of contamination with the High Burn-up Structure (HBS), also called rim, particles, the CORE sample was prepared from powder taken from the centre of the pellet and up to 3 mm in radial position. Within the limitations of the available tools the drilling was continued until shortly before the cladding (4 mm). Then the cladding was compressed with the help of a manual press in order to detach the external region of the SNF, named OUT, leaving the cladding almost free of SNF. Due to the different preparation methods, i.e. friction vs. fracturing, the CORE sample was finer and more homogeneous in size than the OUT sample. Because of this, OUT samples were milled with a Universal Mill (M20 KA, IKA GmbH, Germany). In order to obtain the specific particle size range, i.e., between 50 and 100 µm, both powder samples, CORE and OUT, were sieved (Vibrax-VXR, IKA GmbH, Germany). Figure 15.1 shows macroscopic views of both powder fractions. Finally, in order to remove fines attached to the grain surface, especially in the CORE sample, both powder SNF fractions were washed twice in the selected corrosion media.

Figure 15.1: SNF powders taken from the centre, CORE (left), and the periphery of the fuel, OUT (right).

Table 15.1 shows the sample characteristics, the experimentally set S/V ratio and some important irradiation history parameters. These parameters are reported because of the importance they will have for later comparisons with experiments carried out by other partners within the frame of the project.

Table 15.1: SNF samples selected for the corrosion experiments. Segment (42BWR) and powders (42CORE and 42OUT).

Sample	42BWR	42CORE	42OUT
Length (mm)	2.8±0.1	-	-
Diameter without cladding (mm)	9.4±0.1	-	-
Segment weight with cladding (g)	2.3614±0.0001	-	-
Segment weight without cladding (g)	1.7947±0.0001[a]	-	-
Powder sample weight (g)	-	0.1098±0.0001	0.1167±0.0001
Powder particle size (µm)[b]	-	50±20	46±20
Surface area (mm^2)	800±10[c]	4652±2000[d]	5381±2800[d]
S/V (m^{-1})	18±1	105±50	119±60
FGR (%)	2.3	-	-
Burnup (GWd/t_{HM})[e]	45	-	-
Av. Linear Power Rate (W/cm)	215	-	-

[a] The weight of the pellet without cladding was estimated from the ratio pellet to cladding obtained after complete dissolution of a pellet for inventory determination.

[b] The mean particle size was measured following the same procedure as in (Serrano-Purroy et al., 2012).

[c] The surface area exposed was calculated taking into account the total surface in contact with solution (both faces of the slice and the lateral surface) and a roughness factor of 3.5 (Iglesias et al., 2008).

[d] Specific surface areas were estimated for each SNF fraction assuming a spherical shape of the particles and a roughness factor of 3.5 (Iglesias et al., 2008).

[e] Estimated from gamma scanning and cutting plan positions. It will also be experimentally determined during the course of the project.

Experimental set-up

The experimental set-up used for the pellet is described in (Serrano-Purroy et al., 2013).

Corrosion experiments with powders were carried out under oxidizing conditions (in air) at hot-cell temperature (25°C ± 5). Samples were leached in the same aqueous medium used for the pellet (19 mmol/L NaCl + 1 mmol/L $NaHCO_3$).

The powder samples were placed in 50 mL borosilicate glass test tubes of (150 X 25) mm with plastic screw cap (Schütt Labortechnik GmbH, Göttingen, Germany). The tubes were placed on a rotator stirrer to avoid concentration gradients. The rotator stirrer was made up of polymethacrylate over a stainless steel framework, using an electrical motor with a nominal speed of 30 rpm; this speed is enough to ensure that the SNF is kept in suspension and to assure a perfect contact with the solution.

The solution (50 ± 1 mL) is replaced completely at pre-set intervals, shorter at the beginning of the experiment. Non-filtered solutions are further diluted in 1 mol/L nitric acid according to the levels of concentration and dose rate suitable for subsequent analysis by ICP-MS. Based on these results, the behavior of U as matrix main component and of some selected IRF elements (Cs, Rb, Sr, Mo and Tc) is discussed.

Data treatment

The total released or cumulative moles in solution for element *i*, moles(*i*), is calculated considering the total amount of radionuclide *i* removed in each sampling (Equation 15.1):

$$moles(i)=\sum_0^n moles_{sample}(n,i) \qquad (15.1)$$

where $moles_{sample}$ (n,*i*) correspond to the moles in solution before each complete replenishment *n*. The cumulative concentration $cumC_i$ of an element i in moles/L is then calculated as:

$$cumC_i=\frac{moles(i)}{V} \qquad (15.2)$$

where V corresponds to the average volume of the dissolution samples in dm^{-3}. The Cumulative Fraction of Inventory of an element *i* released in the Aqueous Phase (CumFIAP*i*) is given by Equation 15.3.

$$CumFIAP_i=\frac{m_{i,aq}}{m_{i,SNF}}=\frac{c_iV_{aq}}{m_{SNF}H_i} \qquad (15.3)$$

where $m_{i,aq}$ is the mass of element *i* in the aqueous phase in g, $m_{i,SNF}$ the mass of element *i* in the SNF sample in g, m_{SNF} the mass of SNF used in the experiment in g, H_i corresponds to the fraction of inventory for the nuclide *i* in g/g, c_i is the cumulative concentration of element *i* in solution in g/mL and V_{aq} is the volume of solution in mL. The Fractional Release for an element *i* Normalised to the total Surface area (FNS_i) is:

$$FNS_i=\frac{FIAP_i}{S} \qquad (15.4)$$

where S is the total surface area (m^2). The Fractional Release Normalised to Uranium for an element *i* (FNU*i*) is given by Equation 15.5.

$$FNU_i=\frac{FIAP_i}{FIAP_U} \qquad (15.5)$$

where $FIAP_i$ and $FIAP_U$ are the FIAP of element *i* and uranium, respectively. Finally, the Instant Release Fraction for an element *i* (IRF_i) is given by Equation 15.6.

$$IRF_i=FIAP_i-FIAP_U \qquad (15.6)$$

Speciation

Calculations of uranium speciation in bicarbonate water (19 mmol/L NaCl + 1 mmol/L $NaHCO_3$) were assessed using the geochemical code CHESS (Van der Lee et al., 2002). Most internationally known databases have been made available for use with CHESS, here among the EQ3/6, MINTEQ, PHREEQC and NEA databases. The CHESS default database is based on a moderated version of the LLNL's EQ3/6. For this study, both the default CHESS and the NEA databases were used. All calculations were based assuming the local equilibrium, oxidising conditions and a temperature of 25°C.

15.4 Results and discussion

pH evolution

During the experiments pH measurements were carried out at each sampling time. The pH decreases from initially 8.4 at short contact times down to 7.6 at the end of both experiments. This might be a result of the uranium oxide surface buffer capacity.

Concentration in solution (speciation and secondary phase formation studies)

Table 15.3 to Table 15.5 in Annex 1 show measured concentrations in solution at each contact sampling time for the three experiments.

Figure 15.2 shows the uranium speciation under the experimental conditions. As an example, measured uranium concentrations at each sample for the experiment with the pellet are also included. Analysed uranium concentration in solution at each solution renewal is below uranium secondary phase formation levels. Similar studies were made with the experiments carried out with powders

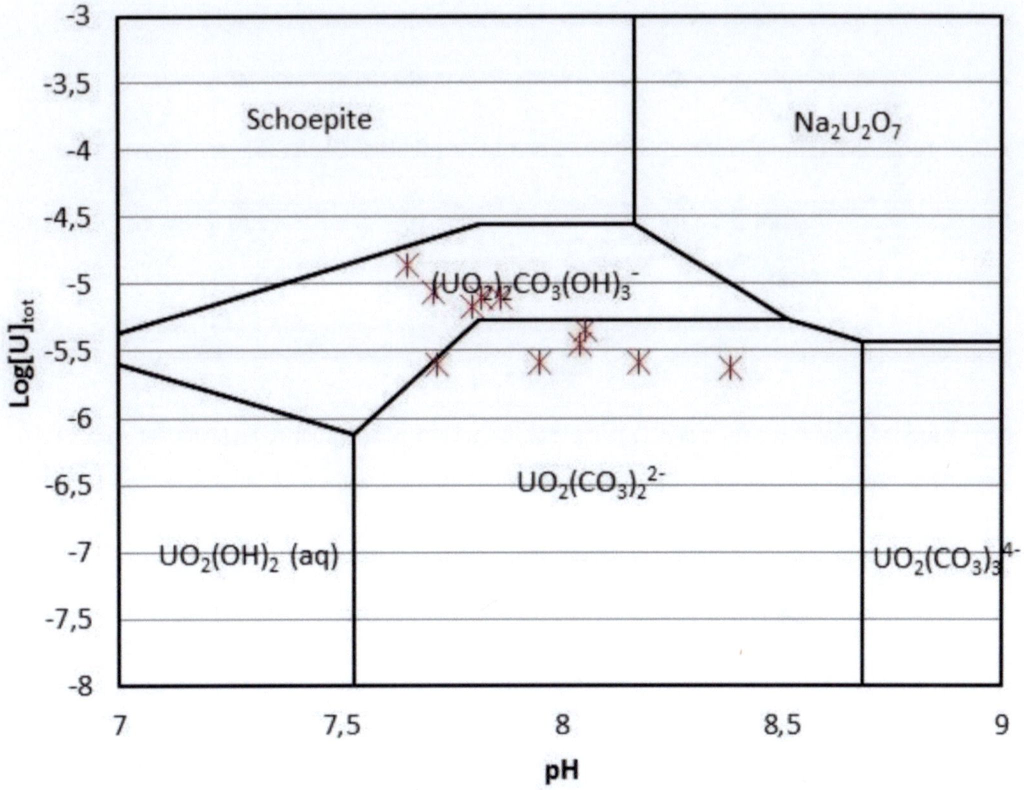

Figure 15.2: Speciation diagram of uranium in the studied experimental conditions (NEA and CHESS databases were used for the calculation). Measured uranium concentrations for each sample in the experiments with the pellet are also included.

In the coming weeks the samples, both cladded segment and powders, will be analysed by SEM in order to corroborate these findings. To properly follow the evolution of the experiments, the cumulative moles in solution as a function of leaching time were calculated according to Equation 15.1, see Figure 15.3 to 5.

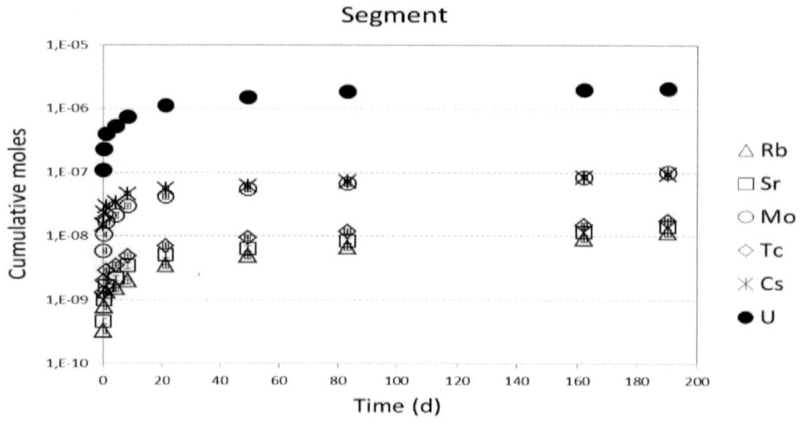

Figure 15.3: Elemental cumulative moles vs time. Experiment carried out with segment.

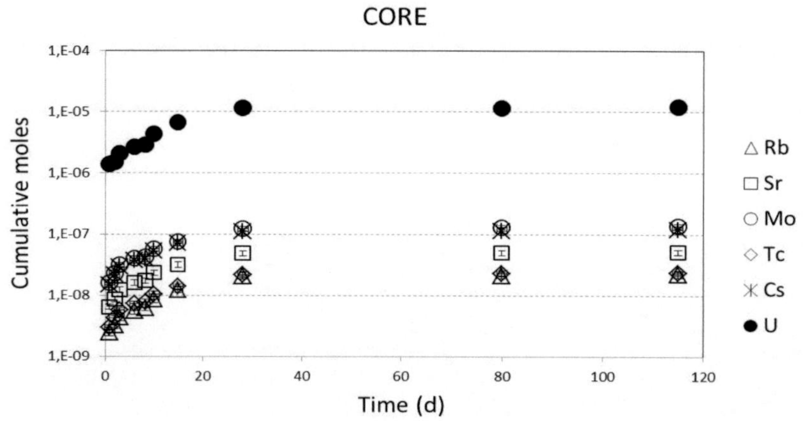

Figure 15.4: Elemental cumulative moles vs time. Experiment carried out with powder from the center (CORE) of the SNF.

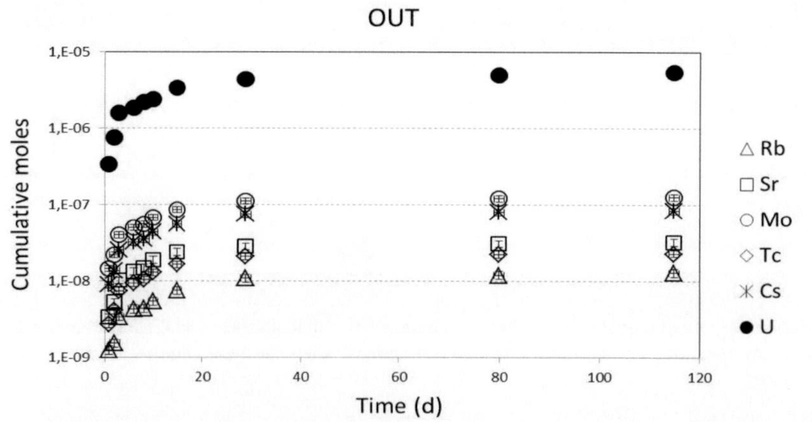

Figure 15.5: Elemental cumulative moles vs time. Experiment carried out with powder from the periphery (OUT) of the SNF.

The dissolution release of both uranium, considered as a tracer of the matrix dissolution, and other elements considered to be Instant Release Fraction (Cs, Mo, Tc, Rb and Sr) is faster at the beginning of the experiment. In the case of uranium the initial higher release suggests a surface peroxidation. However, for the other elements this is attributed to the fraction immediately available for dissolution after contacting the sample with the aqueous solution, i.e. segregated fraction at the gap and the cracks. The second dissolution mechanism, slower and long-term, is attributed to matrix dissolution in the case of uranium and to internal grain boundary IRF dissolution for the other elements.

Cumulative FIAP(%) and FNU

In addition to cumulative moles, and taking into account the experimental inventory, cumulative FIAP values were also calculated, see Figure 15.6 to Figure 15.8.

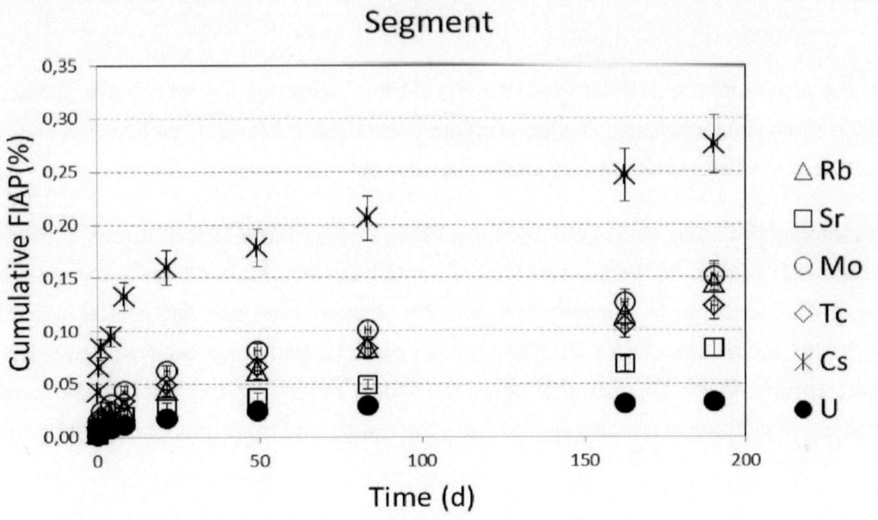

Figure 15.6: Cumulative FIAP(%) vs time. Experiment carried out with the segment.

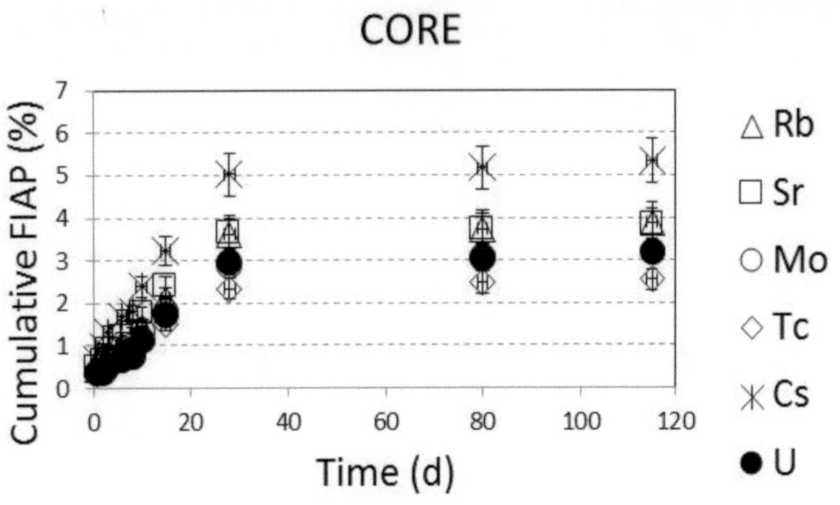

Figure 15.7: Cumulative FIAP(%) vs time. Experiment carried out with powder from the centre (CORE) of the SNF.

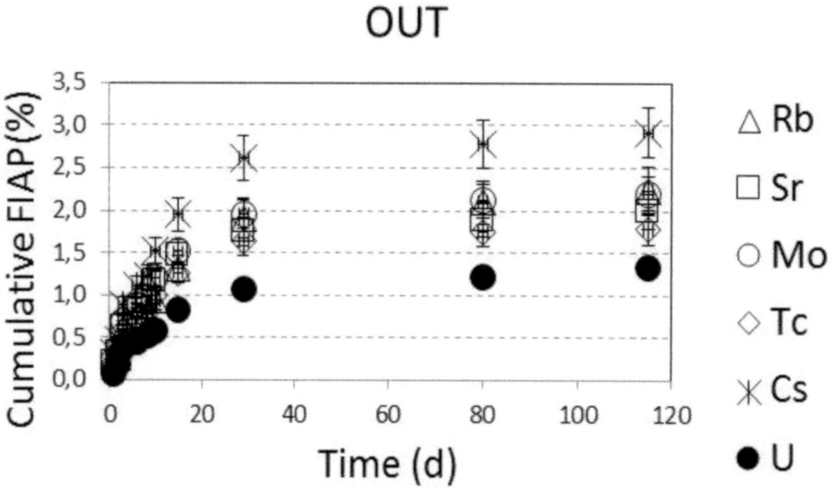

Figure 15.8: Cumulative FIAP(%) vs time. Experiment carried out with powder from the periphery (OUT) of the SNF.

In the case of the pellet and the OUT sample, Cs is the element with the biggest release, followed by Mo, Tc, Rb and Sr which show similar release. Finally, uranium presented the lowest release. However, in the CORE sample Mo and Tc presented a release very similar to uranium.

FNU at each sampling time was also calculated; see Table 15.6 to Table 15.8 in Annex 1. In the case of the segment and the OUT sample all the studied IRF elements have FNU higher than 1, being especially significant the case of Cs. The longer the experiment runs the smaller is the calculated FNU indicating that with enough time the IRF will be completely dissolved and all these elements will approach matrix dissolution. In the experiment with the CORE, Cs, Rb and Sr present a similar behaviour, but Mo has FNU close to 1 and Tc below 1, indicating that they are not segregated from the matrix and therefore don't belong to the IRF.

IRF(%)

Finally, IRF(%) was calculated taking into account cumulative FIAPs. The evolution of the IRF with time is shown in Figure 15.9 to Figure 15.11.

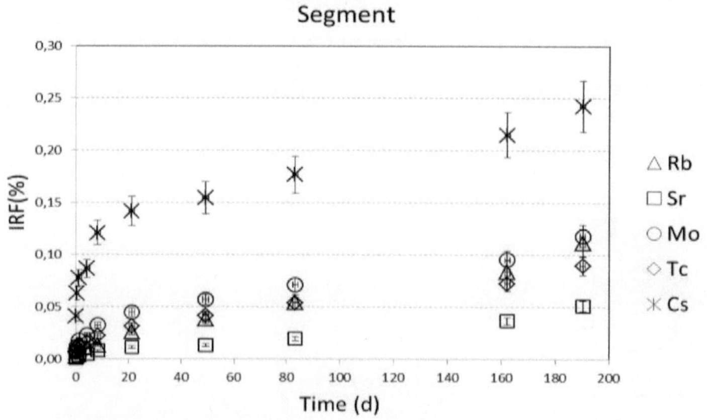

Figure 15.9: IRF(%) vs. time. Experiment carried out with the segment.

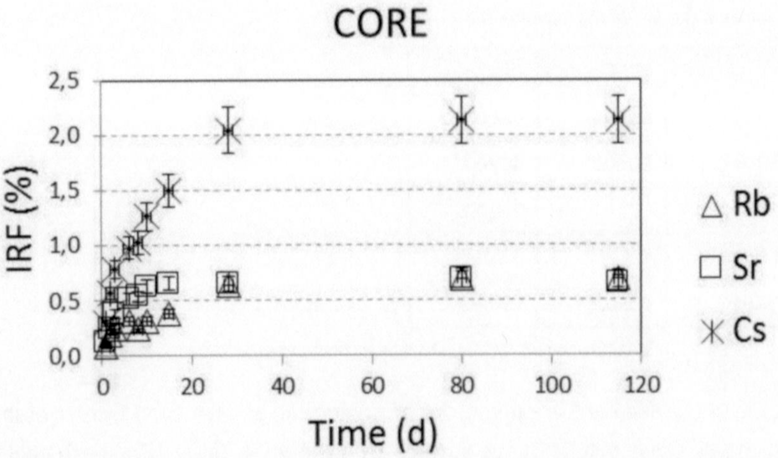

Figure 15.10: IRF(%) vs. time. Experiment carried out with powder from the centre (CORE) of the SNF.

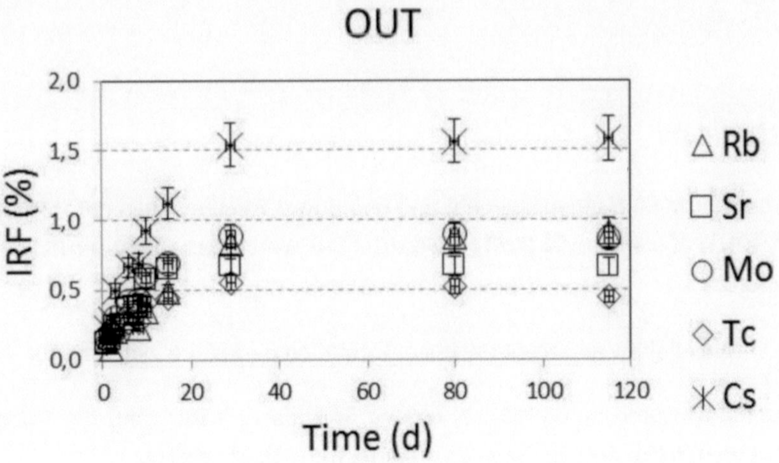

Figure 15.11: IRF(%) vs. time. Experiment carried out with powder from the periphery (OUT) of the SNF.

In all cases, higher release is found at the beginning of the experiment: it is assumed to be that coming from the gap and the cracks in the case of the segment, and from the grain boundaries initially available to water corrosion in the case of the powder specimens. A second IRF contribution over longer duration and characterized by slower dissolution rates is also detected. This is assumed to arise from the fraction segregated at the internal grain boundaries, not immediately accessible to water at the beginning of the experiment. IRF (%) values at the end of the experiments (ca. 200 days for the pellet and 120 days for the powders) are shown in Table 15.2.

At the end of the experiments the highest and most significant IRF (%) corresponds to Cs. In the case of Cs, Rb and Sr no significant differences were found between the powder fractions. The differences found in the case of Mo and Tc might come from different speciation in the centre (higher fractions dissolved in the fuel matrix) and in the periphery. Further studies will be performed in order to clarify these results.

Table 15.2: IRF(%) values at the end of the experiments.

IRF (%)	Pellet	CORE	OUT
Rb	0.11±0.01	0.8±0.1	0.9±0.1
Sr	0.05±0.01	0.7±0.1	0.7±0.1
Mo	0.12±0.01	-	0.9±0.1
Tc	0.09±0.01	-	0.5±0.1
Cs	0.24±0.01	2.1±0.2	1.6±0.2

IRF values in the case of the segment are one order of magnitude smaller than those obtained with powder. However, if we normalise these results by the surface area shown in Table 15.1, both pellet and powder IRF results are in good agreement.

Careful comparison with the literature and the other partner's findings will be carried out at the end of the project by the WP leader.

15.5 Conclusions

Corrosion experiments using a cladded segment and two power fractions selected from the centre (CORE) and the periphery (OUT) of commercial BWR (42GWd/t_U) SNF were successfully carried out and the respective results are reported.

Uranium speciation studies show no secondary phase formation during the experiments.

Among the elements measured by ICPMS, Cs, Rb and Sr dissolve faster than the matrix (uranium) and therefore, are attributed to IRF. Among them, Cs presents the highest release.

Mo and Tc can also be considered IRF in the experiments carried out with the segment and the powder OUT samples. However, in the experiment with the CORE sample, Mo dissolves congruently with uranium while Tc dissolves even slower. Further studies will be dedicated to the clarification of this effect.

After normalising the results by the available surface area, no significant differences were found between the experiment carried out with cladded segment and with the powders.

15.6 Acknowledgement

The research leading to these results has received funding from ENRESA Under ENRESA/ITU/CTM 31698 Agreement and the European Union's European Atomic Energy Community's (Euratom) Seventh Framework Programme FP7/2007-2011 under grant agreement n° 295722 (FIRST-Nuclides project).

The authors want to thank A. Komlan for the technical support and S. Van Winckel for the ICPMS determinations.

15.7 References

Iglesias, E., Quiñones, J. (2008). Analogous Materials for Studying Spent Nuclear Fuel: The Influence of Particle Size Distribution on the Specific Surface Area of Irradiated Nuclear Fuel. Applied Surface Science, 254(21), 6890-6896.

Van der Lee, J. De Windt, L. (2002). CHESS Tutorial and Cookbook (version 3.0). Ecole des Mines de Paris, CIG Fontainebleau.

Serrano-Purroy, D., Aldave de las Heras, L., Glatz, J.-P., Rondinella, V.V., Sureda, R. (2013). WP3. Dissolution Based Release. 2nd Annual Workshop Proceedings of the 7th EC FP CP FIRST-Nuclides project (eds. Kienzler et al.). KIT Scientific Reports 7676, 87-96.

Serrano-Purroy, D., Clarens, F., González-Robles, E., Glatz, J.-P., Wegen, D.H., de Pablo, J., Casas, I., Giménez, J., Martínez-Esparza, A. (2012). Instant Release Fraction and Matrix Release of High Burn-Up UO_2 Spent Nuclear Fuel: Effect of High Burn-Up Structure and Leaching Solution Composition. Journal of Nuclear Materials, 427, 249–258.

15.8 Annex

Table 15.3: Elemental concentration in solution (moles dm^{-3}) at the end of each sampling contact time (not to be confused with the total experimental time which corresponds to the cumulative sampling time and that will be used in some of the later discussions). Experiment carried out with the cladded segment.

Time (d)	0.08	0.17	1.00	3.00	4.00	13.00	28.00	34.00	79.00	28.00
Rb	7.1E-09	1.0E-08	1.1E-08	4.0E-09	1.0E-08	3.1E-08	3.1E-08	3.6E-08	5.0E-08	2.2E-07
Sr	1.0E-08	1.2E-08	1.4E-08	1.1E-08	2.5E-08	3.4E-08	2.9E-08	4.2E-08	6.5E-08	3.7E-07
Mo	1.3E-07	1.0E-07	1.3E-07	9.4E-08	1.8E-07	2.5E-07	2.6E-07	2.7E-07	3.5E-07	1.9E-06
Tc	2.8E-08	1.5E-08	1.8E-08	1.3E-08	2.9E-08	4.5E-08	4.8E-08	5.2E-08	6.0E-08	4.8E-07
Cs	3.2E-07	1.8E-07	1.3E-07	7.8E-08	2.7E-07	1.9E-07	1.4E-07	2.0E-07	2.8E-07	1.6E-06
U	2.4E-06	2.6E-06	3.5E-06	2.6E-06	4.5E-06	8.0E-06	8.0E-06	6.9E-06	2.5E-06	8.6E-06

* Note that isotopic concentrations can be calculated from the inventory determination reported in WP1. These numbers are reported with the appropriate number of significant digits.

Table 15.4: Elemental concentration in solution (moles dm^{-3}) at the end of each sampling contact time (not to be confused with the total experimental time which corresponds to the cumulative sampling time and that will be used in some of the later discussions). Experiment carried out with powder from the centre (CORE) of the SNF.

Time (d)	0.86	1.03	0.98	2.98	2.11	1.84	4.99	13.18	51.96	34.98
Rb	5.8E-08	1.8E-08	2.6E-08	2.8E-08	1.6E-08	5.3E-08	8.1E-08	1.9E-07	1.6E-08	2.0E-08
Sr	1.5E-07	5.1E-08	8.3E-08	8.9E-08	2.7E-08	1.4E-07	1.8E-07	3.7E-07	2.8E-08	3.9E-08
Mo	3.6E-07	1.6E-07	1.9E-07	2.0E-07	6.4E-08	3.2E-07	4.0E-07	1.1E-06	1.7E-07	1.2E-07
Tc	6.7E-08	2.9E-08	3.3E-08	3.7E-08	1.2E-08	5.9E-08	7.5E-08	1.8E-07	2.9E-08	1.2E-08
Cs	3.3E-07	1.6E-07	1.8E-07	1.8E-07	5.7E-08	3.0E-07	4.0E-07	8.9E-07	7.4E-08	8.7E-08
U	3.1E-05	2.8E-06	1.2E-05	1.4E-05	5.7E-06	3.1E-05	5.0E-05	1.1E-04	5.0E-06	1.5E-05

* Note that isotopic concentrations can be calculated from the inventory determination reported in WP1. These numbers are reported with the appropriate number of significant digits.

Table 15.5: Elemental concentration in solution (moles dm^{-3}) at the end of each sampling contact time (not to be confused with the total experimental time which corresponds to the cumulative sampling time and that will be used in some of the later discussions). Experiment carried out with powder from the periphery (OUT) of the SNF.

Time (d)	0.86	1.12	0.89	2.97	2.13	1.82	5.01	14.00	51.13	34.99
Rb	2.9E-08	7.0E-09	4.1E-08	2.2E-08	1.8E-09	2.2E-08	4.8E-08	7.6E-08	2.5E-08	1.9E-08
Sr	7.8E-08	4.4E-08	1.1E-07	7.6E-08	3.5E-08	8.8E-08	1.1E-07	1.0E-07	5.1E-08	3.9E-08
Mo	3.3E-07	1.7E-07	4.0E-07	2.5E-07	1.2E-07	2.8E-07	4.2E-07	5.7E-07	2.0E-07	1.2E-07
Tc	6.2E-08	3.2E-08	7.4E-08	4.8E-08	2.3E-08	5.5E-08	7.7E-08	1.1E-07	3.3E-08	1.2E-08
Cs	2.0E-07	1.1E-07	2.6E-07	1.6E-07	7.3E-08	1.8E-07	2.8E-07	4.3E-07	1.1E-07	8.7E-08
U	8.4E-06	9.4E-06	1.8E-05	6.7E-06	7.2E-06	5.0E-06	2.2E-05	2.3E-05	1.3E-05	1.1E-05

* Note that isotopic concentrations can be calculated from the inventory determination reported in WP1. These numbers are reported with the appropriate number of significant digits.

Table 15.6: FNU at each sampling solution versus cumulative sampling time. Experiment carried out with the cladded segment.

Time (d)	0.08	0.25	1.25	4.25	8.25	21.25	49.25	83.25	162.25	190.25
Rb	2	3	3	2	2	2	3	3	4	4
Sr	2	2	2	2	2	2	2	2	2	3
Mo	5	4	4	4	4	3	3	3	4	5
Tc	5	4	3	3	3	3	3	3	3	4
Cs	25	18	13	11	11	9	7	7	8	8

Table 15.7: FNU at each sampling solution versus cumulative sampling time. Experiment carried out with powder from the centre (CORE) of the SNF.

Time (d)	0.86	1.98	2.87	5.84	7.97	9.79	14.81	28.81	79.94	114.93
Rb	1.2	1.5	1.5	1.4	1.5	1.4	1.3	1.3	1.3	1.3
Sr	1.3	1.6	1.7	1.7	1.7	1.6	1.4	1.2	1.2	1.2
Mo	1.1	1.4	1.4	1.4	1.3	1.2	1.0	1.0	1.0	1.0
Tc	0.9	1.2	1.1	1.1	1.1	1.0	0.9	0.8	0.8	0.8
Cs	1.8	2.4	2.4	2.4	2.3	2.1	1.9	1.7	1.7	1.7

Table 15.8: FNU at each sampling solution versus cumulative sampling time. Experiment carried out with powder from the periphery (OUT) of the SNF.

Time (d)	0.86	1.98	2.87	5.84	7.97	9.79	14.81	28.81	79.94	114.93
Rb	2.6	1.5	1.5	1.7	1.4	1.6	1.6	1.8	1.7	1.7
Sr	2.5	1.8	1.6	1.8	1.7	2.0	1.8	1.6	1.5	1.5
Mo	3.0	2.1	1.8	1.9	1.8	2.0	1.8	1.8	1.7	1.6
Tc	2.5	1.7	1.5	1.6	1.5	1.7	1.5	1.5	1.4	1.3
Cs	3.7	2.6	2.3	2.5	2.3	2.6	2.4	2.4	2.3	2.2

16 Results of leaching experiments on spent UO_2 and MOX nuclear fuel carried out at PSI

Enzo Curti, Ines Günther-Leopold, Hans-Peter Linder and Nenad Milosavljevic*

Nuclear Energy and Safety Department (NES), Paul Scherrer Institut (CH)
** Corresponding author: enzo.curti@psi.ch*

16.1 Abstract

The results of aqueous leaching experiments carried out under oxidizing conditions on cladded or bare high burn-up UO_2/MOX spent fuel from the Gösgen (PWR) and Leibstadt (BWR) nuclear power plants are presented. So far, data are available for ^{137}Cs and ^{129}I up to 182 days leaching time. In general, the calculated Fractions of Inventory in the Aqueous Phase (FIAP) are comparable to those determined in an earlier project on similar fuel samples. The results indicate stronger release from cladded fuel samples compared to bare fuel samples. In the case of the Leibstadt samples, the "cladding contribution" to the total release of ^{129}I is even dominant.

16.2 Introduction

Previous studies have shown that release of radionuclides from uranium oxide (UOX) and mixed uranium-plutonium oxide (MOX) spent nuclear fuel (SNF) is controlled by two mechanisms: the rapid dissolution of soluble elements, such as Cs, I and Cl (Johnson and Tait, 1997; Grambow et al., 2008; Carbol et al., 2009; Johnson et al., 2012) and the slow dissolution rate of the UO_2 grains. In the long-term safety assessment terminology, the rapid release is often referred to as the instant release fraction (IRF) and is thought to include: (1) the release of soluble nuclides from the fuel/cladding gap which occurs in the first weeks to months of contact with aqueous solution and (2) the release of soluble nuclides segregated at fuel grain boundaries accessible to water.

The IRF is a critical parameter in safety assessments, because it is dominated by long-lived, geochemically mobile nuclides such as ^{14}C, ^{36}Cl, ^{90}Sr, ^{99}Tc, ^{129}I, and ^{135}Cs (Nagra, 2002) and may significantly contribute to the total calculated dose. It is well-known that reactor operating conditions have a strong impact on fuel microstructure and on the in-pile segregation of some radionuclides from the UOX/MOX matrix. Fission gas release (FGR) is a particularly important indicator as it is found to correlate – although not in a trivial way - with reactor operating conditions such as linear power rating and fuel burn-up.

Considerable research has been carried out to study the correlation between reactor operation parameters and chemico-physical fuel parameters controlling radionuclide release, such as diffusion coefficients and restructuring of the fuel grains. For instance, at burn-ups exceeding about 50 GWd/t_{iHM} a microstructure develops in the rim of the fuel pellet, which is finer-grained and apparently more porous than the initial fuel structure (Noirot et al., 2008). This led to the hypothesis that labile radionuclides could be released more easily from such restructured rims into aqueous media. Although some data indicate enhanced release of

Cs from the rim zone (see e.g. Fig. 8 in Johnson et al., 2012) the same authors conclude that "the data reported here do not support the hypothesis that the radionuclides segregated at grain boundaries are easily leached". Therefore, the question whether IRF nuclides are preferentially leached from the restructured periphery of high-burnup SNF has not yet been resolved conclusively.

Available data show that the short-term release of IRF nuclides tends to correlated with reactor operational parameters, such as FGR and linear power (Stroes-Gascoyne et al., 1987; Johnson et al., 2004). In principle, such correlations would be very useful "extrapolation tools", since direct experimental IRF determinations are very expensive and time consuming. Thus, a well-assessed empirical relationship between IRF and reactor parameters would allow estimating reliably the potential releases from SNF to be disposed of in a geological repository. The goal of FIRST-Nuclides is to provide new data and eventually a robust database for such correlations.

Currently, there is a significant trend for many reactors towards increasing fuel burn-up, implying that more IRF data from high-burnup SNF are urgently needed to improve the database. Whereas the data obtained in the previous PSI study, documented in Johnson et al. (2012), focused on the different contributions of rim and central pellet region, as well as on the effect of fuel fragmentation, the present series of experiments aims at separating fuel and cladding contributions to the rapid release of radionuclides during aqueous leaching. As in the previous project, the release of ^{137}Cs, ^{129}I, ^{79}Se and ^{14}C will be investigated.

We define as bulk fuel contribution the direct release of radionuclides from bare fuel samples containing core and rim material in representative proportions. In principle, this contribution can be determined if the leached pellet section was completely separated from the cladding and recovered to 100%. This goal was achieved for most fuel samples. In two cases, 30%-40% of the fuel was fine-grained residual material resulting from sample preparation, which was not used in the leaching experiments to avoid undesired variations in surface area. Although unlikely, one cannot exclude a bias towards rim or core material for these two samples.

The definition of release contributions from cladding samples is trickier. Leaching of cladding fragments encompassing the whole cross-section of the zircaloy annulus (with the internal porous zirconia layer) will include, in addition to radionuclides formed in the cladding, part of the gap contribution, i.e. nuclides segregated from the fuel and deposited on the inner side of the cladding. We define this sum as bulk cladding contribution. In contrast, we define as intrinsic cladding contribution the release of "native" IRF radionuclides generated in the cladding itself. Only activation products such as ^{14}C are expected from this source. Because some of our samples were prepared by cutting tangentially only the external part of the cladding without the inner zirconia layer (see section 1.1.1) eventually it will be possible to distinguish between these two types of cladding contributions. The experiments with pure cladding have been started with strong delay and it will therefore not be possible to obtain data on the intrinsic cladding contribution before the end of FIRST-Nuclides.

16.3 Leaching Experiments

Experimental Methods

Sample preparation

PSI selected fuel rods from Leibstadt (BWR) and Gösgen (PWR) nuclear power plants for investigations within the project FIRST-Nuclides. UO_2 spent fuel with standard initial enrichment of ^{235}U and average burnups of 57.5 (Leibstadt) and 56.6 GWd/t_{iHM} (Gösgen), was used for the leaching experiments. Additional leach samples were prepared from a Gösgen MOX fuel rod with 63.0 GWd/t_{iHM} burnup.

Based on the experience made during analogous earlier experiments (Johnson et al., 2012) fuel rod specimens of 20 mm length were used for all leach samples, whereas 10 mm length samples were prepared for the determination of fuel burnup. The cutting was carried out in "Hot Cell 2" at the PSI hot laboratory, using a LECO cutting machine with a diamond plate saw. Some samples were broken into two halves by cutting the cladding on opposite sides without cutting the fuel itself (see also remarks at the bottom of Table 16.1).

Since the handling of the leaching equipment had to be performed with manipulators in a shielded dissolution box, the experimental setup was kept as simple as possible and followed the same design used in Johnson et al. (2012). The fuel and cladding samples were introduced in glass columns (total volume approx. 250 mL) and filled with 30 mL of leaching solution (19 mM NaCl + 1 mM $NaHCO_3$, pH ~ 7.4). After 7 days of "pre-leaching", the entire solution volume was sampled for analysis through a sealed outlet cock with an integrated glass filter (to prevent clogging by solid particles). The columns were immediately refilled with 115 mL of fresh leaching solution. Aliquots of 15 mL each were then taken sequentially after total leaching times of 28, 56 and 182 days ("Leaching", no refill). A last sampling will be performed after 364 days leaching time. The used equipment is shown in Figure 16.1.

Analytical methods

The analytical methods used to obtain aqueous concentrations of the nuclides of interest are described in detail in Deliverable 1.2, pp. 39-42. Therefore, only a very short account is given here. ^{137}Cs and ^{129}I activities were determined by gamma spectrometry. The determination of ^{129}I required previous separation of Cs via multiple precipitation steps and a corresponding loss correction (3% per step). ^{79}Se, as well as the full U, Pu, Nd isotope vectors for the burnup determination, will be measured with a high-resolution inductively coupled plasma mass spectrometry (ICP-MS) system. The same system will be used to determine the isotopic composition of Cs, after chromatographic Ba separation. Finally, ^{14}C will be determined by liquid scintillation counting (LSC) based on the extraction method described by Stroes-Gascoyne et al. (1994).

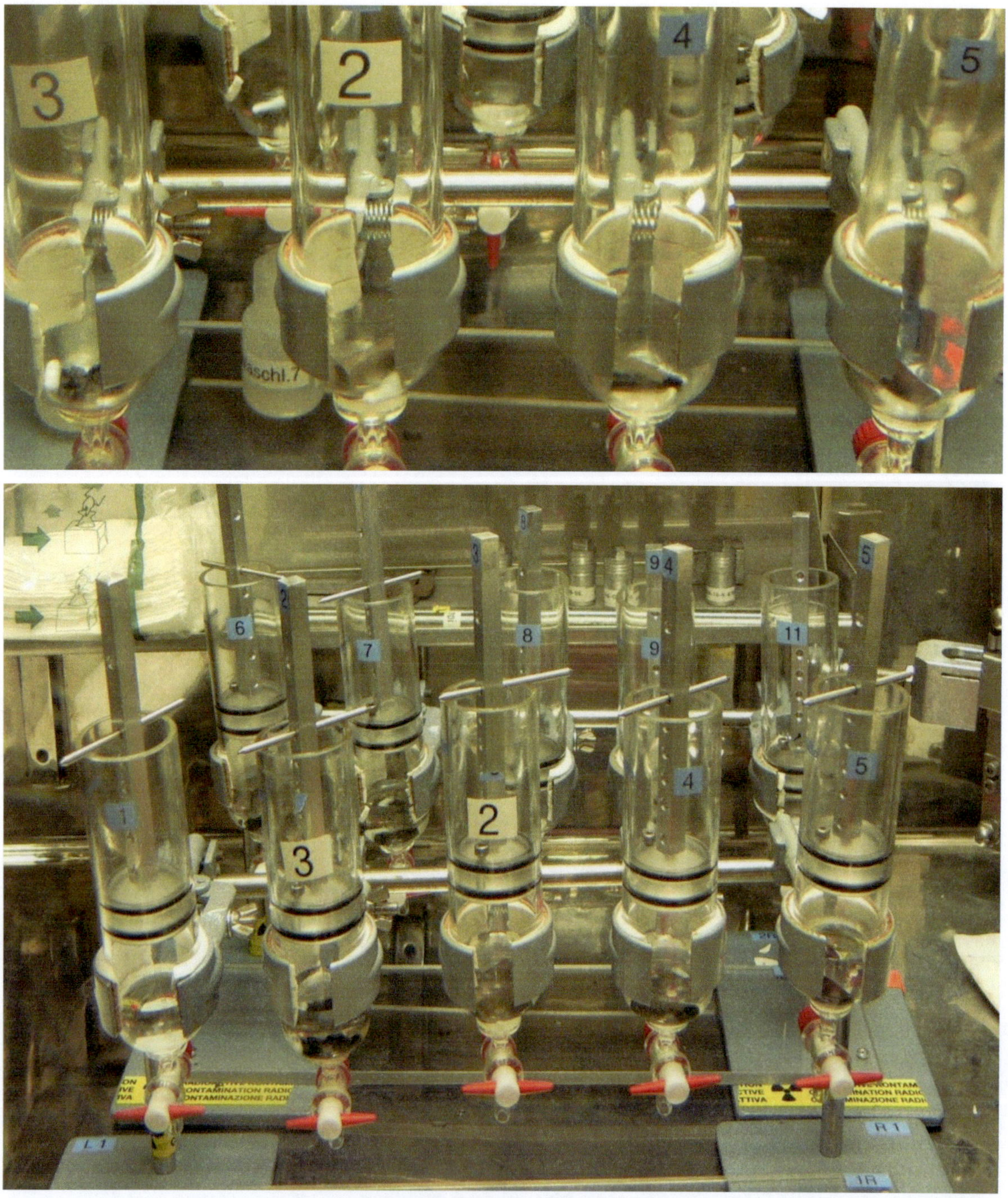

Figure 16.1: Assembly of glass columns used for the PSI leaching experiments. Top: SNF samples before introducing the leaching solution. Bottom: columns "in operation", after filling with leaching solution.

Experimental matrix and status of the experiments

At the date of this reporting, only ^{137}Cs and ^{129}I data were available for 6 out of 9 samples, up to 182 days leaching time. Due to delays in the supply of a special saw for the tangential cutting of fuel-free cladding fragments, no data are currently available for experiments KKL-UO2-C, KKG-UO2-C and KKG-MOX-C. These experiments will be started as soon as possible. All outstanding data gathered after the end of the FIRST-Nuclides project (December 31th, 2014) will be documented in separate publications.

Table 16.1: Matrix of PSI leach experiments. After the first sampling (pre-leaching, 30 mL), the columns were refilled with 115 mL of fresh solution, from which 15 mL were drawn at the indicated reaction times. Black crosses indicate currently available data. Grey crosses indicate outstanding sampling or analyses.

Sample	**Column**	**Remarks**	**Pre-Leaching (30 mL)**	**Leaching (refill: 115 mL, 15 mL samples)**			
			7 d	28 d	56 d	182 d	364 d
Blank	1	(I)	X	-	X	-	X
KKG-UO2-CF	2	(II)	X	X	X	X	X
KKG-UO2-F	3	(III)	X	X	X	X	X
KKG-UO2-CR	4	(IV)	X	X	X	X	X
KKG-UO2-C	8	(V)	X	X	X	X	X
KKL-UO2-CF	5	(II)	X	X	X	X	X
KKL-UO2-F	6	(III)	X	X	X	X	X
KKL-UO2-C	9	(V)	X	X	X	X	X
KKG-MOX-C	11	(V)	X	X	X	X	X
KKG-MOX-F	7	(III)	X	X	X	X	X

Key for sample identification: KKG=Gösgen, KKL=Leibstadt, UO_2=uranium oxide fuel, MOX=mixed oxide fuel, C=cladding, F=fuel, R=fuel residues. Remarks: (I) fuel-free test, (II) intact cladded segment, (III) fuel fragments (obtained by crushing after detaching the cladding), (IV) almost fuel-free cladding ring remaining after preparation of column 3 sample, (V) fragments obtained by tangential cuts of cladding, without fuel residues.

Data reduction and results

Calculation of FIAP

The cumulative "Fraction of Inventory in the Aqueous Phase" (FIAP) as a function of leaching time was calculated from measurements of the radionuclide activity (gamma-spectrometry) or isotope concentration (ICP-MS) in the probed leaching solutions, after normalization to a reference inventory. The reference inventories (expressed as activities) were extrapolated from model data (Appendix B in McGinnes (2002)) referring to the different fuel and reactor types. All activities were extrapolated to the starting time of the experiments and corrected for the local burnup. For the latter correction, deviations from the average burnup were determined by comparing the results of gamma scans along the fuel rod with sample position in the rod. A correction of + 15% (Gösgen samples) and + 5% (Leibstadt samples) was applied to the inventory data obtained from average burnups.

The FIAP calculation adopted here follows the definition of Roudil et al. (2007). Accordingly, the FIAP increment for a nuclide X during a time interval Δt is given by:

$$\Delta \mathrm{FIAP}_X(\Delta t) = \frac{\Delta \bar{A}(X)_{SOL}\ m_{SOL}}{f_U\ \bar{A}(X)_{FUEL}\ m^0_{(U,Pu)O_2}} \quad (16.1)$$

where:

$\Delta\bar{A}(X)_{SOL}$ is the increment in specific activity of nuclide *X* (Bq/g solution) in the sampled solution, corrected for decay to a reference time t_{ref} (start of experiments, 16.09.2013)

$\bar{A}(X)_{FUEL}$ is the specific activity of nuclide *X* in the fuel at t_{ref} (Bq/g_{iHM})

m_{SOL} is the mass of solution in the column (g) during the time interval Δt

$m^0_{(U,Pu)O_2}$ is the initial mass of UO_2 or $(U,Pu)O_2$ fuel in the column (g)

f_U is a mass conversion factor accounting for the transformation of leached inventories from (Bq/g fuel) to (Bq/g_{iHM}), defined as:

$$f_U=(1-x)\frac{MW(U)}{MW(UO_2)}+x\frac{MW(Pu)}{MW(PuO_2)} \tag{16.2}$$

In Equation 16.2 *MW(X)* indicates molar weights (mol/g) and *x* is the enrichment fraction of Pu in the initial fuel (0 for pure UO_2 fuel). Note that the activities in Equation 16.1 can be interchanged with normalized molar amounts, $\bar{n}$ (*X*), or weights, $\bar{m}$(*X*). The cumulative FIAP is obtained by summing up the individual time increments:

$$FIAP_X=\sum_{i=1}^{N}\Delta FIAP_X \tag{16.3}$$

Results may be either presented as fractions (FIAP) or percentages (FIAP% = 100 x FIAP).

The uncertainty of FIAP values was estimated from the square root of the summed up variances of each single quantity affecting Equations 1 and 3, applying standard propagation error formulae (Bevington and Robinson, 2003). The covariance terms were neglected. It turned out that the overall uncertainty of ^{137}Cs and ^{129}I FIAP values is dominated by the uncertainties of gamma activity measurements in the leaching solution, $\bar{A}(X)_{SOL}$, and of the reference inventory, $\bar{A}(X)_{FUEL}$. The statistical uncertainty of ^{137}Cs measurements varied between 1.3% and 3.7% (95% confidence level, 1σ), whereas the uncertainty of ^{129}I measurements was in general higher (2.7% - 37.3%) due to low counting rates. The uncertainty of the reference inventories was estimated to be about 2% - 3% for both nuclides after applying the local burnup correction. The so determined FIAP uncertainties are reported numerically in Table 16.3 and 4 and shown as error bars in Figure 16.2 and Figure 16.3, provided they exceed the symbol size.

16.4 Results

Table 16.2 shows a list of the initial experimental weights of solid introduced in each column (Sample mass = fuel ± cladding, rightmost column) compared to calculated values for the masses of fuel, cladding and total sample, calculated from geometry and dimensions of the pins. The values under the column "Fuel mass (g)" were used for the quantity $m^0(U,Pu)O_2$ in Equation 16.1. The values under the columns denoted "Fuel fraction" and "Cladding fraction" indicate how much of the theoretical full weight of each sample was effectively present in each column. Values of 1 for both fuel and cladding fractions indicate full fuel and cladding integrity in correspondence to the cutting plans, i.e. cylindrical samples with a height of exactly 20 mm. Values below 1 indicate that smaller amounts are present. Values of less than 1 may reflect intentional exclusion of fine-grained material produced during sample preparation or the presence of fuel residues in cladding samples. For instance, the calculated fuel fraction is only 3% in sample KKG-UO2-CR because the intention was to remove the fuel from the cladding as completely as possible.

Table 16.2: Experimental matrix showing calculated masses of fuel and cladding in each leach experiment, compared with weighed total sample masses. See text for explanations.

Sample	Column	Fuel fraction	Cladding fraction	Fuel mass (g) calculated	Cladding mass (g) calculated	Sample mass (g) calculated	Sample mass (g) measured
KKG-UO2-CF	2	0.98	0.98	13.91	2.65	16.57	16.57
KKG-UO2-F	3	0.70	0	9.98	0	9.98	9.98
KKG-UO2-CR	4	0.03	1	0.43	2.71	3.13	3.13
KKG-UO2-C	8						n.d.
KKL-UO2-CF	5	0.99	0.99	12.62	2.67	15.29	15.29
KKL-UO2-F	6	0.97	0	12.31	0	12.31	12.31
KKL-UO2-C	9						n.d.
KKG-MOX-C	11						n.d.
KKG-MOX-F	7	0.56	0	8.00	0	8.00	8.00

Key for sample identification: KKG=Gösgen, KKL=Leibstadt, UO_2=uranium oxide fuel, MOX=mixed oxide fuel, C=cladding, F=fuel, R=fuel residues. n.d.=not yet determined

The "Fuel fraction" and "Cladding fraction" values were determined by fitting them to the experimental total sample weights, using the available knowledge on the samples. For instance, since the samples denoted '-CF' include the full cladding annulus, the additional constraint that Fuel and Cladding Fractions must be equal could be used. In the case of pure fuel samples, obviously the "Cladding fraction" was set to zero. The fitting was carried out with the help of a spreadsheet by varying the fractions until agreement between calculated and measured sample mass was obtained. The Fuel and Cladding fractions will be used after completion of all experiments to quantify *bulk fuel*, *bulk cladding* and *intrinsic* cladding contributions to the total release.

Table 16.3 and 4 show the experimentally determined cumulative FIAP% values for ^{137}Cs and ^{129}I, calculated using Equations 1, 2 and 3, with the corresponding uncertainties. Note that the fourth solution sample was drawn from column #5 (KKL-UO2-CF) already after 148 days instead of 182 days. The anticipated sampling was prompted leaking of the solution from the column, probably caused by the radiation-induced rupture of the Teflon liner sealing the outlet cock. After this event, the column was refilled and the outlet cocks of all columns were replaced with new ones.

Figure 16.2 and Figure 16.3 show the experimentally determined cumulative FIAP% values for ^{137}Cs and ^{129}I up to 182 days leaching time, for experiments with bare and fully cladded spent fuel. Data calculated for sample KKG-UO2-CR, involving cladding with some fuel residues, are not shown in the plots since the calculated FIAP% values (due to the very small fuel mass) are strongly affected by the dominant contribution of the bulk cladding and therefore not representative. These values are greyed out in Table 16.3 and 4.

Table 16.3: Matrix of PSI leach experiments with currently available cumulative FIAP% values for ^{137}Cs.

Sample	Column	7 d	28 d	56 d	148 d	182 d
KKG-UO2-CF	2	4.77 ± 0.23	5.63 ± 0.30	5.74 ± 0.29	-	5.96 ± 0.32
KKG-UO2-F	3	3.01 ± 0.15	3.40 ± 0.16	3.51 ± 0.17	-	3.94 ± 0.21
KKG-UO2-CR	4	20.7 ± 1.0	22.7 ± 1.0	23.0 ± 1.1	-	24.1 ± 1.3
KKL-UO2-CF	5	1.88 ± 0.09	2.44 ± 0.12	2.43 ± 0.12	2.48 ± 0.13	-
KKL-UO2-F	6	1.79 ± 0.09	1.93 ± 0.09	1.96 ± 0.09	-	2.10 ± 0.11
KKG-MOX-F	7	6.32 ± 0.30	6.74 ± 0.31	6.80 ± 0.31	-	7.17 ± 0.38

Key for sample identification: KKG= Gösgen, KKL=Leibstadt, UO2= uranium oxide fuel, MOX= mixed oxide fuel, C=cladding, F=fuel, R=fuel residues

Table 16.4: Matrix of PSI leach experiments with currently available cumulative FIAP% values for ^{129}I.

Sample	Column	7 d	28 d	56 d	148 d	182 d
KKG-UO2-CF	2	6.16 ± 0.41	9.01 ± 0.65	8.50 ± 0.72	-	8.87 ± 0.92
KKG-UO2-F	3	3.40 ± 0.27	3.92 ± 0.94	3.84 ± 1.4	-	3.52 ± 0.28
KKG-UO2-CR	4	21.5 ± 1.6	21.5 ± 1.0	21.5 ± 1.0	-	21.5 ± 1.7
KKL-UO2-CF	5	0	2.05 ± 0.19	2.89 ± 0.32	2.67 ± 0.26	-
KKL-UO2-F	6	0.23 ± 0.03	0.23 ± 0.01	0.23 ± 0.01	-	0.23 ± 0.02
KKG-MOX-F	7	11.05 ± 0.72	11.05 ± 0.51	11.52 ± 2.70	-	11.12 ± 0.88

Key for sample identification: KKG= Gösgen, KKL=Leibstadt, UO2= uranium oxide fuel, MOX= mixed oxide fuel, C=cladding, F=fuel, R=fuel residues

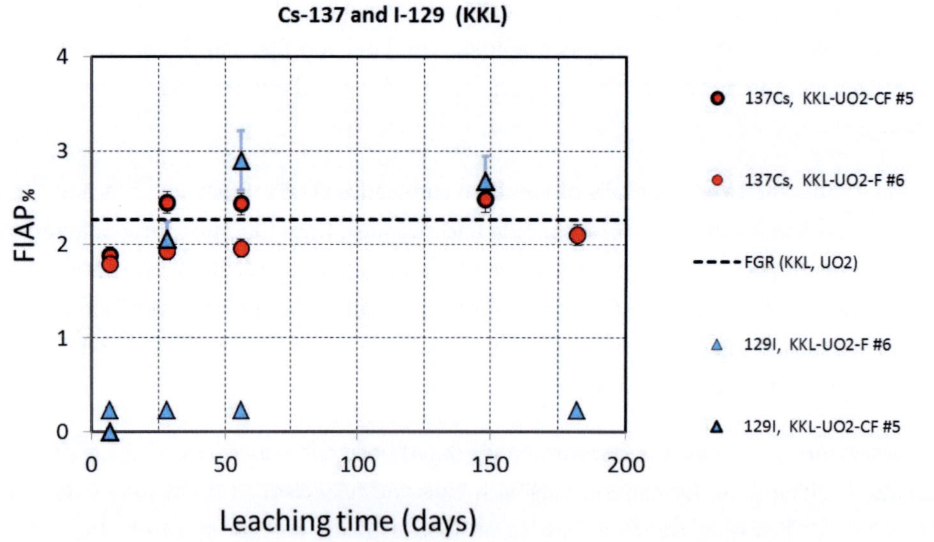

Figure 16.2: Cumulative percentages of ^{137}Cs and ^{129}I inventories in the aqueous phase (FIAP%) as a function of leaching time for UO_2 fuel samples (± cladding) from the Leibstadt BWR, compared with fission gas release (FGR).

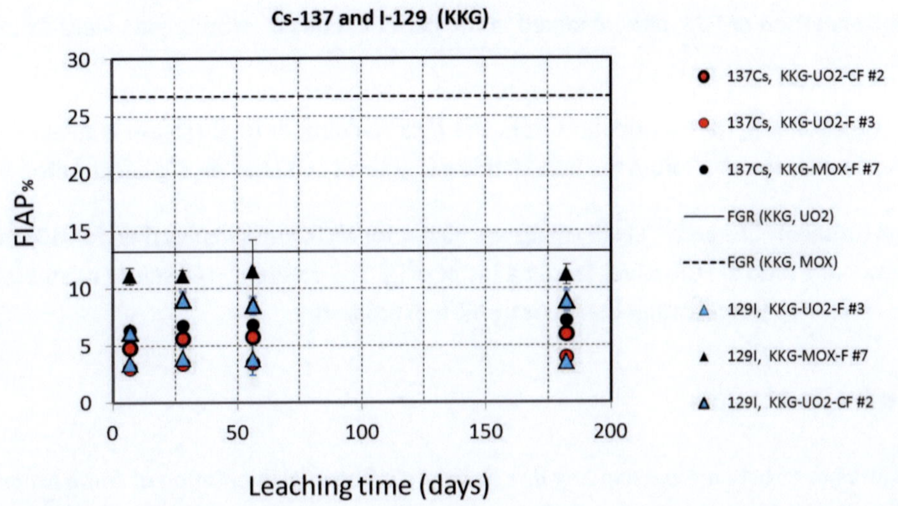

Figure 16.3: Cumulative percentages of ^{137}Cs and ^{129}I inventories in the aqueous phase (FIAP%) as a function of leaching time for UO_2 and MOX fuel samples (± cladding) from the Gösgen PWR, compared with fission gas release (FGR).

16.5 Discussion

The present leach experiments are similar, but not identical, to those described in Johnson et al. (2012). Whereas in the latter study sample preparation was optimized to detect differences in the release from rim vs. central pellet region, or to study the effect of exposed surface area, the present experiments, carried out with fuel from the same reactors, aim at distinguishing release contributions from the fuel and the cladding. A preliminary differentiation of contributions from bulk cladding and UO_2 fuel can already be made thanks to the fact that experiments were carried out both with bare and cladded fuel. However, a differentiation between *bulk* and *intrinsic* cladding contributions will be possible only after completion of the experiments with tangentially cut cladding (KKG-UO2-C and KKL-UO2-C in Table 16.1).

In general, FIAP values were found to be roughly similar to earlier determinations, with maxima in the order of ~ 2-3% for both ^{137}Cs and ^{129}I in BWR (Leibstadt) samples, and ~ 4-7% for ^{137}Cs, up to ~ 9-11% for ^{129}I in PWR (Gösgen) samples.

The following more specific observations can be made from the comparison of Figure 16.2 and Figure 16.3 in this report with Figures 8 and 10 from Johnson et al. (2012):

BWR samples (Leibstadt): In the current experiments, the release of ^{137}Cs and ^{129}I from the BWR SNF samples is significantly faster than in the analogous experiments described in Johnson et al. (2012), where FIAP values increased gradually over 98 days. In the present study "plateau" values are reached already after 7/28 days. At leaching times of 98 days or larger, similar cumulative FIAPs are reached in the earlier and current experiments for both ^{137}Cs and ^{129}I.

PWR samples (Gösgen): ^{137}Cs FIAP% values for cladded UO_2 fuel appear to be systematically higher in the current series of experiments compared with the earlier ones (5 - 6% vs. 4%). ^{129}I FIAP% values for cladded UO_2 fuel appear to be similar in both series of experiments (5 - 10%).

The internal comparison of FIAP data obtained in the current series of experiments leads to the following conclusions:

BWR samples (Leibstadt): The contribution from the bulk cladding to total release is minor for ^{137}Cs, but dominant for ^{129}I. Release of ^{129}I from the bulk cladding is gradual and reaches a "plateau" after 56 days.

PWR samples (Gösgen) ^{137}Cs and ^{129}I FIAP values are higher for MOX fuel samples than for UO_2 fuel samples, approximately in the ratio of FGR values (about a factor of 2). The different FIAP values are therefore mostly related to in-reactor operation rather than to intrinsic fuel properties.

Status of the experiments

Currently, a number of data are outstanding due to delays in the implementation of some experiments and analyses. Sampling of the leaching solutions is complete for 6 out of 9 experiments, but the 364 days solutions (collected on September 16, 2014) are not yet analysed.

High-resolution ICP-MS analyses for the determination of ^{79}Se and fuel burnup are currently underway, whereas ^{14}C analyses are outstanding. Experiments with bare cladding are still waiting to be started. The delay is due to difficulties in installing and testing the special equipment required to operate precise tangential cuts of the cladding.

16.6 Acknowledgement

The research leading to these results has received funding from the European Union's European Atomic Energy Community's (Euratom) Seventh Framework Programme FP7/2007-2011 under grant agreement n° 295722 (FIRST-Nuclides project).

16.7 References

Bevington, Ph.R., Robinson, D.K. (2003). Data Reduction and Error Analysis for the Physical Sciences. 3rd edition, McGraw-Hill, New York.

Carbol, P., Fors, P., Goulder, T., Spahiu, K. (2009). Hydrogen Suppresses UO_2 Corrosion. Geochimica and Cosmochimica Acta, 73, 4366–4375.

Grambow, B., Lemmens, K., Minet, Y., Poinssot, C., Spahiu, K., Bosbach, D., Cachoir, C., Casas, I., Clarens, F., Christiansen, B., de Pablo, J., Ferry, C., Giménez, J., Gin, S., Glatz, J.P., Gago, J.A., Gonzalez-Robles, E., Hyatt, N.C., Iglesias, E., Kienzler, B., Luckscheiter, B., Martinez-Esparza, A., Metz, V., Ödegaard-Jensen, A., Ollila, K., Quiñones, J., Rey, A., Ribet, S., Rondinella, V.V., Skarnemark, G., Wegen, D., Serrano-Purroy, D., Wiss, T. (2008). Final Synthesis Report RTD Component 1: Dissolution and Release from the Waste Matrix. European Commission report, Deliverable D1.6.3, NF-PRO (Contract Number: FI6W-CT-2003-02389).

Johnson, L.H., Tait, J.C. (1997). Release of Segregated Radionuclides from Spent Fuel. SKB Technical Report TR-97-18.

Johnson, L., Poinssot, C., Ferry, C., Lovera, P. (2004). Estimates of the Instant Release Fraction for UO_2 and MOX Fuel at t = 0. Nagra Technical Report, NTB 04-08.

Johnson, L., Günther-Leopold, I., Kobler Waldis, J., Linder, H.P., Low, J., Cui, D., Ekeroth, E., Spahiu, K., Evins, L.Z. (2012). Rapid Aqueous Release of Fission Products from High Burn-Up LWR Fuel: Experimental Results and Correlations with Fission Gas Release. Journal of Nuclear Materials, 420, 54–62.

McGinnes, D.F. (2002). Model Radioactive Waste Inventory for Reprocessing Waste and Spent Fuel. Nagra Technical Report, NTB 01-01.

Nagra (2002). Project Opalinus Clay – Safety Report. Demonstration of Disposal Feasibility for Spent Fuel Vitrified High-Level Waste and Long-Lived Intermediate-Level Waste (Entsorgungsnachweis). Nagra Technical Report, NTB 02-05.

Noirot, J., Desgranges, L., Lamontagne, J. (2008). Detailed Characterisations of High Burn-Up Structures in Oxide Fuels. Journal of Nuclear Materials, 372, 318–339.

Roudil, D., Jégou, C., Broudic, V., Muzeau, B., Peuget, S., Deschanels, X. (2007). Gap and Grain Boundaries Inventories from Pressurized Water Reactor Spent Fuels. Journal of Nuclear Materials, 362, 411-415.

Stroes-Gascoyne, S., Johnson, L.H., Sellinger, D.M. (1987). The Relationship between Gap Inventories of Stable Xenon, ^{137}Cs and ^{129}I in Used CANDU Fuel. Nuclear Technology, 77(3), 320–330.

Stroes-Gascoyne, S., Tait, J.C., Porth, R.J., McConnell, J.L., Lincoln, W.J. (1994). Release of ^{14}C from the Gap and Grain-Boundary Regions of Used CANDU Fuels to Aqueous Solutions. Waste Management, 14(5), 385-392.

17 X-ray absorption spectroscopy of selenium in high-burnup UO_2 spent fuel from the Leibstadt and Oskarshamn-3 reactors

E. Curti[1], A. Froideval-Zumbiehl[1], M. Martin[1], A. Bullemer[1], I. Günther-Leopold[1], A. Puranen[2], D. Jädernäs[2], O. Roth[2], D. Grolimund[3], C.N. Borca[3] and A. Velea[3]*

[1] *Nuclear Energy and Safety Department (NES), Paul Scherrer Institut (CH)*
[2] *Studsvik Nuclear AB (SE)*
[3] *Swiss Light Source (SLS), Paul Scherrer Institut (CH)*
* *Corresponding author: enzo.curti@psi.ch*

17.1 Abstract

In this study, we investigated oxidation state and coordination environment of Se in high-burnup UO_2 spent nuclear fuel (SNF) samples from the Leibstadt and Oskarshamn-3 reactors by means of micro X-ray absorption near-edge spectroscopy (µ-XANES) and micro X-ray fluorescence (µ-XRF). In spite of the low Se concentrations (about 100 ppm), the heavy UO_2 matrix and the extremely small sample size, we were able to collect XANES spectra from the core and rim regions of the fuel pellets.

Sample preparation was carried out at PSI (Leibstadt SNF microparticles) and Studsvik Nuclear AB, where two ~ $(30 \cdot 15 \cdot 5)\mu m^3$ chips were prepared from an Oskarshamn-3 SNF pellet by Focused Ion Beam (FIB) milling. All SNF spectra proved to be strikingly similar independently of the reactor origin and location in the pellet, suggesting a homogeneous chemical state of Se in the matrix of SNF from boiling water reactors. The XANES data were first evaluated by comparing the SNF spectra with those of reference compounds representing different oxidation states of Se (-II, -I, 0, IV, VI). The comparison suggests that Se occurs in the SNF either as a mixture of Se(0) and Se(IV), or entirely as Se(-II). Based on crystal-chemical arguments, the former option is considered to be unlikely. Occurrence as selenide is also supported by *ab initio* XANES calculations indicating that Se(-II) may replace oxygen sites in the UO_2 lattice.

17.2 Introduction

The present contribution focuses on the fate of ^{79}Se, a long-lived safety-relevant nuclide (half-life $3.3 \cdot 10^5$ a) which is traditionally considered as a potentially significant Instant Release Fraction (IRF) contributor owing to the appreciable volatility in the Se(0) state of this element and the high solubility of oxidized selenium species (Scheinost et al., 2008; Curti et al., 2013). This view is based on the assumption that most of the selenium is released as oxidized Se(IV) or Se(VI) from the fuel/cladding gap or easily accessible grain boundaries during early oxidative UO_2 dissolution. However, experimental studies have shown that oxidative conditions will be suppressed in a deep repository as long as molecular hydrogen, generated by the corrosion of iron or steel used in overpack construction, is present (Carbol et al., 2009). So far, no data are available on the chemical state of selenium in spent UO_2 fuel.

Selenium may occur in different oxidation states. The oxidized species Se(IV) and Se(VI) form soluble salts, whereas the reduced forms Se(0) and Se(-II) are sparingly soluble (Scheinost et al., 2008; Curti et al., 2013) and have thus limited Se mobility in aqueous solutions. Recent leaching experiments on high burnup UO_2 fuel from two Swiss power plants (Johnson et al., 2012) failed to detect dissolved ^{79}Se after leaching in aqueous solutions during 98 days, in spite of the low detection limit (0.5 ppb) of the experimental setup, suggesting that selenium in spent fuel is not as mobile as commonly assumed.

In order to determine the chemical nature of Se in SNF, we carried out X-ray absorption spectroscopy (XAS) experiments on high-burnup spent fuel microsamples from the Leibstadt (Switzerland) and Oskarshamn-3 (Sweden) boiling water reactors. The main objective was to assess oxidation state and coordination environment of selenium in the irradiated fuel. Such information is expected to give clues for understanding the behavior of ^{79}Se during aqueous leaching of spent fuel. The feasibility of XAS measurements at Se-K edge on spent fuel was successfully tested in preliminary measurements carried out on non-irradiated UO_2 containing trace Se concentrations. The details are documented in the 2nd Annual Workshop Proceedings (Curti et al., 2014).

17.3 Experimental Methods

Sample preparation

During the first and second beamtimes at the MicroXAS beamline (SLS) in November 2012 and May 2013, microsamples from a high-burnup spent fuel pellet (average 78.7 GWd/t_U) unloaded in August 2004 after 9 fuel cycles from the Leibstadt boiling water reactor (Switzerland) were investigated. Due to the radiation protection limits enforced at the SLS and current limitations in the preparation techniques at PSI, only tiny dispersed SNF particles could be measured. The sample was prepared by the "peeling method" as described in detail by Degueldre et al. (2011). Briefly, a cross-cut section of a fuel pellet impregnated with resin was polished using a silicon carbide disc and sand paper, which loosened micrometer-sized fuel particles. The surface of the so-prepared fuel section was then pressed on adhesive Kapton tape, allowing loosened spent fuel micro-particles (10-15 μm max.) to stick on the tape. The particles were then immobilized by covering them with a second Kapton tape strip. The resulting sample shows a "halo" of the pellet with a few scattered spent fuel particles (Figure 17.1) presumably conserving the original radial distance in the pellet. A sub-sampling was necessary in order to reduce the activity below the limit of 100 LA allowed for a single sample at the MicroXAS beamline (rectangles in Figure 17.1). For each of the two allocated beamtimes two sub-samples from the same pellet were produced, one from the rim region and one from the internal part of the pellet. Figure 17.1 shows the sub-samples prepared for the May 2013 campaign.

For the third beam time at the MicroXAS beamline (June 2014) spent fuel microsamples from the Oskarshamn-3 reactor (Sweden) were prepared by Focused Ion Beam (FIB) milling at Studsvik Nuclear AB, using a Zeiss® equipment with a Ga ion source. Both samples originate from a "matchstick"-sized strip of the same pellet (average burn-up 62.9 GWd/t_U) encompassing the entire diameter of the pellet (from clad to clad) and passing through the center.

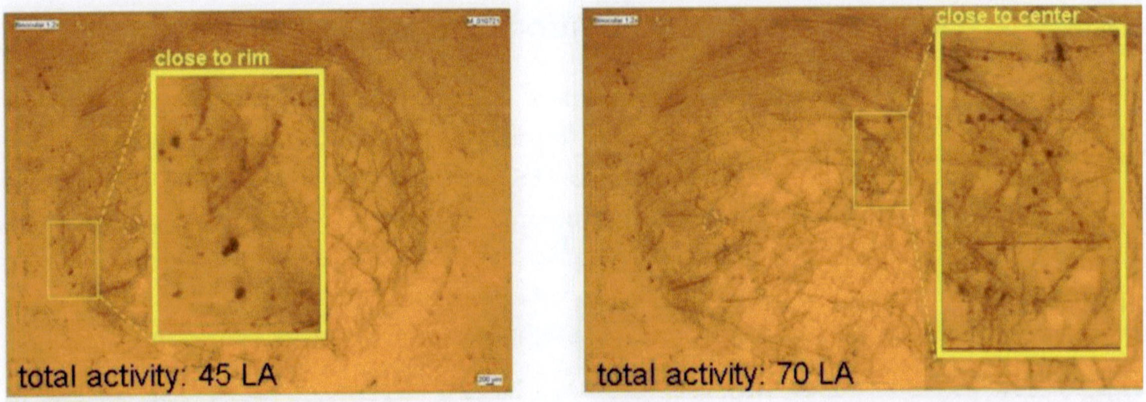

Figure 17.1: Sample preparation for XANES measurements of the Leibstadt spent fuel. See text for explanation.

Two polished microsamples of approximately (L · W · H) = (30·15·5) μm^3 were produced: *"RIM" sample* from the rim region, including the outer boundary of the pellet, i.e. the fuel-gap interface (Figure 17.2). *"CEN" sample* from the center of the pellet.

The RIM sample was brought with a special tip onto an Omniprobe® Lift-Out Grid, using a carbon source as welding material (instead of Pt) to avoid interferences with the emission spectra of selenium. The CEN sample detached from the grid during preparation, but it could be recovered and fixed with a strip of Kapton tape.

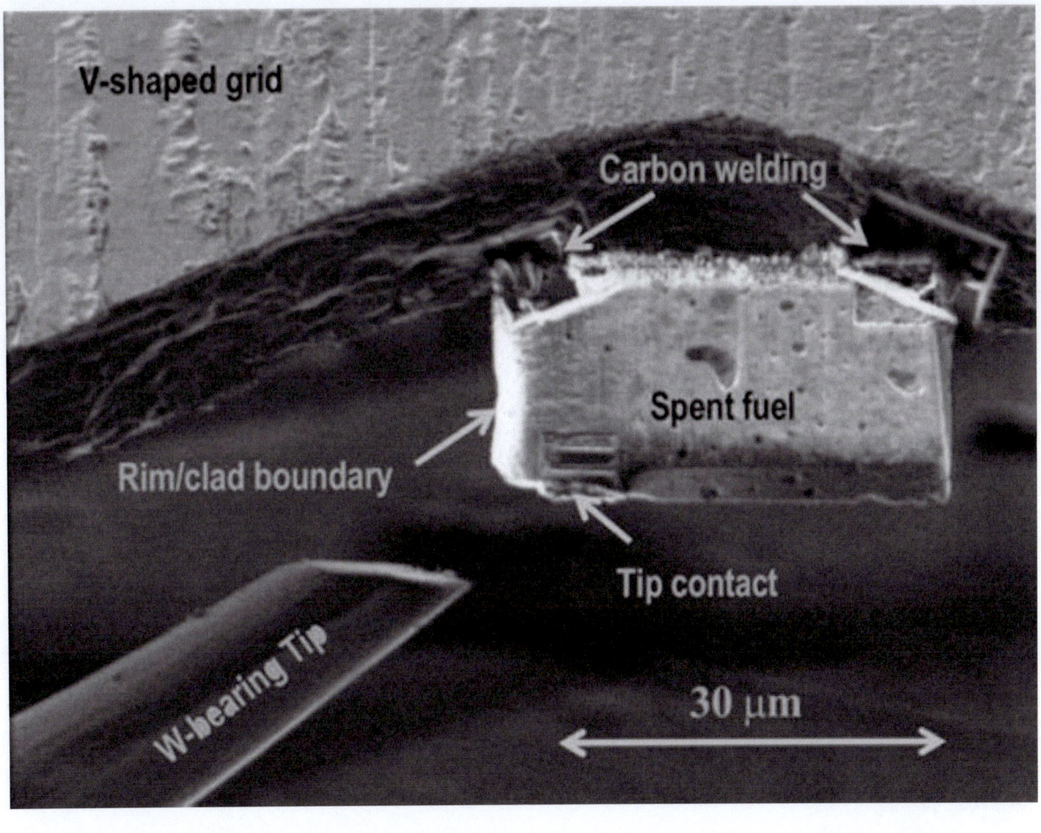

Figure 17.2: FIB Sample ("RIM") of Oskarshamn spent fuel for XANES measurements (courtesy Studsvik Nuclear AB).

17.4 XAS/XRF experimental conditions and data treatment

All measurements of absorption and emission spectra on spent fuel obtained at the MicroXAS beamline were carried out in fluorescence mode, using a micro-focused monochromatic X-ray beam of about 2 - 3 μm (vertical)·5 μm (horizontal) size in the energy ranges 12.6 - 13.2 keV (Se-K edge and Se-K_α fluorescence regions) and 17.2 - 17.3 keV (U-L_α fluorescence). The absorption spectra of Se reference compounds where measured simultaneously in fluorescence and transmission. A Ketek® single element Si detector was used to measure energy-dependent fluorescence intensities, while the transmitted signal was detected with a photodiode. Energy calibration was carried out using a Se(0) reference.

Micro X-ray Fluorescence (μ-XRF) maps of the Se-K_α, U-L_α and the transmitted signals were obtained by scanning the sample surface with the micro-focused X-ray beam at the appropriate fixed energy, while collecting "on the fly" the emitted fluorescence intensity.

XANES data normalization was carried out using the ATHENA package (Ravel and Newville, 2005). Due to the unfavorable signal/noise ratio of the spectra collected on Leibstadt spent fuel, the in-built ATHENA interpolative smoothing procedure was applied to allow a careful determination of the edge position.

17.5 Results

XAS/XRF on spent fuel from the Leibstadt boiling water reactor

XRF maps of spent fuel samples

Figure 17.3 a-c shows XRF maps of the U-L_α, Se-K_α and transmitted signals of a 1.3 mm·1.0 mm section from the intermediate region of the pellet (between rim and center). Figure 17.3b also shows the locations (letters) at which Se-K edge XANES spectra were successfully recorded. The maps were recorded at an X-ray energy of 17.3 keV (above the U-L_{III} edge) allowing us to identify unequivocally the location of the scattered, tiny spent fuel particles dispersed across the "imprint" of the pellet on the Kapton tape. With this information locations with sufficiently high Se concentrations could be identified for collecting absorption spectra. The Se fluorescence signal was, as expected, very weak and rarely exceeded a few hundreds counts per second.

Micro-XANES

Due to the low signal/noise ratio of the Se-K_α fluorescence emitted from the SNF sample, an interpolative smoothing procedure was applied to allow for a careful determination of the edge position. The in-built utility in the ATHENA software (Ravel and Newville, 2005) was used to this purpose. Figure 17.4 shows a typical example of raw and the corresponding smoothed micro-XANES spectrum obtained from a single scan on a selected spent fuel particle. Moreover, because all XANES spectra collected from the spent fuel samples proved to be very similar, the single smoothed spectra were merged; giving rise to an averaged spectrum of all Leibstadt SNF data plus two limiting spectra representing the upper and lower standard deviation (see Figure 17.5).

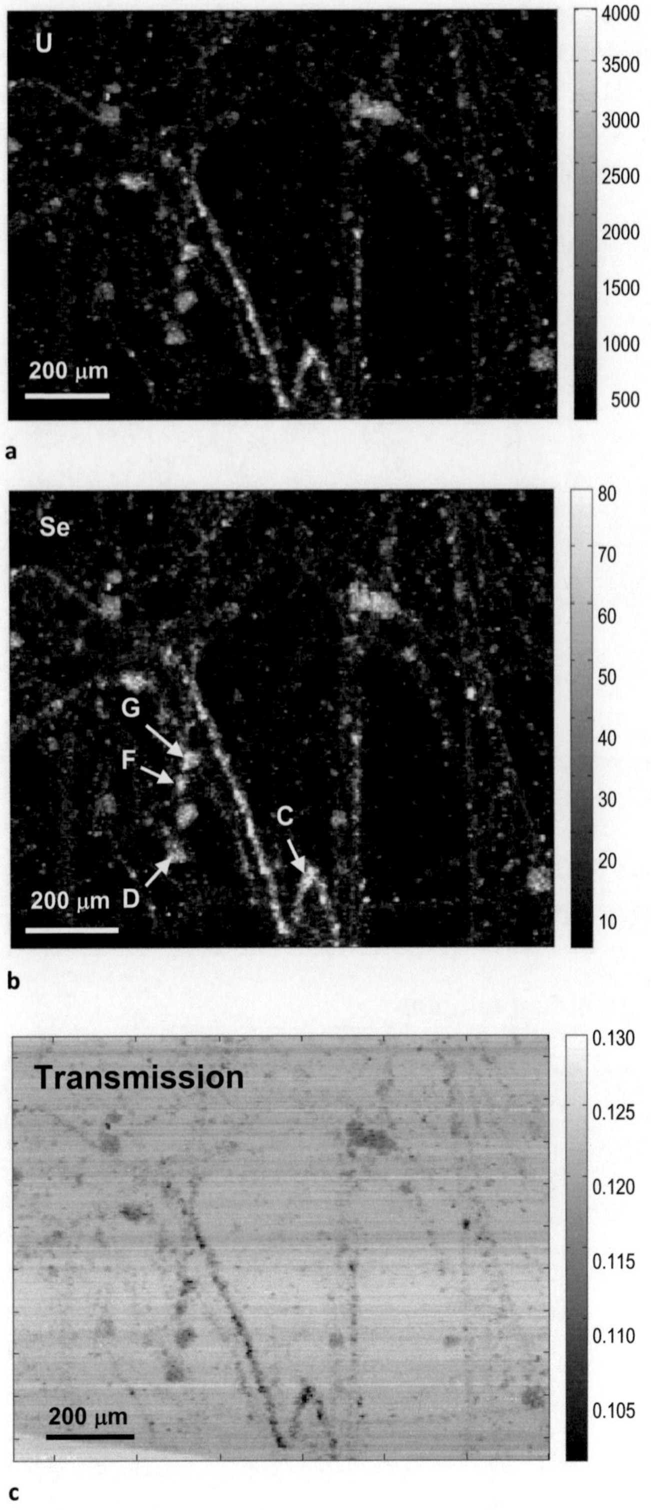

Figure 17.3: Micro-XRF maps of U (a) and Se (b) fluorescence signals, and transmitted intensity map (c) obtained from a section of a dispersed Leibstadt UO_2 spent fuel sample. Letters and arrows in Figure 17.3 b denote fuel particles where Se K-edge micro-XANES spectra were recorded. Grey-coded scales represent detector counts (a, b) and voltages of the diode used to measure the transmitted signal (c).

This data treatment was also carried out using ATHENA in-built routines. The comparison of the spectra obtained from the SNF sample with those of several reference compounds representing different oxidation states of selenium clearly shows a striking similarity between the spectra of SNF and of two Se(-II) compounds, namely ZnSe and Na_2Se. The spectral features of all other compounds, including FeSe(-II), clearly differ from those of SNF, allowing us to rule them out as sources of the signals recorded from the SNF samples.

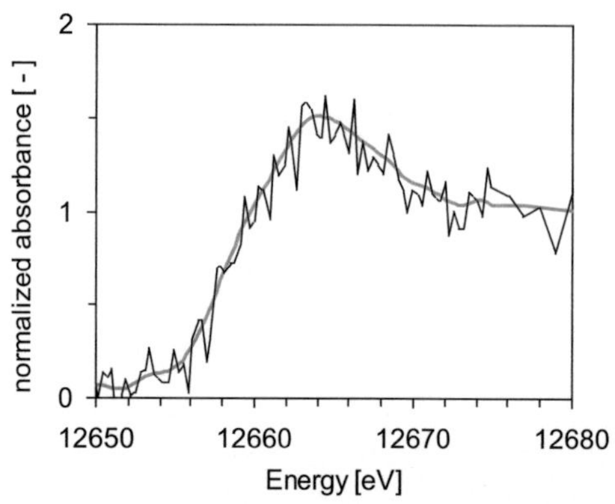

Figure 17.4: Typical example of a raw (black thin line) and smoothed (grey thick line) Se-K edge XANES spectrum from the Leibstadt spent fuel sample (location C, see Figure 17.3 b).

17.6 XAS/XRF on spent fuel from the Oskarshamn-3 boiling water reactor

XRF element maps of spent fuel samples

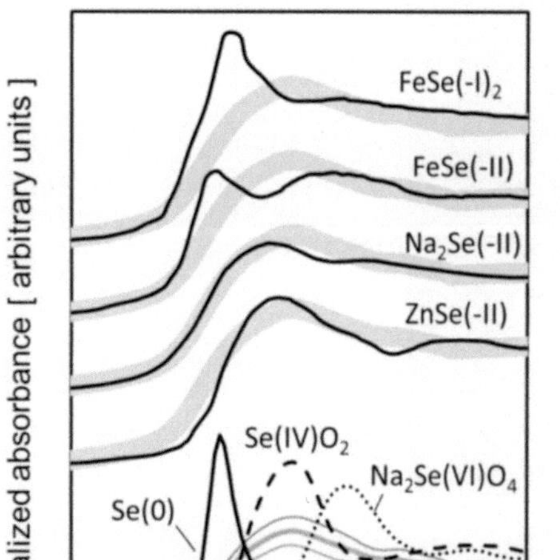

Several elements (U, Nd, Se) were mapped over the limited available surface of both FIB SNF samples from Oskarshamn-3 (CEN and RIM samples). Within the precision of the method, the element distributions (including Se) were found to be uniform. Variations in sample thickness and edge effects make it difficult to detect presumably small concentration gradients across a maximum distance of about 30-35 µm (length of the FIB samples). Se-K_{α} fluorescence signals were found to be more intense than in the case of the Leibstadt samples, probably due to the better quality of the FIB samples (larger thickness and flat, polished surface).

Figure 17.5: Average smoothed Se K-edge µ-XANES spectra from all measurements obtained from the spent fuel particles (SNF, thick grey line) and envelope defined by the standard deviation of all individual spectra (thin grey lines or shading) compared with the spectra of reference compounds.

Micro-XANES

Micro XANES Se K-edge spectra were recorded on both CEN and RIM samples (Figure 17.2) originating from the core and periphery, respectively, of a fuel pellet from Oskarshamn-3. As in the case of the Leibstadt samples, the recorded spectra are all very similar. The Oskarshamn-3 and Leibstadt XANES are practically indistinguishable since the average spectra from the Oskarshamn-3 samples fall within the envelope defined by the standard deviation of 15 spectra measured on Leibstadt spent fuel (Figure 17.6).

Nevertheless, very small systematic differences have been observed between the spectra of the two samples from Oskarshamn: Compared to those from the rim region (RIM sample), the normalized spectra from the center of the pellet (CEN sample) are shifted by 0.5 eV to lower edge energies and have significantly larger edge crest heights. These differences are minor and lie within the standard deviation envelope defined by the spectra recorded on Leibstadt. At the current status of data analysis, we cannot exclude that they might be due to instrumental bias.

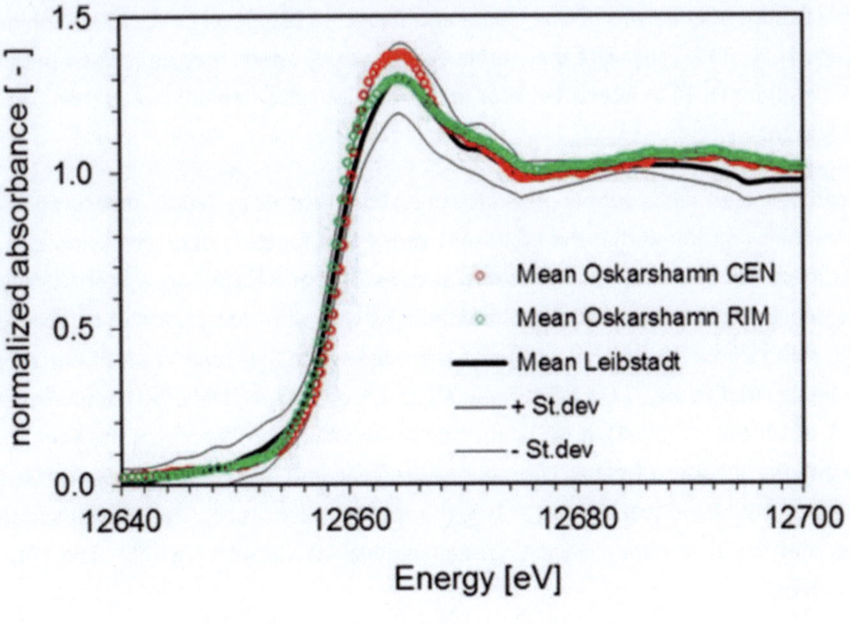

Figure 17.6: Averaged Se K-edge micro-XANES spectra from the central and peripheral regions of a spent fuel pellet from the Oskasrhamn-3 reactor compared to the averaged spectrum (± standard deviation) measured on spent fuel from the Leibstadt reactor.

17.7 Interpretation of XANES data

Comparison with reference compounds (fingerprinting method)

The comparison of the spectra obtained from spent fuel with the experimental spectra of Se reference compounds having different oxidation states (-II, -I, 0, IV and VI, see Figure 17.5) suggests that selenium may occur in the fuel as Se(-II) (selenide). However, Figure 17.5 (bottom plot) also suggests that the spent fuel XANES spectra could also be reconstructed through linear combinations of the Se(0) and $Se(IV)O_2$ reference spectra, whereas contributions from the $Na_2Se(VI)O_4$ spectrum can be excluded.

Indeed, a quantitative linear combination fit analysis confirmed this anticipation (see Curti et al. (2014) for details). The results of the fit indicate that the experimental spectra obtained on the spent fuel samples may be acceptably represented by a linear combination of the Se(0) and Se(IV)O_2 spectral components with about equal weights (i.e. about 50% - 50% mixture on atomic basis). This means that the spectroscopic data may be explained alternatively assuming that Se occurs as a mixture of Se(0) and Se(IV) in the investigated fuel.

The occurrence of two Se species, Se(0) and Se(IV), at almost constant ratio over the entire location of the SNF samples appears however quite unlikely. The overall similarity among all recorded Se-K XANES spectra rather points to a unique chemical state of Se in the investigated spent fuel sample.

Formation of Se(-II)-Zn or Se(-II)-Na bonds, as suggested by the similarity of the SNF spectra to those of ZnSe and Na_2Se, can be practically excluded. Because of the very low fission yield, selenium will remain a trace component even in high burn-up fuel and will hardly form concentrated Se phases. The concentrations of Na and Zn in SNF are expected to be even less due to negligible fission yields. Therefore, dissemination of Se in UO_2 lattice is a more probable option. The existence of a large number of stable U(IV) selenide compounds, notably various polymorphs of USe_2 (Beck and Dausch, 1989; Noel et al., 1996; Kohlmann and Beck, 1997 and Bugaris et al., 2011) suggests that stable Se(-II) – U(IV) bonds may be formed in UO_2 SNF, e.g. via substitution of Se(-II) for O(-II) or interstitial sites in the UO_2 lattice. Selenium would then form a dilute solid solution of USe_2 in UO_2.

Further spectroscopic arguments support this interpretation. Lenz et al. (2008) measured a large set of Se reference compounds and found that the edge shift varies as a function of crystal symmetry. For instance, the absorption spectra of cubic Se(-II) compounds, such as ZnSe and Na_2Se, are systematically shifted by 1 - 3 eV to higher energies with respect to Se(-II) compounds with different crystal structures, e.g. hexagonal FeSe. The edge shift relative to trigonal Se(0) and the edge-white line separation of our spent fuel XANES spectra were determined to be -0.1 ± 0.5 eV and 4.6 ± 0.5 eV, respectively. By comparison with the data given in Table 2 of Lenz et al. (2008), it turns out that these values closely match the edge shifts and edge-white line separations of cubic reference compounds like ZnSe and particularly Na_2Se. The latter compound has cubic symmetry with same space group as UO_2 (fluorite type structure). This provides further support to the hypothesis that Se(-II) occupies oxygen sites or equivalent vacancies within the UO_2 lattice in the studied SNF samples.

17.8 Ab initio XANES calculations

In order to test this hypothesis, *ab initio* calculations of Se-K XANES absorption spectra have been carried out using the FDMNES code (Joly, 2001) with the support of the FitIt software (Smolentsev and Soldatov, 2007) which allows for geometrical optimization of the spectra. This package allows one to perform series of FDMNES calculations for preselected "boundary" geometries of the atomic clusters. Geometrical transformations may include variations of interatomic distances and/or angles for predefined groups of atoms. The calculated spectra are then used by an in-built routine as nodes to construct multidimensional interpolation polynomials, with the help of which XANES spectra can be calculated for any cluster geometry comprised between the predefined geometrical boundaries. This allows a smooth fitting of the experimental spectra. Following strategy was pursued:

1) An initial cluster (cluster 1) was defined in which a single Se central absorber is surrounded by four U and six O atoms with geometry identical to the coordination environment of oxygen in UO_2, but with all interatomic distances increased by 14%. This simulates an arbitrary minimum value for the lattice expansion induced by the Se(-II) for O(-II) substitution, leading to Se-U distances of 2.70 Å and Se-O distances of 3.12 Å.

2) Starting from the latter geometrical configuration, two additional clusters were defined by independently increasing all Se-U or all Se-O distances by further 20%. This generates two additional "limiting" coordination environments, characterized by Se-U distances of 3.24 Å with Se-O at 3.12 Å (cluster 2) and Se-O distances of 3.74 Å with Se-U at 2.70 Å (cluster 3).

3) Using FitIt, FDMNES theoretical Se K-edge XANES spectra were calculated for the three clusters described above and six other intermediate geometries. This yielded a total of nine nodes for the interpolation of intermediate cluster geometries. Once the interpolation procedure is set up, FitIt is capable to approximate the theoretical spectra of all possible geometrical configurations comprised within the predefined limiting cluster geometries (in this case three). It is then possible to display graphically and smoothly the effect of such geometrical variations on the XANES spectra and fit the experimental spectra.

The results of the theoretical analysis are illustrated in Figure 17.7, which shows the best fits of two spectra from Leibstadt SNF, and Figure 17.8, which pictures the optimized Se-U-O cluster corresponding to the best fit spectra. The geometrical optimization indicates Se-U distances very similar to those encountered in USe_2 compounds, which is a further indication in favor of the hypothetical selenide for oxygen substitution in the SNF UO_2 lattice.

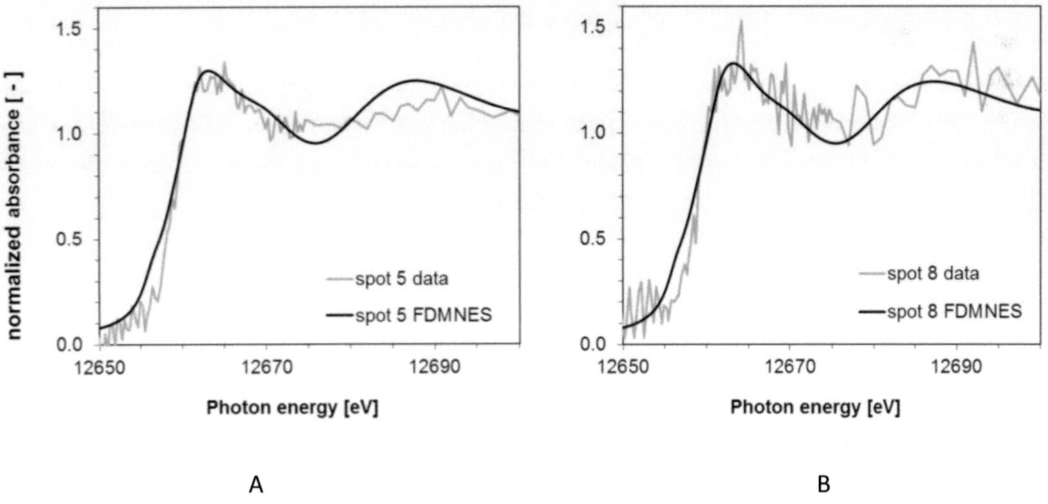

Figure 17.7: Fits of Se K-edge spectra collected at two distinct spots in the Leibstadt spent fuel sample through geometrical optimization of theoretical XANES.

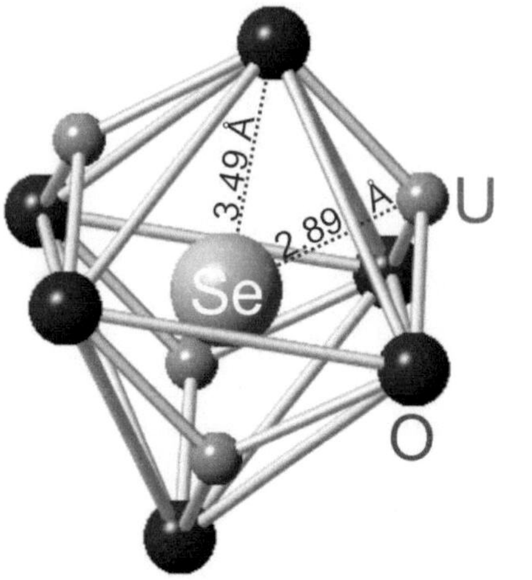

Figure 17.8: Polyhedral representation of Se-coordination environment resulting from geometrical optimization of a Leibstadt SNF spectrum ("spot 5") obtained via FDMNES / FitIt calculation.

17.9 Conclusions and Future work

X-ray spectroscopic studies on fission products in spent nuclear fuel are very challenging due to severe technical and radioprotection constraints. Nevertheless, this pilot study shows that valuable data can be obtained even on low-yield fission products using state-of-the-art X-ray sources (intense, stable micro-beams) and advanced preparation methods (FIB milling). The data obtained in this study indicate that a substantial part of ^{79}Se and other Se fission products are present in UO_2 SNF in a homogeneous chemical form, probably as dispersed Se(-II) replacing oxygen sites in the UO_2 lattice. More details on the spectroscopic work carried out on Leibstadt SNF can be found in Curti et al. (2014).

Future work (to be carried out outside the FIRST-Nuclides project) will involve improvements in the fluorescence detection techniques to obtain EXAFS quality absorption spectra and micro-diffraction for the identification of minor SNF phases. Further, it is planned to extend spectroscopic studies to other fission products, as well as to investigate SNF samples after aqueous leaching.

17.10 Acknowledgement

The Se spectroscopy work documented in this chapter has been conducted by PSI in close collaboration with the FIRST-Nuclides partners KIT-INE and Studsvik Nuclear AB. KIT-INE provided free access to the INE Beamline at ANKA for assessing the feasibility of Se-spectroscopy measurements in UO_2 spent fuel (November 2012). Studsvik Nuclear AB prepared two spent fuel microsamples from the Oskarshamn-3 reactor by Focussed Ion Beam (FIB) milling and provided assistance during the joint measurement campaign at the MicroXAS beamline (June 2014). The collaboration between Studsvik Nuclear AB and PSI was supported by a TALISMAN proposal, which covered part of the fuel samples transportation costs. J. Rothe and K. Dardenne from KIT-INE are warmly thanked for their invaluable assistance during preliminary measurements at the INE-beamline (ANKA). G. Smolentsev (PSI) supported us in the *ab initio* calculations. The Swedish Nuclear Fuel and Management Co (SKB) is also acknowledged for financial support of the work performed at Studsvik Nuclear AB.

The research leading to these results has received funding from the European Union's European Atomic Energy Community's (Euratom) Seventh Framework Programme FP7/2007-2011 under grant agreement n° 295722 (FIRST-Nuclides project).

17.11 References

Beck, H.P., Dausch, W. (1989). The Refinement of -USe_2, Twinning in a $SrBr_2$-Type Structure. Journal of Solid State Chemistry, 80, 32-39.

Bugaris, D.E., Choi, S.E., Copping, R., Glans, P.-E., Minasian, S.G., Tyliszczak, T., Stosh, A.K., Shuh, D.K., Ibers, J.A. (2011). Pentavalent and Tetravalent Uranium Selenides, $Tl_3Cu_4USe_6$ and $Tl_2Ag_2USe_4$: Syntheses, Characterization, and Structural Comparison to Other Layered Actinide Chalcogenide Compounds. Inorganic Chemistry, 50, 6656-6666.

Carbol, P., Fors, P., Gouder, Th., Spahiu, K. (2009). Hydrogen Suppresses UO_2 Corrosion. Geochimica and Cosmochimica Acta, 73, 4366–4375.

Curti, E., Aimoz, L., Kitamura, A. (2013). Selenium Uptake onto Natural Pyrite. Journal of Radioanalytical and Nuclear Chemistry, 295, 1655–1665.

Curti, E., Froideval Zumbiehl, A., Günther-Leopold, I., Martin, M., Bullemer, A., Linder, H.P., Borca, C.N., Grolimund, D., Rothe, J., Dardenne, K. (2013). X-Ray Absorption Spectroscopy of Selenium in Non-Irradiated Doped UO_2 and High Burn-Up Spent UO_2 Fuel.

Curti, E., Froideval-Zumbiehl, A., Günther-Leopold, I., Martin, M., Bullemer, A., Linder, H.P., Borca, C.N., Grolimund, D. (2014). Selenium Redox Speciation and Coordination in High-Burnup UO_2 Fuel: Consequences for the Release of ^{79}Se in a Deep Underground Repository. Journal of Nuclear Materials, 453(1-3), 98-106. DOI: 10.1016/j.jnucmat.2014.07.003.

Degueldre, C., Martin, M., Kuri, G., Grolimund, D., Borca, C. (2011). Plutonium–Uranium Mixed Oxide Characterization by Coupling Micro-X-Ray Diffraction and Absorption Investigations. Journal of Nuclear Materials, 416, 142–150.

Johnson, L., Günther-Leopold, I., Kobler Waldis, J., Linder, H.P., Low, J., Cui, D., Ekeroth, E., Spahiu, K., Evins, L.Z. (2012). Rapid Aqueous Release of Fission Products from High Burn-Up LWR Fuel: Experimental Results and Correlations with Fission Gas Release. Journal of Nuclear Materials, 420, 54-62.

Joly, Y. (2001). X-ray Absorption Near-Edge Structure Calculations beyond the Muffin-Tin Approximation. Physical Review B, 63, 125120-1 –125120-10.

Kohlmann, H., Beck, H.P. (1997). Synthesis and Crystal Structure of the -Modifications of US_2 and USe_2. Zeitschrift für Anorganische und Allgemeine Chemie, 623, 785-790.

Lenz, M., Van Hullenbusch, E.D., Farges, F., Nikitenko, S., Borca, C.N., Grolimund, D., Lens, P.N.L. (2008). Selenium Speciation Assessed by X-Ray Absorption Spectroscopy of Sequentially Extracted Anaerobic Biofilms. Environmental Science & Technology, 42, 7587–7593.

Noel, H., Potel, M., Troc, R., Shlyk, L. (1996). Crystal Structure and Physical Properties of USe_2 and $USe_{2-x}Te_x$ (x = 0.24 and 0.72). Journal of Solid State Chemistry, 126, 22-26.

Ravel, B., Newville, M. (2005). ATHENA, ARTEMIS, HEPHAESTUS: Data Analysis for X-Ray Absorption Spectroscopy Using IFEFFIT. Journal of Synchrotron Radiation, 12, 537-541.

Scheinost, A.C., Kirsch, R., Banerjee, D., Fernandez-Martinez, A., Zaenker, H., Funke, H., Charlet, L. (2008). X-Ray Absorption and Photoelectron Spectroscopy Investigation of Selenite Reduction by Feii-Bearing Minerals. Journal of Contaminant Hydrology, 102, 228–245.

Smolentsev, G., Soldatov, A.V. (2007). FitIt: New Software to Extract Structural Information on the Basis of XANES Fitting. Computational Materials Science, 39, 569-574.

18 Charcterisation and leaching test for the experimental determination of IRF radionuclides from belgian high-burnup spent nuclear fuel

Thierry Mennecart, Karel Lemmens and Christelle Cachoir*

Belgian Nuclear Research Centre, SCK•CEN (BE)
** Corresponding author: thierry.mennecart@sckcen.be*

18.1 Abstract

In the framework of the FIRST-Nuclides program, SCK•CEN has conducted leaching experiments on a PWR UOX fuel, taken from the Belgian Tihange 1 reactor, with a rod average burnup of 50.5 MWd/kg_{HM} in order to determine the rapid release of some of the most critical radionuclides. This paper describes how the leaching experiments were conducted, and presents results of the solution analysis. Information about the irradiation history was also provided.

The preliminary results show that the IRF is higher for the declad segment than for the clad segment, and that the IRF for iodine is close to the FGR for the declad segment.

18.2 Introduction

As one of the partners of FIRST-Nuclides with a hot-cell infrastructure and the required analytical laboratories, SCK•CEN has performed leaching tests on spent fuel samples with a relatively high burn-up. For this purpose, a fuel rod was selected from the spent fuel stock available at SCK•CEN for which the characteristics are known and can be made public. The selected fuel samples originate from a PWR UOX fuel from the Belgian Tihange 1 reactor with a calculated (operation-based) average burnup of 50.5 MWd/kg_{HM} (Govers et al., 2014).

The results obtained from leaching tests depend on how the fuel samples are prepared before leaching commences. For instance, clad fuel segments and declad fuel fragments, or fragments from the centre or the periphery of the pellet will exhibit different leaching behaviours. A combination of differently prepared samples provides the most complete information, SCK•CEN thus performed one test with a clad fuel segment and one test with declad fuel fragments.

In the following sections, we describe the sample preparation, the sample pre-test characterization, the experimental setup, the leachate composition, the solution analyses and the preliminary results, before ending with our conclusions.

18.3 Sample preparation

Two fuel segments, about 2.4 cm long, including one whole and two half pellets, were taken from the central part of the fuel rod (Govers et al., 2014), and prepared as samples for leaching test. The samples were prepared in hot cells under ambient atmosphere, implying that we can expect the formation of a thin oxidized $UO_{2.33}$ layer at the fuel surface by taking up oxygen at the interstitial sites in the UO_2 lattice (Johnson et al., 1988).

The first segment was used as it was after cutting, without any further treatment. For a second test with the declad sample, the rod was cut in three parts with similar length (Figure 18.1). For each piece, the cladding was removed by pressure using a hydraulic pump with a piston which has a diameter close to the inner diameter of the cladding. Due to the strong exerted pressure during this operation, the fuel pellets were broken in fragments and powder of different sizes. In order to compare the two leaching tests, the three parts of the cladding and all the different fuel fragments and the fuel powder were leached together. Like this, we got the equivalent sample for the two leaching tests. Only the accessibility of the spent fuel for the leaching solution differs between these two leaching tests.

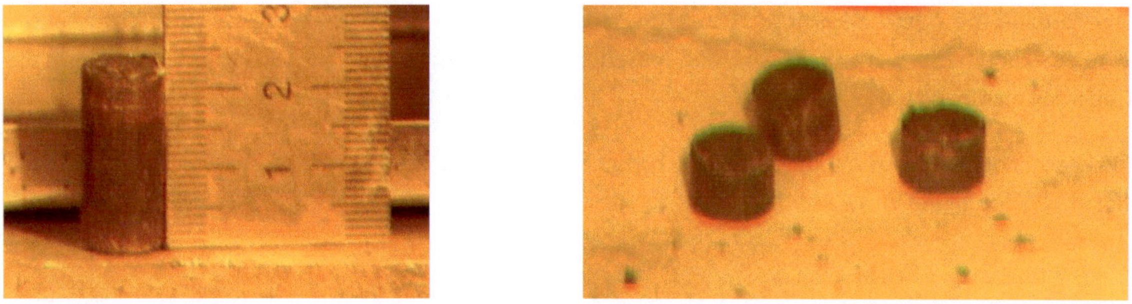

Sample of SNF for leaching test 1
'Intact sample'

The three pieces of the sample prepared for leaching test 2
Declad sample'

Figure 18.1: Samples prepared for the two leaching tests with spent nuclear fuel.

A third, adjacent segment was cut in order to determine the burnup by means of radiochemical analysis. A fourth shorter segment was cut for the microstructural characterization of the fuel before leaching. That same sample was also analysed by electron-probe microanalysis (EPMA) to determine the local concentration of various elements.

A rotary disk using a cooling solution (diamond saw) was used to cut the sample set aside for the determination of the burnup and the optical measurements. The aim of the "FIRST-Nuclides" program is to study the release of the most soluble radionuclides when the spent nuclear fuel is in contact with a solution for the first time. Consequently, the classic method using a cooling solution was not conceivable to cut the rod segments for the leaching test. Hence, a commercial tube cutter was used (tool of plumbing) to cut these segments under dry conditions. Only the cladding was cut with this tool. The fuel pellet was manually broken at the cutting plane.

18.4 Sample pre-test characterization

Although fabrication data and detailed irradiation history are available, a cross-check between these estimations and experimental observations of the state of the irradiated fuel is needed in order to reduce uncertainties on e.g. available free surfaces for leaching and on fission product inventory and location at the end of the irradiation. Good agreement between experiments and theoretical models for the behaviour of "observable" fission products should allow us to make firm hypotheses about the location of the few isotopes for which the experimental sensitivity is too low.

Burnup analyses

The average pellet burnup was determined by radiochemical analysis of a small fuel segment (equivalent of 2 pellets) cut in the vicinity of the test samples. The proposed cutting scheme specifies a cutting from mid-pellet to mid-pellet in order to keep a relevant inventory of Cs, bearing in mind the relocation of Cs to the pellet-pellet interface observed upon γ-scanning. The burnup determination is based on the measurement of various burnup indicators, mainly Nd isotopes, by Thermal Ionization Mass Spectrometry (TIMS), and confirmed by ^{137}Cs using ⍰-spectrometry. Usually, ^{144}Ce is also used for the confirmation of the value, but due to its short half-life (284.9 days) and the long cooling period, it could not be detected anymore in the fuel. A local burnup was calculated equal to 54.6 MWd/kg_{HM} (Adriaensen et al., 2014). This value is close to the calculated value, 54.3 MWd/kg_{HM} (Govers et al., 2014).

Other relevant isotopes in terms of the fast / instant release fraction were also measured, within the range permitted by the technique. A comparison between the calculated values - and the experimental concentrations is shown on the Figure 18.2. For most of the element, a good correlation is found, only for ^{242}Cm and ^{109}Ag, a discrepancy higher than 25% was found.

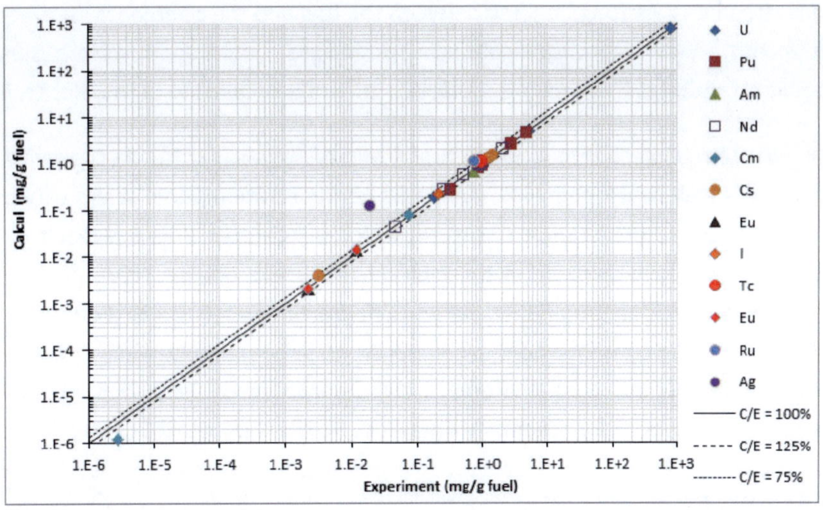

Figure 18.2: Comparison between the calculated radionuclides concentrations in the fuel and their measured concentrations by radiochemical analysis.

Microstructure analysis – Electron-probe microanalysis

A segment of the fuel rod was examined using different surface analysis techniques. After polishing, optical microscopy and Scanning Electron Microscopy (SEM) were used to observe the surface state and the different zones (Figure 18.3). One observes seven radial cracks which is typical for a medium linear rating fuel (Verwerft et al., 2014). Higher magnification images on various regions on the surface were taken from the inner part of the pellet to the outer part and the gap between the cladding and the pellet.

Figure 18.3: Optical micrograph of the microstructure of the spent fuel sample (the large dark part is due to a missing part of the fuel lost during the sample preparation).

Most of the gap is closed and a strong bonding between the fuel and the cladding is observed. In the centre of the pellet, and with a very high magnification of the images, bright spots were observed due to the presence of metallic precipitates ("five metals particles", generally composed of Mo, Ru, Tc, Rh and Pd).

The same sample was also subjected to Electron-Probe MicroAnalysis (EPMA). At the pellet periphery, the high burnup structure is well-developed, with numerous restructuration pores at the very periphery of the pellet. In the outer 30 µm, Xe depletion is observed, which is interpreted as the result of clustering of a large fraction of the locally generated Xe in the high burnup pores (Figure 18.4). In this layer, very fine precipitates of five-metal particles (Mo, Ru, Tc, Rh and Pd) are also found. Up to a depth of more than 100 µm into the pellet, elevated Cs concentrations are found on grain boundaries, while the same grain boundaries are depleted in Xe. Deeper into the fuel around 1,200 µm from the pellet periphery, thermally activated clustering of Xe and of the metallic elements is observed. At the very centre of the fuel, Xe depletion is almost complete and large metallic precipitates are found. At the very centre, one also finds ceramic precipitates, likely to be related to the perovskite $BaZrO_3$, with possible substitution on both Ba and Zr sites by other elements.

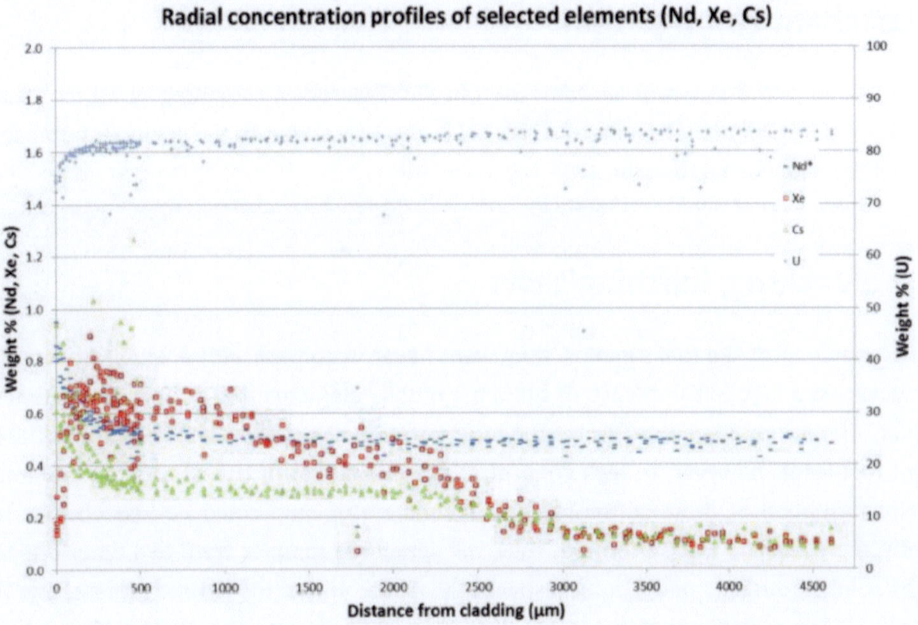

Figure 18.4: Example of a radial quantitative scans of the spent fuel sample by EPMA.

18.5 Leaching tests - Experimental setup

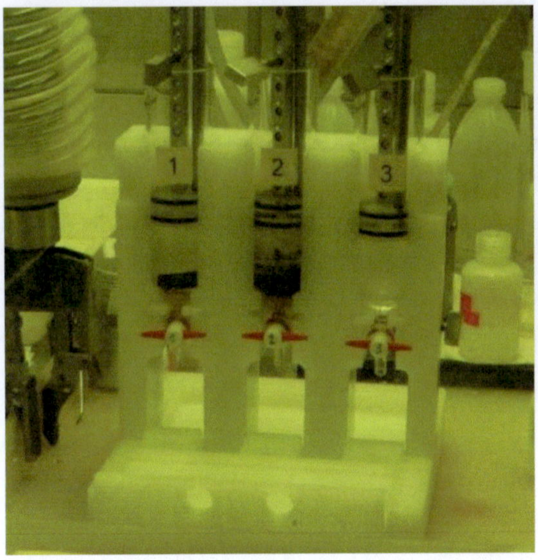

The experiments were performed inside a hot cell under an air atmosphere. The experimental setup is based on previous work performed at the Paul Scherrer Institute – PSI. The leaching device (Figure 18.5) was already described in a previous paper (Mennecart et al. 2013). Because there will be little further exchange between the solution in the vessel and the air in the hot cell, fuel oxidation by atmospheric oxygen should be limited during the leaching test.

Figure 18.5: Experimental setup used for the leaching of the spent fuel with an intact segment (1) and a declad segment (2), the column 3 is a blank test.

Two parallel leaching tests were conducted: one leaching unit contained the intact fuel segment, a second unit contained the 'declad' (fragments of different sizes + powder + cladding pieces) sample. A third unit was only filled with the leachate, as a blank test to estimate the cross contamination.

18.6 Leachant composition

The test were done with a standard leachant used in FIRST-Nuclides, consisting of 19 mM NaCl + 1 mM $NaHCO_3$. The pH of this solution is around 7.4. No pH buffer was added to the solution, because important pH changes were not expected to occur.

18.7 Preleaching, leaching tests

During the preparation of the fuel samples, they were never in contact with a liquid. At contact with the leachant, we expect a high initial release of the most mobile elements and a rapid dissolution of the oxidized uranium. This information is crucial for the interpretation of the experiments. If the initial dissolution of the spent fuel leads, however, to high concentrations in solution of the considered radionuclides, the smaller further variation of these concentrations in function of time could be less obvious to observe. Consequently, a preleaching was performed. Thus, the spent fuel samples were two times preleached with 50 mL of the solution during 3 and 4 days, respectively. At the end of the second preleaching, the column was filled with 150 mL of fresh solution to cover the whole test duration. After 7, 35, 182 and 350 days (end of the experiment) samplings of 20 mL were taken.

All (pre-)leached solutions were analyzed to determine the concentrations of the radionuclides by various techniques (see section on solution analyses). The same procedure was also applied for test 3 (blank), but only solutions after 7 and 350 days were analyzed. The other samples are kept in the hot-cell for analysis in case of questionable results. In the case of the test with the intact spent fuel sample, the solution looked translucent and any detached fuel particles were retained in the column thanks to the frit on the bottom of the column. In the case of the test with the declad spent fuel, the solution was cloudy due to the fine particles of spent fuel in suspension. During the replacements of the solution for the preleaching and the further sampling, some of the finest powder particles have passed through the frit into the solution for analysis. To estimate the loss of material during these steps, the solutions were filtered with a previously tarred paper filter. After drying, the filter was weighed and the mass of spent fuel was estimated. The amount of powder on the filter represents less than 1% of the total mass of the spent fuel used per test, and it is considered as negligible for the further calculations.

18.8 Solution analyses

Only the most relevant radionuclides were measured, using either mass-spectrometry or radioanalytical techniques. The concentration of ^{238}U was used to estimate the dissolution rate of the fuel and was measured by means of ICP-MS. This technique also gave the concentrations of $^{233\text{-}236}U$, $^{239\text{-}242,244}Pu$, ^{129}I, ^{99}Tc, $^{106,108}Pd$, $^{90,91,94}Zr$, $^{95\text{-}98}Mo$, $^{100\text{-}102,104}Ru$, $^{111,112,114}Cd$, ^{93}Nb, ^{103}Rh, ^{126}Te, $^{107,109}Ag$, $^{118,120}Sn$ and ^{133}Cs in the leachates. Some of these radionuclides (especially Pd, Ag, Ru, Sn and Tc) have concentrations close to the detection limit.

The concentrations of ^{94}Nb and $^{134,137}Cs$ were measured by γ-spectrometry. The concentrations of ^{14}C, ^{63}Ni and ^{90}Sr were determined using liquid scintillation counting (LSC). In some cases, prior to the analyses, a separation was required to avoid interference with others radionuclides. This is the case for ^{14}C, ^{63}Ni, ^{94}Nb and ^{90}Sr.

Although the measurement of the ^{14}C-, ^{63}Ni- and ^{94}Nb-content of waste matrices is performed routinely at SCK•CEN, their analysis in spent fuel and/or leachates has not been performed at SCK•CEN previously. An evaluation and optimization of existing methodologies was required. Analyses of ^{79}Se was not performed.

18.9 Results

After analysis of the leachates (preleaching and sampling) and the determination of the initial inventory of the spent fuel, the data were treated to estimate the total release of the radionuclides during each step, the cumulative Fraction of the Inventory in the Aqueous Phase (CumFIAP) and the (cumulative) Instant Release Fraction (IRF). The CumFIAP in function of the time is reported in the Figure 18.6 for the intact and the declad segment. To calculate the IRF, the CumFIAP(U) is subtracted from the CumFIAP. In this way, the radionuclide release due to matrix dissolution is eliminated in the IRF.

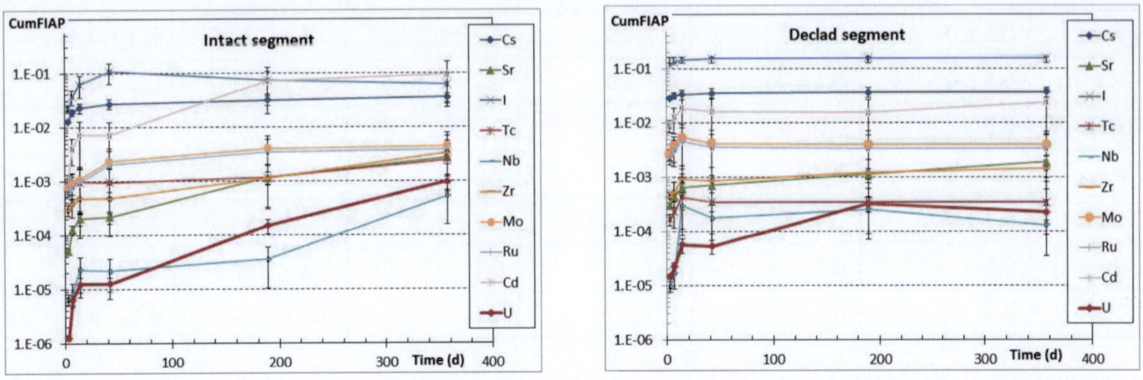

Figure 18.6: Cumulative FIAP in function of the time for the experiment with the intact segment (left) and the declad segment (right).

In Figure 18.6, only part of the radionuclides are plotted. Usually, the behavior of one specific RN was independent of the isotopic form. For this reason, only one isotope of each RN is presented. Moreover, for some elements, the analytical results gave values below the detection limit of the method. Their FIAP values are not plotted, as the ^{14}C and the Pu. Nevertheless, the calculations were done and an estimation of the IRF is provided as upper limit and are reported on the Table 18.1.

In the two experiments and whatever the radionuclides, we observe a high initial release of the RN in solution during the preleaching period and the first sampling. In general, the FIAP values in the experiment with the intact segment are slightly lower than the values estimated for the experiment with the declad segment. In both experiments, the iodine release in solution is the biggest followed by the caesium. In the case of the intact segment, the IRF of iodine reached a value close to 10% after 42 days (cumulative duration) and decreased slowly during the rest of the experiment. The reason for this decrease is not clear at this moment. It's different in the experiment with the declad segment, where iodine is still the RN with the biggest release in solution but the release increases continuously to reach a plateau at 15.6%.

Table 18.1: Cumulative IRF (%) values for the experiment with an intact segment (Int) and with a declad segment (Dec) in function of time (in days). Values in grey are upper values using the detection limit of the analytical method, in this case, the uncertainties are omitted for the whole series, " -- " indicates negative values.

Int	**3**	**7**	**14**	**42**	**189**	**357**
Cs	1.3 ± 0.2	1.9 ± 0.3	2.3 ± 0.4	2.7 ± 0.4	3.1 ± 0.5	3.4 ± 0.6
Sr	0 ± 0	0 ± 0	0 ± 0	0 ± 0	0.1 ± 0	0.2 ± 0
C	0.8	1.0	1.2	1.2	1.5	1.5
Pu	0.0	0.0	0.0	0.0	--	--
I	2.2 ± 0.5	3.7 ± 0.9	6.2 ± 1.5	10.8 ± 2.3	7.3 ± 1.9	6 ± 2.3
Tc	0.1 ± 0	0.1 ± 0	0.1 ± 0	0.1 ± 0	0.1 ± 0	0.1 ± 0.1
Pd	0.0	0.0	0.0	0.0	0.0	--
Nb	0.0	0.0	0.0	0.0	--	--
Zr	0.0	0.0	0.0	0.0	0.1	0.2
Mo	0.1 ± 0	0.1 ± 0	0.1 ± 0	0.2 ± 0.1	0.3 ± 0.1	0.3 ± 0.1
Ru	0.1 ± 0	0.1 ± 0	0.1 ± 0	0.2 ± 0.1	0.3 ± 0.1	0.3 ± 0.1
Rh	0.0	0.0	0.0	0.0	0.0	--
Ag	2.0 ± 0.7	2.0	2.0	2.0	2.1	2.4
Cd	0.1 ± 0	0.4 ± 0.1	0.7 ± 0.2	0.7 ± 0.2	7.2 ± 2.5	9.2 ± 3.2
Sn	0.1	0.1	0.1	0.1	0.1	0.2
Te	0.1	0.1	0.1	0.1	0.1	0.1

Dec	**3**	**7**	**14**	**42**	**189**	**357**
Cs	2.8 ± 0.5	3.2 ± 0.5	3.4 ± 0.6	3.5 ± 0.6	3.6 ± 0.6	3.7 ± 0.6
Sr	0 ± 0	0 ± 0	0.1 ± 0	0.1 ± 0	0.1 ± 0	0.2 ± 0
C	0.8	0.9	0.9	1.0	1.0	1.0
Pu	0.0	0.0	0.0	0.0	--	--
I	11.8 ± 2.9	13.4 ± 3.3	14.4 ± 3.5	15.6 ± 3.8	15.6 ± 3.8	15.6 ± 5
Tc	0 ± 0	0 ± 0	0 ± 0	0 ± 0	0 ± 0	0 ± 0
Pd	0.0	0.0	0.0	0.0	--	0.0
Nb	--	--	0.0	0.0	--	--
Zr	0.0	0.1	0.1	0.1	0.1	0.1
Mo	0.2 ± 0.1	0.3 ± 0.1	0.5 ± 0.2	0.4 ± 0.1	0.3 ± 0.1	0.3 ± 0.1
Ru	0.2 ± 0.1	0.3 ± 0.1	0.5 ± 0.2	0.3 ± 0.1	0.3 ± 0.1	0.3 ± 0.1
Rh	0.0	0.0	0.0	0.0	--	--
Ag	0.2	0.2	0.6	0.4	0.3	0.3
Cd	0.9 ± 0.3	1.3 ± 0.4	1.8 ± 0.6	1.6 ± 0.5	1.5 ± 0.5	2.4 ± 0.8
Sn	0.1	0.1	0.1	0.1	0.1	0.1
Te	0.1	0.1	0.2	0.1	0.1	0.1

Usually, iodine is considered as a volatile element, well related to the FGR (Johnson et al., 2012). The IRF value for iodine has to be compared with the 14% and 11% for the measured and calculated FGR, respectively (Govers et al, 2014).

Unfortunately, the values of the ^{14}C concentrations are always below the DL, only one value is reliable. We can only conclude that the IRF for the ^{14}C is below 1.5%.

18.10 Conclusions

The experiments performed by SCK•CEN with clad and declad UOX PWR fuel segments have allowed the determination of the fast release in solution of the easily soluble radionuclides. After the preparation of the spent fuel samples in dry conditions, the samples were leached with a standard solution and samplings were performed for the analysis of the main radionuclides. The experiments were stopped after one year and the results were treated to estimate the FIAP and IRF for each radionuclides analyzable in our institute. The preliminary results show that the IRF is higher for the declad segment than for the clad segment, and that the IRF for iodine is close to the FGR for the declad segment. The IRF for ^{14}C is below 1.5%. In addition to the leaching experiments, a post-irradiation examination was performed. In comparison with the results of these investigations, most of results (sample inventory, rod FGR) matched well with the theoretical values based on the fuel fabrication and irradiation characteristics.

18.11 Acknowledgement

The research leading to these results has received funding from the European Union's European Atomic Energy Community's (Euratom) Seventh Framework Programme FP7/2007-2011 under grant agreement n° 295722 (FIRST-Nuclides project).

This work was performed as part of the programme of the Belgian Agency for Radioactive Waste and Enriched Fissile Materials (NIRAS/ONDRAF) on the geological disposal of high level/long-lived radioactive waste. The authors gratefully acknowledge the technical support from Ben Gielen and Herman Van Eyck, the analytical service of the SCK•CEN as well as AREVA, Electrabel and Tractebel for the data related to the fuel rod.

18.12 References

Adriaensen, L., Dobney, A., VanBree, P. (2014). Radiochemical Burnup Analysis of Fuel Sample FT1X57-D05-BU1. Restricted Contract Report, SCK•CEN-R-5640.

Govers, K., Verwerft, M. (2014). Fuel Fabrication Characteristics. Databook for Rod D05 & E12 Extracted from Fuel Assembly FT1X57, Tihange 1 NPP: Data on the Fuel Initial Composition, Irradiation History and In-Pile Behavior. Restricted Contract Report, SCK•CEN-R-5514.

Johnson, L.H., Shoesmith, D.W. (1988). Radioactive Waste Forms for the Future. Eds. W. Lutze and R.C. Ewing, North Holland.

Johnson, L., Günther-Leopold, I., Kobler Waldis, J., Linder, H.P., Low, J., Cui, D., Ekeroth, E., Spahiu, K., Evins, L.Z. (2012). Rapid Aqueous Release of Fission Products from High Burn-Up LWR Fuel: Experimental Results and Correlations with Fission Gas Release. Journal of Nuclear Materials, 420, 54-63.

Mennecart, Th., Lemmens, K., Cachoir, C., Adriaensen, L., Dobney, A. (2013). Leaching Tests for the Experimental Determination of IRF Radionuclides from Belgian High-Burnup Spent Nuclear Fuel: Overview of Pre-Test Fuel Characterization, Experimental Set-Up and First Results. 2nd Annual Workshop Proceedings of the 7th EC FP CP FIRST-Nuclides project (eds. Kienzler et al.). KIT Scientific Reports 7676, 123-129.

Verwerft, M., Vos, B., VanDenBerghe, S., Govers, K. (2014). Post-Irradiation Examination Report Rod D05-OM1. Restricted Contract Report, SCK•CEN-R-5579.

19 Corrosion test of commercial UO_2 BWR spent nuclear fuel for fast/instant release studies

Albert Martínez-Torrents[1], Rosa Sureda[1], Joan de Pablo[1,2], Frederic Clarens[1], Daniel Serrano-Purroy[3], Laura Aldave de las Heras[3] and Ignasi Casas[2]*

[1] *Fundació CTM Centre Tecnològic (ES)*

[2] *Department of Chemical Engineering, Universitat Politècnica de Catalunya (ES)*

[3] *European Commission, Joint Research Centre, Institute for Transuranium Elements (EC)*

** Corresponding author: albert.martinez@ctm.com.es*

19.1 Abstract

The release behaviour of commercial UO_2 spent nuclear fuel was studied using batch leaching experiments performed in the hot cell facilities of EC-JRC-ITU (Karlsruhe, Germany) in the framework of the FIRST-Nuclides program. This paper describes the experimental methodology used and results after one-year of experiment.

19.2 Introduction

A standard BWR fuel rod was selected for leaching test investigations. The fuel had an standard ^{235}U enrichment and an average burn-up of 54 GWd/t_{HM}. All the main parameters can be found in previous report D.1.2 for WP1 (Metz et al., 2013). Two types of fuel samples with different morphologies: fuel cladded segments and powder were used in the framework of the project. The powder fraction was prepared from different sections of the pellet. One was obtained from the internal part (CORE) and other from the exterior (OUT), in order to see the differences due to the migration of radionuclides during the irradiation and the effect of the high burn-up structure (HBS).

In the present work, a cladded segment sample, powder from the core of the pellet and powder from the exterior of the pellet, were leached in order to study the Instant Release Fraction (IRF). This concept takes into account the amount of radionuclides that can potentially be released during the initial contact period between fuel and solution, although there is not a clear consensus in the definition. These results will be useful in order to identify the different radionuclide contributions: gap, grain boundary and fuel matrix.

19.3 Experimental

Fuel Pellet Sample

The cladded segment sample was a disk of (2.5 ± 0.1) mm cut at the middle of the pellet avoiding inter-pellets zones as referred in previous report (D.1.2.; Metz et al., 2013). The total weight of the sample including the cladding was (2.0617 ± 0.0001) g with a fuel pellet diameter of (8.7 ± 0.1) mm. Based on the inventory results (report for WP1), the initial mass of the fuel sample can be calculated as approximately 79.4% of the total weight (fuel +cladding), which corresponds to a fuel sample weight of (1.6 ± 0.1) g. An image of the cross sections of the sample is shown in Figure 19.1 where the typical fractures are observed due to the high thermal gradients during reactor operation.

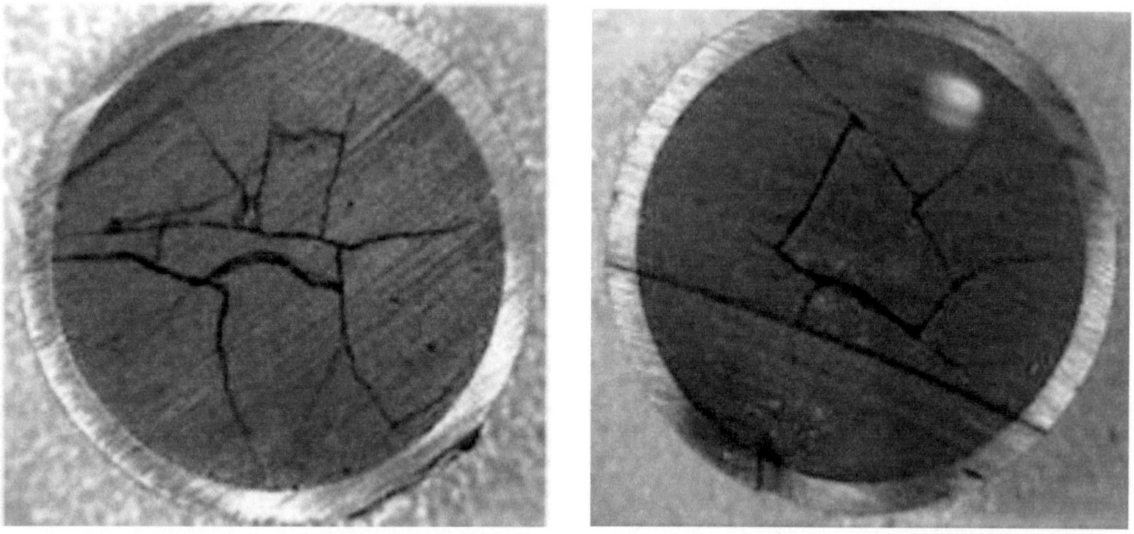

Figure 19.1: Cross sections of spent fuel sample used in leaching experiment.

The surface are of the fuel samples used in our test was not exactly known but minimum values were calculated from direct geometric measurement. Considering the pellet as a cylinder totally exposed to the solution, the surface area was 187 ± 6 mm^2. However, a higher specific surface area for a fractured pellet was assumed with a surface roughness factor of 3.5 (Iglesias et al., 2008), corresponding to a specific surface area of 655 ± 22 mm^2.

Powder Samples.

Powder samples from the CORE region of the pellet were obtained drilling the pellet with a bit of 3 mm (Figure 19.2a). A fraction not used in this project was collected using a bit of 8 mm. The OUT fraction was obtained from the remaining fuel attached in the walls of the rod. It was obtained by pressuring the rod to release the fuel fraction (Figure 19.2b). Each powder was sieved in order to obtain fractions of powder with different particle size. In the experiments particles with a particle size bigger than 50 µm were used. Specifically the mean diameter was of 72.6 ± 28 µm and 90.7 ± 37 µm for the CORE and OUT powder particles respectively (Figure 19.3). Knowing the particle size the specific surface area was estimated from the sample weight considering the particles as spheres and a roughness factor of 3.5 (Iglesias et al., 2008). The specific surface area was of 4,010 ± 1,820 mm^2 for the CORE experiment and 2,090 ± 1,010 mm^2 for the OUT experiment.

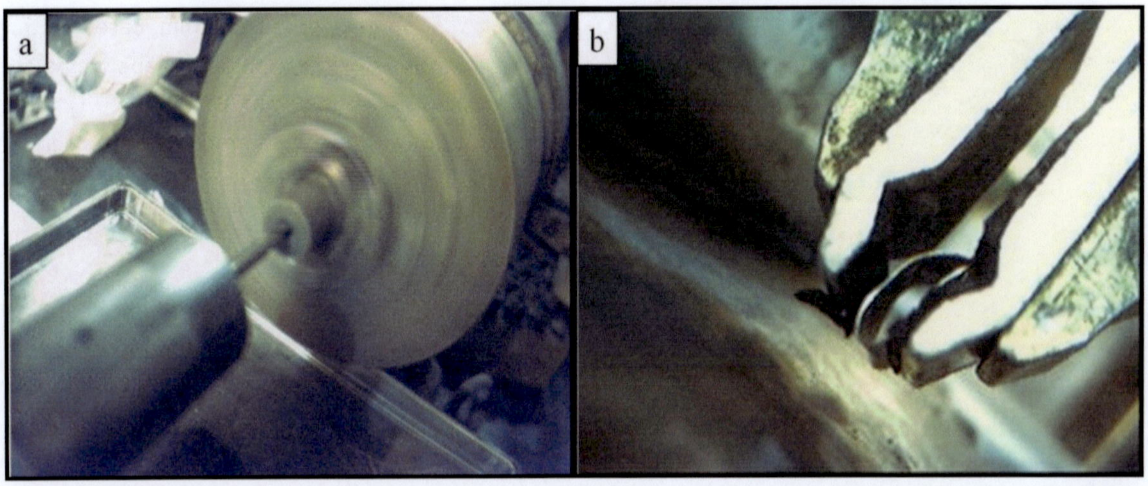

Figure 19.2: Powder samples preparation a: CORE; b: OUT.

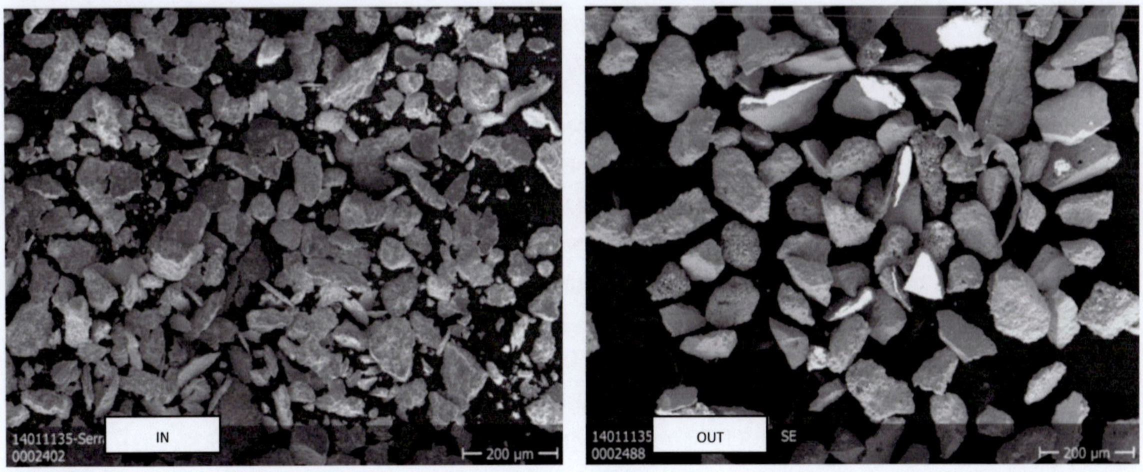

Figure 19.3: Powder particles.

Methodology and analytical methods

The experiments were performed in carbonated solution (1 mM $NaHCO_3$ + 19 mM NaCl, with initial pH of (8.4 ± 0.1) that were initially equilibrated with air under oxic conditions and a normal hot cell temperature (25 ± 5) °C. For all experiments a volume solution of (50 ± 1) mL was used.

The cladded segment samples were suspended into a plastic vessel through a platinum thread. The vessel was placed in to a stirrer to keep it under continuous orbital rotation. The system, see Figure 19.4, ensures full contact within sample-leaching solution as well as the homogenization of the solution at all contact times. No previous washing steps were performed in cladded segment samples to avoid the loss of valuable information concerning the IRF.

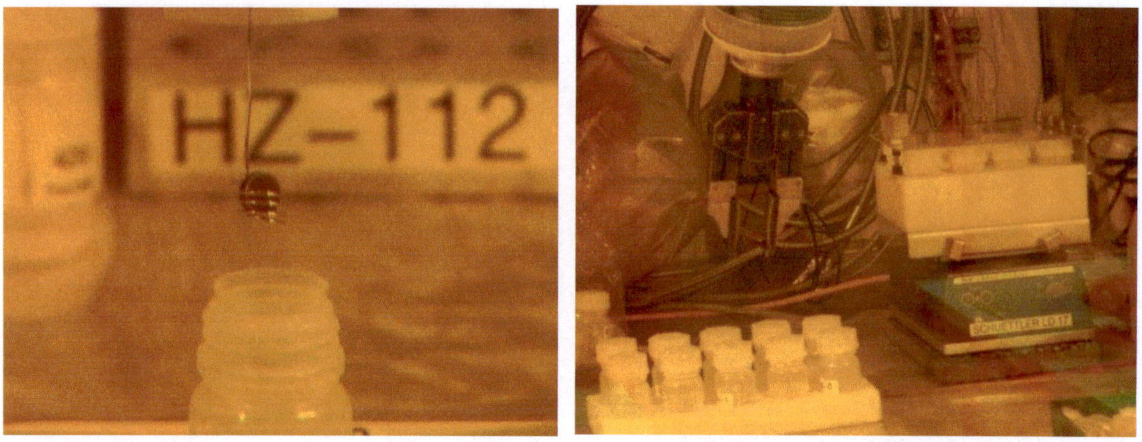

Figure 19.4: Experimental set-up.

Blank tests were also carried out using the same procedure but without the fuel sample to check for possible cross-contamination in the hot cell. The contact periods were: 0.1, 0.2, 1, 4, 8, 21, 49, 83, 162 and 190 days. After each one, the whole solution was renewed.

Powder samples were prepared adding a known quantity of the selected fuel powder to a glass tube and then adding 50 mL of the sample-leaching solution. The glass tubes were continuously stirred. Two tubes with the same solution but without fuel powder were used as hot cell blanks. At each sampling the solution was renewed. The first two samplings were considered as washings to avoid the influence of fines and oxidized layers into the experiment results. The contact periods were: 0.8, 2, 3, 4, 7, 9, 11, 16, 22, 30, 81 and 116 for the CORE and 0.8, 2, 3, 4, 7, 9, 11, 16, 30, 81 and 116 for the OUT. The two last samples will be analysed and reported before the end of the project. Unfiltered and filtered (through a 0.2 micron pore size PVDF membrane) aliquots of fuel solutions were diluted and acidified with 1 M HNO_3. Samples were analysed by ICP-MS using a Thermo ELEMENT 2 instrument (Thermo Electron Corporation, Germany).

Calibration curves were produced using a series of dilutions of certified multi-element standard solutions in the concentration range of the major elements in solution. All the samples were measured with the addition of In, Ho, Co and Th as internal standards. The concentrations of the elements were calculated from the isotopic data. Whenever was possible, isobaric interferences were corrected based on isotopic ratios previously calculated with the ORIGEN code (ORIGEN-ARP, 2000) as reference.

Measurements of pH have been performed during the entire experiment using an Orion 525A+ pH-meter, a gel pH Triode L/M (reference 9107BN) supplied from Thermo-Electron, USA. The pH electrode was calibrated with commercial pH buffer solutions (METLER TOLEDO Inc., USA; pH 4.01 (Ref. 501307069), pH 7.00 (Ref.51302047), pH 9.21 (Ref. 51302070)).

19.4 Static leaching results

In order to analyse the results from leachate of the experiment with fuel 54 BWR, data will be presented focussing on: Cumulative moles, Fraction of Inventory in the Aqueous Phase ($FIAP_x$), FIAPx normalised to the uranium fraction (FNU_x) and Instant Release Fraction (IRF).

Measured Cumulative moles.

The cumulative moles were calculated from the concentration of non-filtered samples due to the changes in the behaviour in some radionuclides like technetium in the filtered samples.

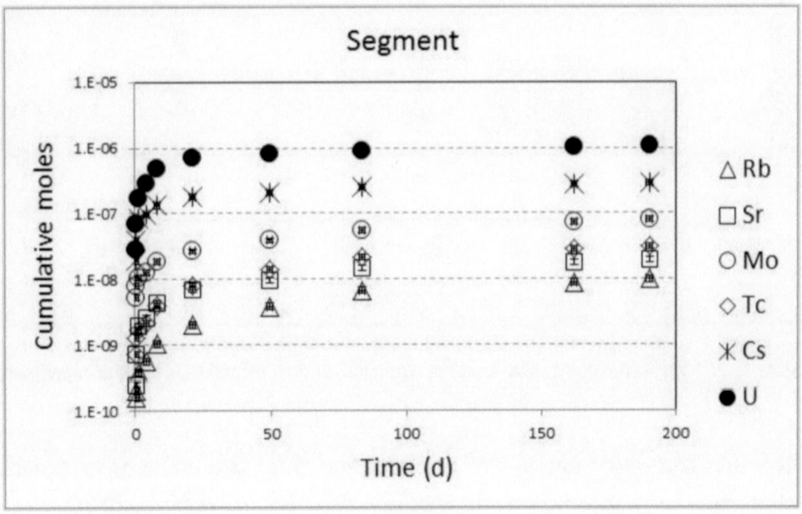

Figure 19.5: Measured cumulative moles as a function of time for the pellet experiment.

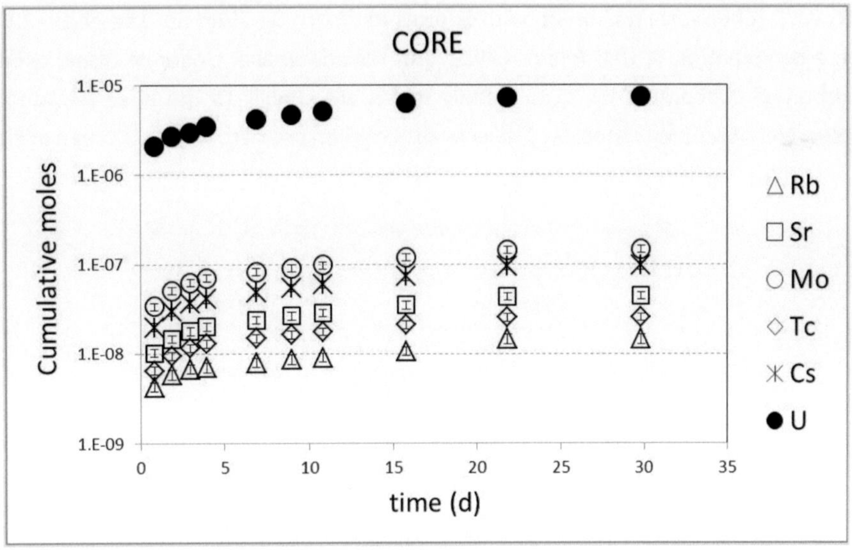

Figure 19.6: Measured cumulative moles as a function of time for the CORE powder experiment.

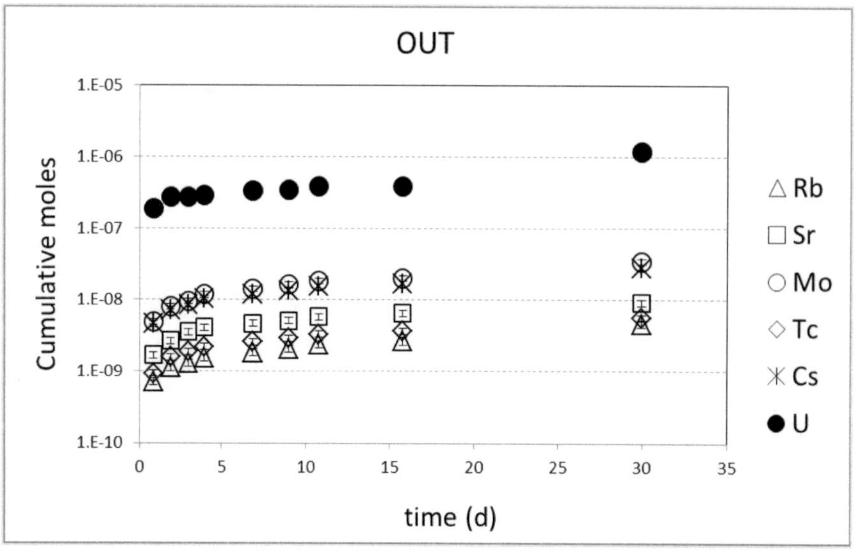

Figure 19.7: Measured cumulative moles as a function of time for the OUT powder experiment.

It is possible to see that the major amount of moles comes from the uranium as expected, followed by cesium and molybdenum.

Predominance diagrams were used to identify which are the main aqueous or solid phases for uranium that may be formed under the experimental conditions of our system. Speciation calculations were performed with the CHESS v.2.5 (Chemical Equilibrium with Species and Surfaces, Van der Lee et al. (2002)) computer code based on a combination of the default CHESS and NEA database. Under oxidising conditions, and in the pH and carbonate concentrations of this study which are similar to the ones found in most natural waters, the predominant aqueous species of hexavalent uranium are carbonates, as seen in Figure 19.5.

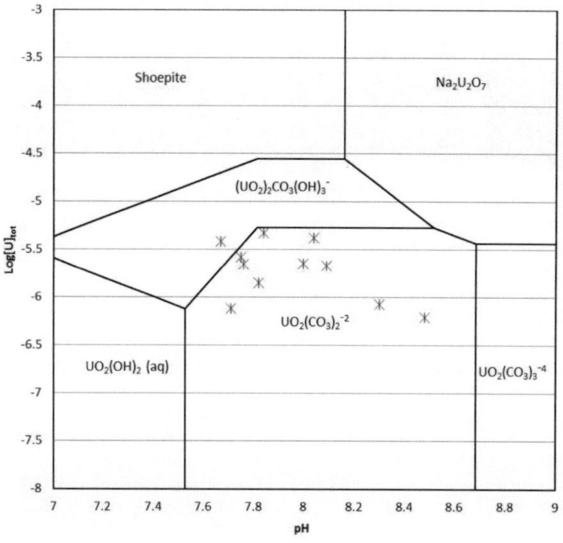

Figure 19.8: Predominance diagram of main uranium species present in our system. Symbols represent our experimental data.

Measured Fraction of the inventory in aqueous phase (FIAP)

In order to better compare the results obtained for different nuclides the FIAP was calculated. In this sense, the fraction of inventory of an element x released in aqueous phase ($FIAP_x$) was calculated for an element x according to Equation 19.1 (Hanson and Stout, 2004):

$$FIAP_x = \frac{m_{x,aq}}{m_{x,SNF}} = \frac{c_x V_{aq}}{m_{x,SNF} H_x} \tag{19.1}$$

where: $m_{x,aq}$ is the mass of element x in the aqueous phase and m_{xSNF} is the mass of element x in the SNF; c_x corresponds to the concentration of element x in solution (g/mL); V_{aq} is the volume of solution (mL), m_{SNF} is the initial mass of SNF sample (g) and H_x represents the fraction of the inventory in the SNF for the element x (g/g).

The cumulative FIAP, FIAPc, of an element x in sample n can be calculated following Equation 19.2:

$$FIAPc(n) = \sum_{i=1}^{n} FIAP_x(i) \tag{19.2}$$

$FIAP_C$ values for Rb, Sr, Mo, Tc, Cs and U are plotted in Figure 19.9, 10 and 11. In general, a higher release during the initial time intervals is observed for uranium and the group of elements: Cs, Mo, Tc, Rb and Sr. This fast initial dissolution observed in uranium can be attributed to the dissolution of the oxidized surface layer present at the beginning of the experiment. For the rest of elements typically considered as a part of the IRF, this fast release is assumed to come mainly from the more accessible and open surfaces (i.e. gap and cracks).

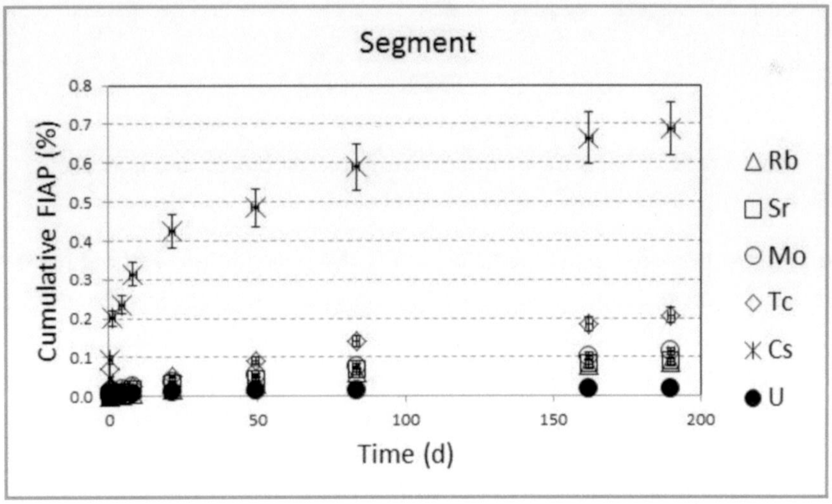

Figure 19.9: Cumulative FIAP as a function of time for the pellet experiment.

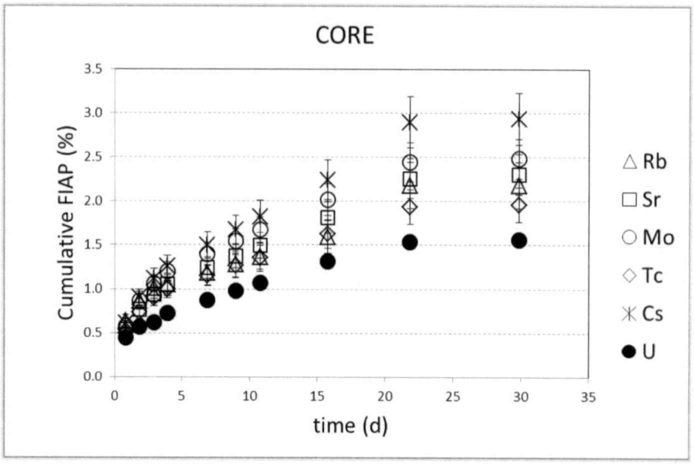

Figure 19.10: Cumulative FIAP as a function of time for the CORE powder experiment.

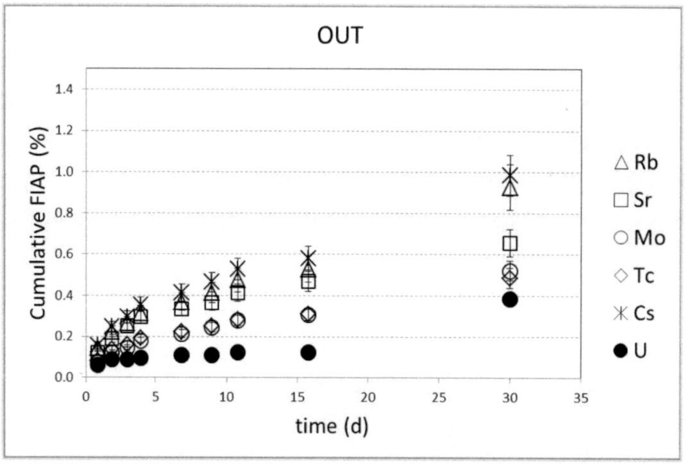

Figure 19.11: Cumulative FIAP as a function of time for the OUT powder experiment.

It is possible to see how in all the experiments cesium is the radionuclide with the highest FIAP. Molybdenum and strontium seem to follow a similar pattern oppositely to technetium and rubidium where in some cases are the second biggest FIAP and in some cases are the second lowest FIAP. It is possible that differences in the speciation of these elements produce difference in their release. Uranium has the lowest FIAP in all the cases. This means that all these radionuclides were dissolved faster than the matrix.

Fractional release Normalized to Uranium (FNU)

FNU values were also calculated in order to compare with uranium in reference to matrix release, see Table 19.1, 2 and 3.

$$FNU_i = \frac{FIAP_i}{FIAP_U} \tag{19.3}$$

$$FNUc_i = \frac{FIAPc_i}{FIAPc_U} \tag{19.4}$$

where FNU_i is the Fractional release Normalized to Uranium for element *i*, $FIAP_i$ is the Fraction of the inventory in aqueous phase for element *i* and $FIAP_U$ is the Fraction of the inventory in aqueous phase for Uranium. $FNUc_i$ is the Cumulative Fractional release Normalized to Uranium for element *i*, $FIAPc_i$ is the Cumulative Fraction of the inventory in aqueous phase for element *i* and $FIAPc_U$ is the Cumulative Fraction of the inventory in aqueous phase for Uranium.

All elements from this group showed higher fractional release compared to U (FNU>1), specially cesium with ratios of one order of magnitude higher than the rest for the pellet experiment. For the CORE powder experiments cesium and molybdenum are higher than 1 after 30 days, and for the OUT powder experiments Rubidium, Strontium and cesium are higher than uranium after 30 days.

Table 19.1: Cumulative FNU values (%) and cumulative time for the pellet experiment.

Time (d)	**0.1**	**0.2**	**1**	**4**	**8**	**21**	**49**	**83**	**162**	**190**
Rb	3	1	1	1	1	1	2	4	4	5
Sr	2	3	3	2	2	2	3	4	5	5
Mo	15	9	5	4	3	3	4	5	6	6
Tc	9	7	4	3	3	4	6	9	10	11
Cs	88	76	65	47	37	33	33	37	37	37

Table 19.2: Cumulative FNU (%) values and cumulative time for the CORE powder experiment.

Time (d)	**0.8**	**2**	**3**	**4**	**7**	**9**	**11**	**16**	**22**	**30**
Rb	1	2	2	1	1	1	1	1	1	1
Sr	1	1	2	1	1	1	1	1	1	1
Mo	1	2	2	2	2	2	2	2	2	2
Tc	1	1	1	1	1	1	1	1	1	1
Cs	1	2	2	2	2	2	2	2	2	2

Table 19.3: Cumulative FNU values (%) and cumulative time for the OUT powder experiment.

Time (d)	**0.8**	**2**	**3**	**4**	**7**	**9**	**11**	**16**	**30**
Rb	2	3	3	3	3	4	4	4	2
Sr	2	2	3	3	3	3	3	4	2
Mo	1	1	2	2	2	2	2	2	1
Tc	1	2	2	2	2	2	2	3	1
Cs	3	3	3	4	4	4	4	5	3

Estimation of the IRF(%)

The IRF determinations are based on FIAP results presented in previous section. Only those elements whose release was faster than uranium were considered as contributors of the IRF. The calculation of IRF does not include the matrix contribution and for this reason FIAP of uranium was subtracted in each case. At this point, IRF was calculated after 190 days of experiment in the case of the pellet experiment and 30 days of experiment for the powder experiments (Figure 19.12, 13 and 14).

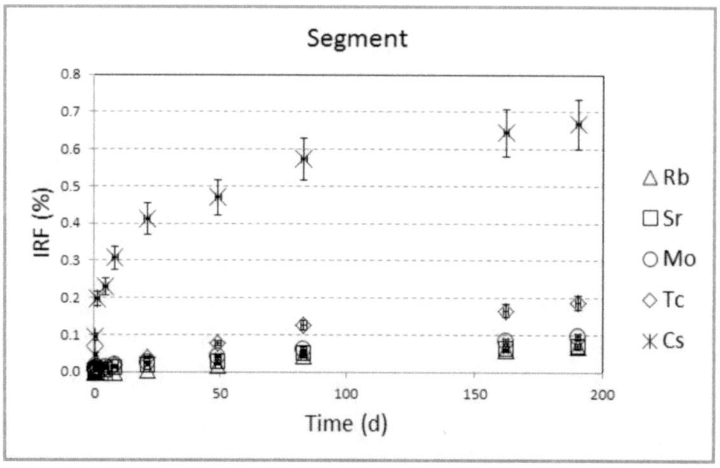

Figure 19.12: IRF as a function of time for the pellet experiment.

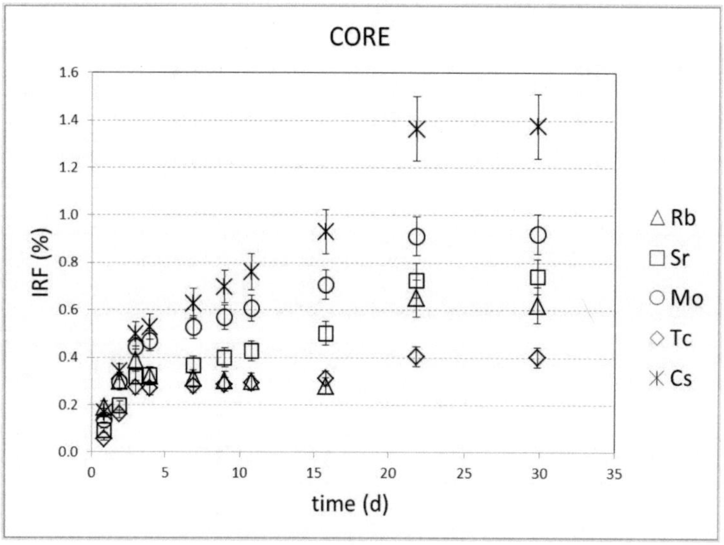

Figure 19.13: IRF as a function of time for the CORE powder experiment.

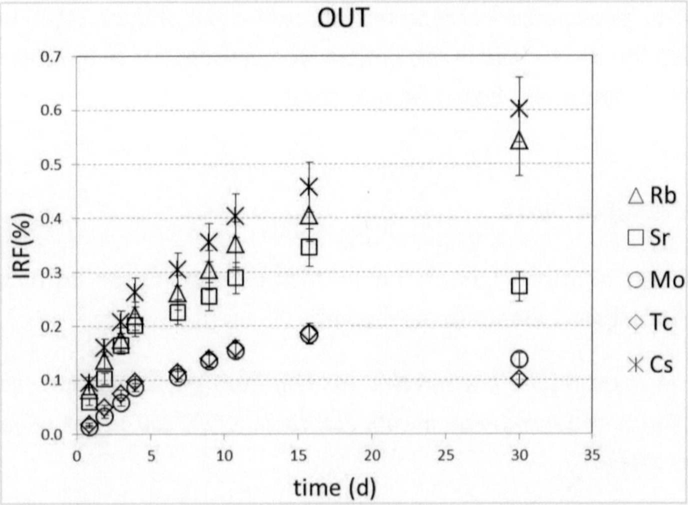

Figure 19.14: IRF as a function of time for the OUT powder experiment.

The higher values are obtained for Cs with a fraction of 1.4% for CORE powder fraction, 0.6% for the OUT powder fraction and 0.7% for the pellet. It has to be taken into account that the last sampling time presented for powder experiments is at 30 days and for the pellet experiment is at 190 days. The IRF(%) at 30 days for Cesium in the pellet experiment is between 0.4 and 0.5%. Considering that the specific surface area is bigger for powder than for the pellet, it is logical to expect a higher release in powder experiments. Looking at the trend of the experimental data, for the powder experiments, it seems that the following samples at 80 and 115 days will give higher releases for cesium and rubidium. It is not so clear what will happen with strontium, technetium and molybdenum of the OUT powder fraction. It seems that the percentage of their instant release fraction is decreasing with time. It can be seen in the experiment with CORE powder (Figure 19.13) that the IRF(%) at sample 21 is higher than what could be expected looking at the trend of the previous samples. This behaviour is attributed to some experimental error, perhaps due to the dissolution of some small piece of fuel. The analysis of the rest of the samples will help in correcting this experimental error.

19.5 Conclusions and Future work

Instant release fraction experiment with cladded fuel segment, powder from the CORE fraction and powder from the OUT fraction were performed. Final results for the pellet experiment and results for the first 30 days of powder experiments are reported. The experimental data show no uranium secondary phase precipitation following the methodology used.

Elements Cs, Mo, Tc, Rb and Sr follow similar trends as observed in other studies with higher fractional release compared to uranium and therefore, are identified as IRF. Specially, cesium clearly presents the higher release compared to the rest. The higher release found at the beginning of the experiment is assumed to come mainly from the gap contribution.

More data is expected to be analysed and reported before the end of the project, for the powder experiments that will complete the work made in this project. Comparisons between similar works performed by different authors in the literature will be made in the future.

19.6 Acknowledgement

The authors also would like to thank all the staff of EU Joint Research Centre Institute of Transuranium for their daily help in the hot cells and the analytical support.

The research leading to these results has received funding from the European Union's European Atomic Energy Community's (Euratom) Seventh Framework Programme FP7/2007-2011 under grant agreement n° 295722 (FIRST-Nuclides project).

19.7 References

Hanson, B., Stout, R.B. (2004). Re-Examining the Dissolution of Spent Fuel: A Comparison of Different Methods for Calculating Rates. Materials. Research Society Symposium Proceedings, 824, 89-94.

Iglesias, E., Quinones, J. (2008). Analogous Materials for Studying Spent Nuclear Fuel: The Influence of Particle Size Distribution on the Specific Surface Area of Irradiated Nuclear Fuel. Applied Surface Science, 254(21), 6890-6896.

ORIGEN-ARP (2000). Automatic Rapid Process for Spent Fuel Depletion, Decay and Source Term Analysis. "NUREG/CR-0200" Revision 6. Volume 1, Section D1, ORNL/NUREG/CSD2/V1/R6.

Van der Lee, J.; Dewindt, L. (2000). CHESS Tutorial (version 3.0). Ecole des Mines de Paris, Fontainebleau, France.

20 Isotopes dissolution during wet storage of damaged and leaking VVER fuel

Emese Slonszki and Zoltán Hózer*

Centre for Energy Research Hungarian Academy of Sciences (HU)
** Corresponding author: emese.slonszki@energia.mta.hu*

20.1 Abstract

Within the FIRST-Nuclides project dissolution rates of several isotopes, like ^{134}Cs, ^{137}Cs, ^{154}Eu, ^{155}Eu, ^{125}Sb and UO_2 from VVER fuel were determined for pH 7 and pH 4 - 4.5 in the coolant based on activity measurements at the Paks NPP. The present report summarizes both the design and operational characteristics of fuels and the calculation methods and dissolution rates of isotopes during and after the incident of Unit 2 of Paks NPP and during the wet storage of a leaking fuel assembly of Paks NPP. The dissolution rates were different in the two evaluated conditions which is attributed to the pH. During the wet storage of the damaged VVER fuel the fitted release rate range of long lived isotopes was $1\cdot10^{-5}$ - $3\cdot10^{-5}$ L/d, while this range was $4\cdot10^{-6}$-$8\cdot10^{-6}$ L/d in case of leaking VVER fuel. The release rates of ^{125}Sb isotope shows good agreement with $8\cdot10^{-6}$ L/d release rate in both investigated event.

20.2 Introduction

The objective of MTA EK contribution to WP1 in the framework of FIRST-Nuclides project is the characterization of VVER fuel (Hózer and Slonszki, 2013), while to WP3 is the determination of dissolution rates for several isotopes from damaged and leaking VVER fuel assemblies stored in water for several years (Slonszki and Hózer, 2013). There were two series of measurements at the Paks NPP that can be used for the evaluation of fuel dissolution in wet environment:

1) Thirty fuel assemblies were damaged at the power plant during a cleaning tank incident in 2003, which were than stored in stored in a special service area of the spent fuel storage pool for almost four years. Based on the continuously measured activity concentrations we could calculate release rates for several isotopes.

2) A leaking fuel assembly was identified at the NPP in 2009. The assembly was removed from the reactor core and the placed in the spent fuel storage pool. A special measurement programme was carried out in the spent fuel storage pool to investigate the activity release from the leaking fuel rod at wet storage conditions. The data from this programme was used for the calculation of dissolution rates.

20.3 Characterisation of damaged and leaking fuel

The main characteristics of the damaged and leaking fuel VVER-440 fuel assemblies have been collected.

There were no special examinations of the fresh fuel assemblies before loading them into the reactor core, for this reason factory data were used to characterise the fuel. The operational parameters were derived using power histories of from the NPP. The calculations were carried out with fuel behaviour codes FUROM and TRANSURANUS.

20.4 1. Characterisation of damaged VVER fuel

On 10 April 2003 severe damage of fuel assemblies took place during an incident at Unit 2 of Paks Nuclear Power Plant in Hungary (Hózer et al., 2009; Hózer et al., 2010; NEA/CSNI/R, 2008; Slonszki et al., 2010). The assemblies were being cleaned in a special tank below the water level of the spent fuel storage pool in order to remove crud buildup. That afternoon, the chemical cleaning of assemblies was completed and the fuel rods were being cooled by circulation of storage pool water. The first sign of fuel failure was the detection of some fission gases released from the cleaning tank during that evening. The cleaning tank cover locks were released after midnight and this operation was followed by a sudden increase in activity concentrations. The visual inspection revealed that all 30 fuel assemblies were severely damaged (Figure 20.1). The first evaluation of the event showed that the severe fuel damage happened due to inadequate coolant circulation within the cleaning tank.

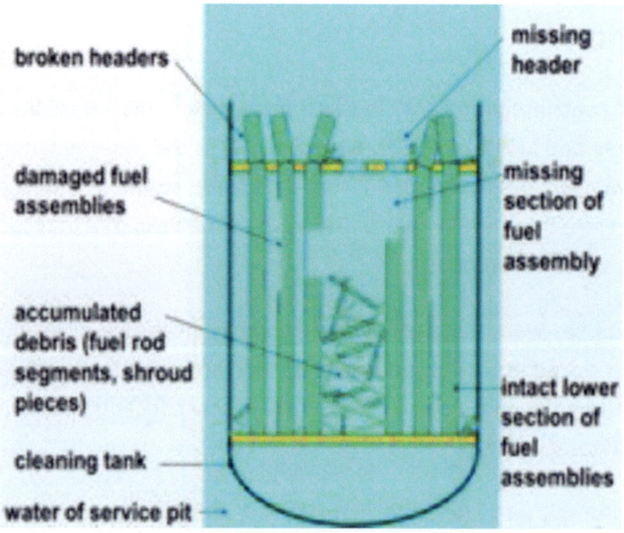

Figure 20.1: Distribution of damaged fuel in the cleaning tank (Slonszki et al., 2010).

The damaged fuel assemblies were removed from the cleaning tank in 2006 and were than stored in special canisters in the spent fuel storage pool of the Paks NPP.

Following several discussions between expert from different countries and international organisations the OECD–IAEA Paks Fuel Project was proposed. The project focused on the numerical simulation of the most important aspects of the incident. This activity helped in the reconstruction of the accidental scenario. A database, which was necessary for the code calculations, was collected and those data were available for the FIRST-Nuclides project.

Design characteristics of VVER-440

In the first step we collected the design characteristics data of VVER-440 in our first report of this project (Hózer and Slonszki, 2012), where the fuel assembly data were collected from open literature (Solonin et al., 1997), that includes the data of both working and follower type VVER-440 assemblies, for both types can be found among the damaged assemblies. Some material properties of E110 (Zr1% Nb) cladding material was also collected from open literature (Smirnov et al., 1997).

Operational data of damaged fuel assemblies

The operational data of fuel assemblies has been collected in order provide burnup specific data for each fuel assembly that was cleaned in the cleaning tank. This part of database collection included several calculations, which were based on real fuel cycles of each assembly. Operational data for Unit 2 and cycles 16 - 19 were provided by the Paks NPP. The decay heat of each node of each assembly was calculated using power distribution and considering the storage time between reactor shutdown and the incident.

The fuel assemblies with similar power histories and of the same type (follower or working) were grouped into 6 groups and only one representative assembly was calculated for each group. The burnup varied between 10.8 and 26.7 MWd/kg_U.

Isotope inventory

The isotope inventories were determined taking into account the real power histories of each fuel assembly for almost one thousand isotopes. For future evaluation only the long lived isotopes will be used. The inventory is important data for the determination of fractional releases.

Table 20.1: Isotope inventory of all 30 damaged fuel assemblies at the reactor shutdown.

Isotope	Activity [Bq]	Isotope	Activity [Bq]	Isotope	Activity [Bq]
^{239}Np	2.30E+18	^{131}I	1.23E+17	^{127}Te	8.90E+15
^{133}I	2.61E+17	^{88}Rb	1.06E+17	^{133m}Xe	8.05E+15
^{133}Xe	2.51E+17	^{131}Te	1.06E+17	^{241}Pu	7.67E+15
^{135}I	2.47E+17	^{105}Ru	1.05E+17	^{137}Cs	7.22E+15
^{140}La	2.45E+17	^{88}Kr	1.04E+17	^{129m}Te	7.11E+15
^{99}Mo	2.37E+17	^{91m}Y	1.03E+17	^{137}Ba	6.90E+15
^{140}Ba	2.35E+17	^{105}Rh	9.78E+16	^{90}Y	6.53E+15
^{95}Zr	2.33E+17	^{147}Nd	8.55E+16	^{90}Sr	5.90E+15
^{95}Nb	2.30E+17	^{135}Xe	7.44E+16	^{134}Cs	5.75E+15

Isotope	Activity [Bq]	Isotope	Activity [Bq]	Isotope	Activity [Bq]
^{97}Nb	2.25E+17	^{237}U	7.12E+16	^{111m}Ag	5.09E+15
^{97}Zr	2.23E+17	^{149}Pm	6.36E+16	^{111}Pd	5.08E+15
^{141}Ce	2.16E+17	^{135m}Xe	5.17E+16	^{111}Ag	5.04E+15
^{141}La	2.15E+17	^{129}Sb	3.78E+16	^{136}Cs	3.08E+15
^{97m}Nb	2.12E+17	^{106}Rh	3.73E+16	^{242}Am	2.48E+15
^{99m}Tc	2.09E+17	^{85m}Kr	3.66E+16	^{244}Am	1.71E+15
^{143}Ce	2.06E+17	^{129}Te	3.58E+16	^{242}Cm	9.69E+14
^{143}Pr	2.00E+17	^{106}Ru	3.20E+16	^{125}Sb	5.77E+14
^{92}Y	1.86E+17	^{153}Sm	3.07E+16	^{181}Hf	4.86E+14
^{92}Sr	1.85E+17	^{105m}Rh	2.99E+16	^{154}Eu	2.33E+14
^{91}Y	1.83E+17	^{109}Pd	2.89E+16	^{155}Eu	1.22E+14
^{132}I	1.81E+17	^{131m}Te	2.25E+16	^{110m}Ag	1.09E+14
^{132}Te	1.78E+17	^{151}Pm	2.22E+16	^{122}Sb	8.19E+13
^{91}Sr	1.78E+17	^{83}Br	1.70E+16	^{180m}Hf	4.37E+13
^{103}Ru	1.72E+17	^{83m}Kr	1.70E+16	^{239}Pu	3.62E+13
^{103m}Rh	1.72E+17	^{147}Pm	1.69E+16	^{124}Sb	3.35E+13
^{89}Sr	1.46E+17	^{238}Np	1.65E+16	^{3}H	3.18E+13
^{145}Pr	1.39E+17	^{148}Pm	1.49E+16	^{240}Pu	3.06E+13
^{93}Y	1.38E+17	^{156}Eu	9.60E+15	^{244}Cm	2.30E+13
^{144}Pr	1.25E+17	^{243}Pu	9.30E+15	^{175}Hf	1.88E+13
^{144}Ce	1.23E+17	^{127}Sb	9.20E+15	^{242}Am	4.68E+12

20.5 Characterisation of the 70873 leaking fuel assembly

The No. 70873 fresh follower fuel assembly was loaded into position No. 33 of cycle 22 of Unit 4 at the Paks NPP. The enrichment of U was 3.82%.

Design characteristics of 70873 leaking fuel assembly

The design characteristics of this assembly are the same as that of damaged fuel described in Chapter 1.1.1.

Operational data of 70873 leaking fuel assembly

The assembly was operated for one year and was removed from the core during the refueling period. For this reason the assembly reached only 14 MWd/kg_U burnup. According to the analyses of primary coolant activities, this assembly had only one single leaking fuel rod, the other 125 rods were intact. The operational history of the assembly was provided by the power plant.

Isotope inventory of 70873 leaking fuel assembly

The isotope inventory of 70873 assembly during the operation and the storage was calculated based on these data by ORIGEN-ARP 5.01 program.

Table 20.2: Isotope inventory of one fuel rod in the leaking 70873 fuel assembly at the reactor shutdown.

Isotope	Mass [g]	Activity [Bq]
^{239}Np	8.31	5.67E+14
^{140}La	0.47	7.86E+13
^{95}Zr	11.77	7.40E+13
^{95}Nb	6.35	7.31E+13
^{141}Ce	8.12	6.80E+13
^{99}Mo	0.42	5.88E+13
^{99m}Tc	0.04	5.32E+13
^{103}Ru	5.54	5.24E+13
^{89}Sr	5.49	4.69E+13
^{144}Ce	36.38	3.41E+13
^{97}Nb	0.00	2.73E+13
^{97}Zr	0.05	2.72E+13
^{106}Ru	7.74	7.61E+12
^{137}Cs	62.21	1.59E+12
^{90}Sr	32.14	1.30E+12
^{134}Cs	2.82	1.08E+12
^{136}Cs	0.04	8.06E+11
^{125}Sb	0.37	1.12E+11
^{242}Cm	0.08	7.74E+10
^{154}Eu	0.51	3.74E+10
^{239}Pu	552.90	2.48E+10
^{155}Eu	0.16	2.35E+10
^{110m}Ag	0.01	1.77E+10
^{240}Pu	97.59	1.61E+10
^{3}H	0.00	1.12E+10
^{122}Sb	0.00	9.68E+09
^{124}Sb	0.00	7.67E+09
^{244}Cm	0.04	1.04E+09
^{241}Am	0.62	6.25E+08

20.6 Determination of dissolution rates for damaged and leaking VVER fuel stored in water

Several isotopes were measured during and after the incident of Unit 2 of Paks Nuclear Power Plant and during the wet storage of No. 70873 leaking fuel assembly of Unit 4 of Paks Nuclear Power Plant. Concentrations of those isotopes were measured regularly and their decay corrected integrated releases were estimated in this work. It involves 11 isotopes and uranium in the first case and 13 isotopes and uranium in the second case. We used two approaches for determination of the release rates:

1) First of all we calculated the corrected integrated releases of every isotope from the measured activity data. After that these data were divided by the total time which belongs to measurement (the elapsed time from the first measurement to the last measurement). The calculated release rate values are given by this method.

2) In the other case linear fitting was applied to the corrected integrated releases data which was resulted the fitted release rates. These procedures are described in the following Chapter 2.1.

20.7 Release rates for damaged VVER fuel

Calculation methods

The activity concentration data of period between 11th April 2003 and 8th January 2007 were provided by the Paks Nuclear Power Plant (Hózer et al., 2009; Hózer et al., 2010; Slonszki et al., 2010). Measurements were available from the pit with the cleaning tank, from the spent fuel storage pool and from the reactor system. Several measurements were taken before and after the filters of the purification system.

It was supposed in the evaluation process, that the measured activity concentration was typical for the total volume of water connected to the cleaning tank with damaged fuel.

Aims of this work to evaluate the activity release of radionuclides first of all for the first two weeks which period describes the incident and the other hand the about 4 year's period which corresponds to the wet storage. The elapsed time since the incident was divided into short periods, which were characterized by stabilized technological conditions allowing us to consider constant release rate during a given period. The number of periods was different for the different isotopes, for example 53 periods of ^{137}Cs isotope were identified. The positions of gates between the reactor pool and the spent fuel storage pool were taken into account.

The operation of water purification system was considered using the flowrate and efficiency values. The real flowrate of the water purification system was corrected by the efficiency of filters for each isotope and for each period:

$$\varepsilon=\frac{C_{before}-C_{after}}{C_{before}} \quad (20.1)$$

ε - efficiency; C_{before} - activity concentration before the filter (Bq/kg); C_{after} - activity concentration after the filter (Bq/kg).

The release rate was calculated for each period and the total release was determined as the result of integration over the total calculated time by three different methods.

1. Linear fitting was applied to those periods, where the activity continuously increased. That was typical for periods without the operation of water purification system.
2. Monotone decreasing or constant activities were typical for periods with operation of water purification system. In this case the following expression was applied for each isotope:

$$\alpha = g_f \frac{C - C_0 e^{g_f * \Delta t/M}}{1 - e^{g_f * \Delta t/M}} \tag{20.2}$$

   α - dissolution rate of the given isotope (Bq/h); g_f - flowrate of water purification system (kg/h); C - activity concentration at the beginning of the period (Bq/kg); C_0 - activity concentration at the end of the period (Bq/kg); Δt - period (h); M - mass of water (kg).

3. In the third case the average activity value was multiplied by the flowrate of the water purification system. This approach was applied when the scatter of the measured points did not allow us to use the above methods. These periods could not be characterized by a constant dissolution rate that is why an average value was considered.

The inventory of radioactive fission products continuously decreased after shutdown and during the storage of spent fuel in the cleaning tank. To compare with the initial inventory and the results of analytical work, corrections were applied for decay. These corrected values were higher than the measured values. The physical meaning of the corrected values relates to the time of the incident, simulating that the total release would have been at that time.

The total release for the considered for 4 years was calculated as the sum of the individual periods. First of all this value was divided by the total time - from the incident until removing of damaged fuel assemblies – giving the "calculated release rate" of the isotope (Table 20.3). Alternatively, a linear fitting was applied to the corrected integrated release data and yielded release rate for every examined isotope.

The activity concentration of uranium was given in µg/l and converted stoichiometrically to a concentration of UO_2 using the ratio of molar mass of UO_2 and ^{238}U.

Finally, the release rates refer to the isotope inventory of incident (Table 20.3 and Table 20.4).

Results

Based on the calculated integrated releases the total measuring period can be divided into two parts. The activity of the first two week describes the incident while the other parts of period correspond to the wet storage.

After the incident the change in the release rate was influenced first of all by the decay of short lived isotopes. However the water chemistry conditions played important role, too. The pH decreased from 7 to 4 – 4.5 due to the increase of the boric acid concentration from 15 g/kg to 21 g/kg and the led to increase of ^{242}Cm and ^{238}Pu activity by two and three orders of magnitude. Similar effect was observed for other isotopes as well at the same time.

Table 20.3: Calculated and fitted release rates of the gamma radiant isotopes.

Isotope	Isotope inventory (Bq)	Calculated release rate (1/d)	Fitted release rate (1/d)	
			Incident pH≈7	Wet storage pH≈4-4.5
^{144}Ce	1.20E+17	1.27E-05	2.00E-05	1.20E-05
^{140}Ba	1.16E+17	2.71E-05	4.15E-05	1.45E-05
^{131}I	4.15E+16	1.38E-04	5.78E-04	2.31E-06
^{137}Cs	7.22E+15	2.10E-05	1.33E-04	1.66E-05
^{134}Cs	5.69E+15	2.76E-05	1.69E-04	2.11E-05
^{125}Sb	5.75E+14	6.75E-06	-	7.85E-06
^{154}Eu	2.32E+14	2.94E-05	-	2.30E-05
^{155}Eu	1.22E+14	3.78E-05	-	2.70E-05

The uranium concentration (μg/L) measurements started in five month after the incident, so there are no uranium data for the time period close to the incident. Using the available data the amount of dissolved UO_2 was calculated and compared to the original mass of the fuel. Using this approach a relative release or dissolution rate was calculated, that is similar to the release rates of the radioactive isotopes. The data shows that 1.8 % (71.5 kg) of the total mass (3,969 kg) of the UO_2 was dissolved during the four years of wet storage after the incident. This means that 19 days were necessary for the dissolution of 1 kg UO_2 into the coolant and the average release rate of UO_2 = 60.8 g/d.

Table 20.4: Calculated and fitted release rates of the alpha radiant isotopes and UO_2.

Isotope	Isotope inventory (Bq)	Calculated release rate (1/d)	Fitted release rate (1/d)	
			pH≈7 (incident+2 months)	pH≈4-4.5
^{242}Cm	9.25E+14	4.42E-05	1.80E-07	5.16E-05
^{238}Pu	7.36E+13	3.62E-05	8.94E-08	4.10E-05
^{244}Cm	2.31E+13	2.05E-05	2.70E-08	1.93E-05
UO_2	3969 kg	1.53E-05	-	1.70E-05

Two fits were made for the measured isotopes (column 4 and 5 in Table 20.3 and Table 20.4), one for the interval of the incident and one for the interval of the wet storage. In case of the long lived isotopes the fitted and calculated release rates during the wet storage gives good agreement. For most isotopes the fitted value is below the release rate which was calculated from the integrated releases (Table 20.3 and Table 20.4).

20.8 Release rates for leaking VVER fuel

Calculation methods

The processed activity concentration data of isotopes were from the measurements between 27th April 2009 (reactor shutdown) and 3rd May 2010. The release rate of isotopes was estimated on the direct measurements from the spent fuel storage pool of Unit 4.

The calculation method is same as calculation method of damaged VVER fuels (Chapter 2.1.1). So the elapsed time since the reactor shutdown was divided into short periods even as before too. Also the number of periods was different for the different isotopes. For example in the case of ^{137}Cs isotope 31 periods were identified.

The method of evaluation of release rates corresponds with the above described calculation. It should be noted that only those releases were used in this evaluation which were characterized by steady-state release rates; the transients (e.g. sipping which technique was used for the identification of leaking fuel assemblies) were not taken account of. The isotope inventory of the reactor shutdown was used to correct the integrated releases.

Results

In case of the No. 70873 leaking fuel assembly, the fitted release rates of isotopes approximate well the calculated release rates (Table 5).

***Table 5:** Calculated and fitted release rates of isotopes.*

Isotope	Isotope inventory (Bq)	Calculated release rate (1/d) pH≈7	Fitted release rate (1/d) pH≈7
^{95}Zr	7.40E+13	3.92E-08	3.79E-08
^{106}Ru	7.61E+12	2.43E-07	2.37E-07
^{137}Cs	1.59E+12	6.24E-06	5.83E-06
^{134}Cs	1.08E+12	6.55E-06	6.09E-06
^{125}Sb	1.12E+11	6.66E-06	6.79E-06
^{154}Eu	3.74E+10	4.53E-06	3.82E-06
^{155}Eu	2.35E+10	1.18E-05	7.89E-06
UO_2	0.955 kg	8.98E-08	7.54E-08

20.9 Conclusions

Our final report summarizes the design and operational characteristics of fuel that were stored in the given periods in the spent fuel storage pool of the power plant, the corresponding isotope inventories in order to facilitate the calculation of fractional release rates and finally calculation methods and calculated dissolution rates of several isotopes from damaged and leaking fuels.

The dissolution rates for VVER fuel stored in water for long periods were determined on the basis of data from the Paks NPP. The two datasets provided information on slightly different conditions: in both cases the water temperature was similar, but in cases of damaged fuel the pH of coolant was significantly lower compared to the leaking fuel. This effect can explain the observed differences in dissolution rates.

During the wet storage of the damaged VVER fuel the fitted release rate range of long lived isotopes was $1 \cdot 10^{-5}$ - $3 \cdot 10^{-5}$ L/d (only the release rate of ^{125}Sb isotope fell slightly short of this range with about $8 \cdot 10^{-6}$ L/d) (Table 20.3). These values are three to six times higher than the release rate of leaking VVER fuel. The release rates of ^{125}Sb isotope show the best agreement of all isotopes when comparing the damaged and leaking fuel under wet storage.

Following a review of the reliability of measured data, the dissolution rates were determined for eleven isotopes in case of damaged fuel and for seven isotopes in case of leaking fuel. Additionally uranium dissolution rates were calculated in both cases. In most of the damaged VVER fuel isotopes the release rates of individual isotopes were comparable to that of uranium, while in case of leaking VVER fuel, dissolution of UO_2 was much slower than most other isotopes.

Our calculated release rates may be regarded as a conservative upper limit compared to the expected releases in the groundwater of deep geological repository since the release from the fuels took place in the spent fuel storage pool with high boric acid concentration.

20.10 References

Hózer, Z., Szabó, E., Pintér, T., Baracska Varjú, I., Bujtás, T., Farkas, G., Vajda, N. (2009). Activity Release from Damaged Fuel during the Paks-2 Cleaning Tank Incident in the Spent Fuel Storage Pool. Journal of Nuclear Materials, 392, 90–94.

Hózer, Z., Aszódi, A., Barnak, M., Boros, I., Fogel, M., Guillard, V., Győri, Cs., Hegyi, G., Horváth, G.L., Nagy, I., Junninen, P., Kobzar, V., Légrádi, G., Molnár, A., Pietarinen, K., Perneczky, L., Makihara, Y., Matejovic, P., Perez-Feró, E., Slonszki, E., Tóth, I., Trambauer, K., Tricot, N., Trosztel, I., Verpoorten, J., Vitanza, C., Voltchek, A., Wagner, K.C., Zvonarev, Y. (2010). Numerical Analyses of an Ex-Core Fuel Incident: Results of the OECD-IAEA Paks Fuel Project. Nuclear Engineering and Design, 240, 538–549.

Hózer, Z., Slonszki, E. (2012). Characterisation of Spent VVER-440 Fuel to Be Used in the FIRST-Nuclides Project (2012). 1st Annual Workshop Proceedings of the 7th EC FP CP FIRST-Nuclides project (eds. Kienzler et al.). Scientific Reports; 7639, 87-101.

NEA/CSNI/R (2008) 2. OECD-IAEA Paks Fuel Project Final Report.

Slonszki, E., Hózer, Z., Pintér, T., Baracska Varjú, I. (2010). Activity Release from the Damaged Spent VVER-Fuel during Long-Term Wet Storage. Radiochimica Acta, 98, 231-236.

Slonszki, E., Hózer, Z., (2013). Determintaion of Dissolution Rates for Damaged and Leaking VVER Fuel Stored in Water. 2nd Annual Workshop Proceedings of the 7th EC FP CP FIRST-Nuclides project (eds. Kienzler et al.). KIT Scientific Reports; 7676, 155-162.

Smirnov, V., Smirnov, B., Kanashov, V., Kuzmin, Y., Bibilashvili, V., Novikov, A., Enin, V., Rozhkov, V., Tzibulya, E. (1997). Behaviour of VVER-440 and VVER-1000 in a Burnup Range of 20-48 MWd/kg_U. Proceedings Second International Seminar on VVER Fuel Performance, Modelling and Experimental Support.

Solonin, M., Bibilashvili, Y., Loltoukhovsky, A., Medvedev, A., Novikov, V., Nikishov, A., Nikulina, A., Sokolov, N., Sokolov, F., Lunin, G., Proselkov, V., Panyushkin, A., Tsiboulia, V., Afanasiev, V., Rozhkov, V., Enin, A., Samoylov, O., Vasilchenko, I., Tsikanov, V., Golovanov, V., Ovchinikov, V., Smirnov, V., Smirnov, A. (1997). VVER Fuel Performance and Material Development for Extended Burnup in Russia. Proceedings Second International Seminar on VVER Fuel Performance, Modelling and Experimental Support.

21 Aqueous leaching of ^{79}Se from spent nuclear fuel

Anders Puranen, Michael Granfors and Olivia Roth*

Studsvik Nuclear AB, Hot Cell Laboratory (SE)
** Corresponding author: anders.puranen@studsvik.se*

21.1 Abstract

The fission product ^{79}Se is one of the concerns in the long term safety assessment of spent nuclear fuel repositories. Data on the aqueous release of ^{79}Se from spent nuclear fuel are however scarce. In this study a dedicated Se analysis campaign was performed by separate subsampling from cumulative leaching studies on fragments from two high burnup UO_2 fuels under aerated conditions (Standard UO_2 and UO_2 Al/Cr-additive fuel irradiated in Oskarshamn 3). The results indicate a ^{79}Se release fraction evolution over 0 - 360 days of leaching (based on six samplings) that approximately follows that of molybdenum. The Se release is slower than rapid release elements such as Cs, but higher that the matrix dissolution rate of U.

21.2 Introduction

One of the concerns in the long term safety assessment of spent nuclear fuel repositories is the fission product ^{79}Se. This is due to the long half-life and the potentially high mobility of dissolved selenium oxyanions, even though ^{79}Se is a very minor fission product.

The very low fission yields of ^{79}Se, in combination with the long half-life beta decay and isobaric interferences makes the analysis of ^{79}Se particularly difficult either by radiometric or mass spectroscopic methods (Jörg et al., 2010). As a result data on aqueous release of ^{79}Se are scarce. (Johnson et al., 2012) performed a spent fuel leaching study under similar conditions to those in this study, reporting an aqueous ^{79}Se release fraction relative to the fuel inventory of <0.22% ($\mu g/\mu g_U$) for up to 98 days of leaching. In this study a dedicated Se analysis campaign was performed by separate subsampling and analysis from an ongoing leaching study with results from up to 360 days of leaching.

21.3 Experimental details

The leaching experiments are performed in a Hot Cell at the Studsvik Hot Cell Laboratory. Leaching of fuel fragments + separated cladding are performed in glass flasks. Each flask contains 200 mL leaching solution (10 mM NaCl + 2 mM $NaHCO_3$). Fuel fragments (~9 g_U) and the separated cladding are placed in glass baskets with filter bottom. The baskets are attached to stop cocks and immersed in the leaching solution. The experiments are performed under aerated, stagnant conditions (no stirring). The leaching is performed in a cumulative way in which the leaching solution is completely exchanged at every sampling interval. Details on the leaching procedure can be found in (Roth et al., 2013). The samples originate from the standard UO_2

sample 5A2 with a local calculated burnup of ~63 MWd/kg_U and from the Al/Cr additive UO_2 fuel sample C1 with a calculated local burnup of ~65 MWd/kg_U from the Oskarshamn 3 boiling water reactor (further details in Roth et al. (2014)).

The analysis was performed using a Perkin Elmer 6100 DRC II Inductively Coupled Plasma Mass Spectrometer adapted for radionuclide analysis. The Dynamic Reaction Cell mode was used with methane as the reaction gas to suppress isobaric interferences. The samples were analysed without any dilution beyond the addition of concentrated nitric acid to 1% and the addition of ^{115}In as the internal standard. Calibration was performed by Se standards of natural isotopic composition (Perkin Elmer Pure Plus). The calibration was verified by running an ^{82}Se enriched standard as a sample (SpectraScan). Br additions showed that any ^{79}Br interference was negligible due to the poor Br response at the present conditions. Following blank subtraction and internal standard normalization, evaluation of the mass range of 74 - 83 (dominated by isotopes of Se, Br and Kr) indicated ~11.2% of the Se was transferred to Se-hydride, SeH^+ under the analytical conditions, in agreement with the observations of Sloth and Larsen (2000). In order to correct for the hydride effect on the Se isotopic composition the method described by Wang et al. (2011) was used, with the addition of corrective equations for ^{79}Se. The ^{79}Se mass response factor was then determined by interpolating the bias factors of ^{78}Se and ^{80}Se. Table 21.1 gives the ICP-DRC-MS instrument settings.

Table 21.1: ICP-DRC-MS instrument parameters and settings for the selenium analysis.

Nebulizer gas flow [L/min]	1
Auxiliary gas flow [L/min]	1.2
Plasma gas flow [L/min]	15
RF power [W]	1400
Spray chamber	Cyclonic
Nebuliser	Quartz
Cones	Platinum
DRC Methane flow [ml/min]	0.50
DRC RPa	0.00
DRC RPq	0.45

Figure 21.1 shows the Origen calculated local pellet average inventory for the mass range 74 - 83 of the investigated fuels (5 y after discharge).

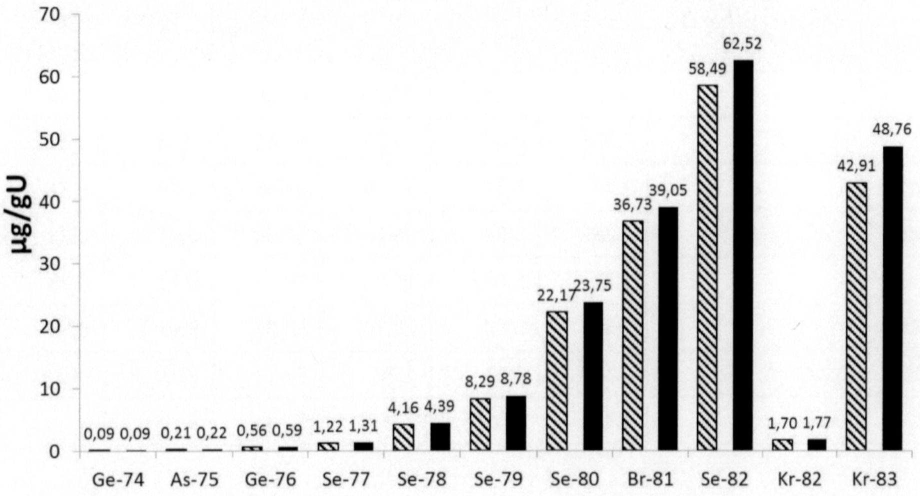

Figure 21.1: Origen calculated inventory for the 5A2 (diagonal) and C1 (solid) fuel samples.

21.4 Results

Table 21.2 shows the nominal and analysed isotopic composition for natural Se, the ^{82}Se reference and the Origen calculated pellet average Se abundance for the two leached fuels 5A2 and C1, as well as examples of the leaching results.

Table 21.2: Examples of natural, calculated and measured Se isotopic abundances from samples and references with measured ^{79}Se and total Se concentrations.

Sample	^{74}Se [%]	^{76}Se [%]	^{77}Se [%]	^{78}Se [%]	^{80}Se [%]	^{82}Se [%]	^{79}Se [%]	^{79}Se [ppb]	Se tot [ppb]
Natural Se	0.89	9.37	7.63	23.77	49.61	8.73	-		
1 ppb Se	0.80	9.12	6.43	24.28	50.91	8.46	(0.08)	<0.01	1.04
5 ppb Se	1.06	9.62	6.88	24.11	49.20	9.14	(0.01)	<0.01	5.07
10 ppb Se	0.87	9.13	7.68	23.80	49.83	8.69	(0.08)	<0.01	10.10
10 ppb Se + 10 ppb Br	0.89	9.02	6.90	23.78	49.82	9.59	(0.04)	<0.01	9.70
Origen C1 Se abundance	-	-	1.30	4.35	23.57	62.05	8.72		
C1-1 (1d)	-	-	3.71	10.03	24.19	53.25	8.83	0.22	2.55
C1-2 (7d)	-	-	2.85	8.08	24.65	54.97	9.45	0.26	2.77
C1-3 (23d)	-	-	2.15	5.51	24.80	58.28	9.26	0.33	3.56
C1-4 (56d)	-	-	1.42	4.45	26.26	58.98	8.90	0.64	7.17
C1-5 (91d)	-	-	1.37	4.37	25.94	59.70	8.63	0.68	7.90
C1-6 (182d)	-	-	1.36	4.37	25.75	59.96	8.57	1.11	12.90
Origen 5A2 Se abundance	-	-	1.29	4.41	23.50	62.01	8.79		

Sample	^{74}Se [%]	^{76}Se [%]	^{77}Se [%]	^{78}Se [%]	^{80}Se [%]	^{82}Se [%]	^{79}Se [%]	^{79}Se [ppb]	Se tot [ppb]
5A2-1 (1d)	-	-	5.19	12.31	21.90	48.68	11.92	0.31	2.56
5A2-2 (7d)	-	-	4.05	8.48	24.25	53.41	9.81	0.20	2.04
5A2-3 (23d)	-	-	2.37	5.61	25.50	57.04	9.48	0.31	3.31
5A2-4 (56d)	-	-	2.26	5.37	25.66	57.63	9.07	0.35	3.89
5A2-5 (91d)	-	-	1.72	5.04	25.75	58.70	8.79	0.38	4.29
5A2-6 (182d)	-	-	1.52	4.60	26.20	59.05	8.63	0.55	6.40
11 ppb ^{82}Se	0.17	0.17	0.16	0.62	1.77	97.12	(0.01)	<0.01	11.40
^{82}Se ref	0.16	0.29	0.33	0.59	1.96	96.66	-		

As can be seen in Table 21.2 the analytical results on the standards agrees fairly well with the nominal Se abundance of the natural and the ^{82}Se enriched standard in the low ppb concentration range. The Se isotopic composition of the leached samples is also quite close to the pellet average Origen calculated Se abundances. Furthermore the isotopic ratios of ^{82}Se/^{77}Se and ^{82}Se/^{79}Se also agrees rather well with radial Se profiles obtained by Laser Ablation ICP-MS on fuel cross sections from the studied fuels (Puranen et al, 2014)

Figure 21.2 shows the cumulative release fractions of the three major fissiogenic Se isotopes from sample 5A2 relative to the ^{135}Cs, ^{100}Mo and ^{238}U results. Figure 21.3 shows the corresponding release rates.

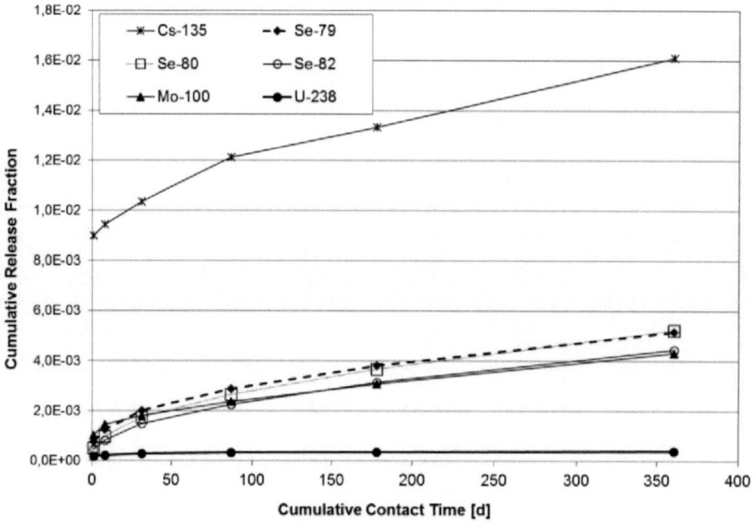

Figure 21.2: Cumulative release fraction of ^{79}Se and selected radionuclides from the fragment leaching of sample 5A2.

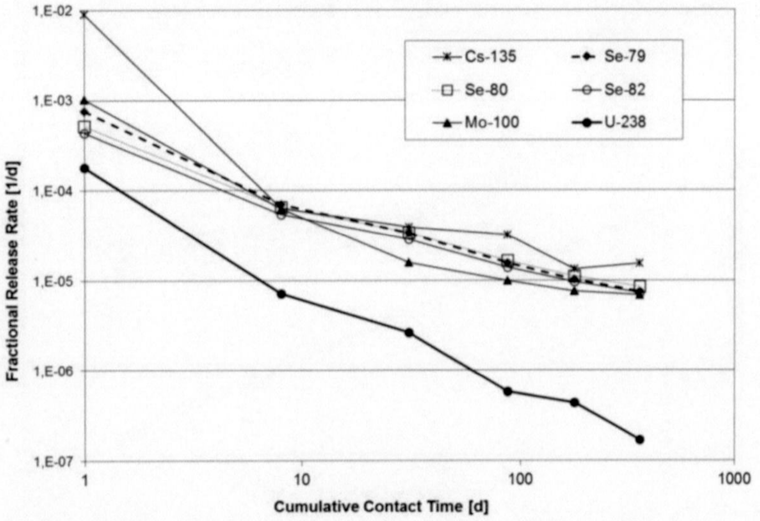

Figure 21.3: Release rates of ^{79}Se and selected radionuclides from the fragment leaching of sample 5A2.

Figure 21.4 shows the cumulative release fractions of the three major fissiogenic Se isotopes from sample C1 relative to the ^{135}Cs, ^{100}Mo and ^{238}U results, with the corresponding release rates in Figure 21.5.

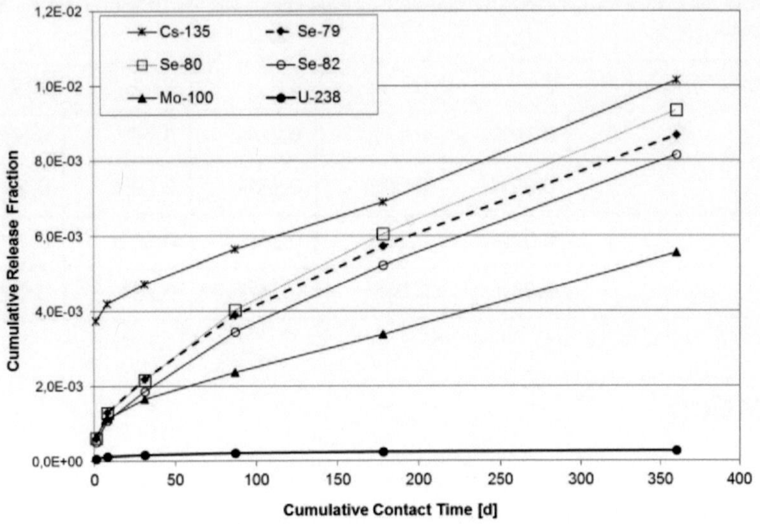

Figure 21.4: Cumulative release fraction of ^{79}Se and selected radionuclides from the fragment leaching of sample C1.

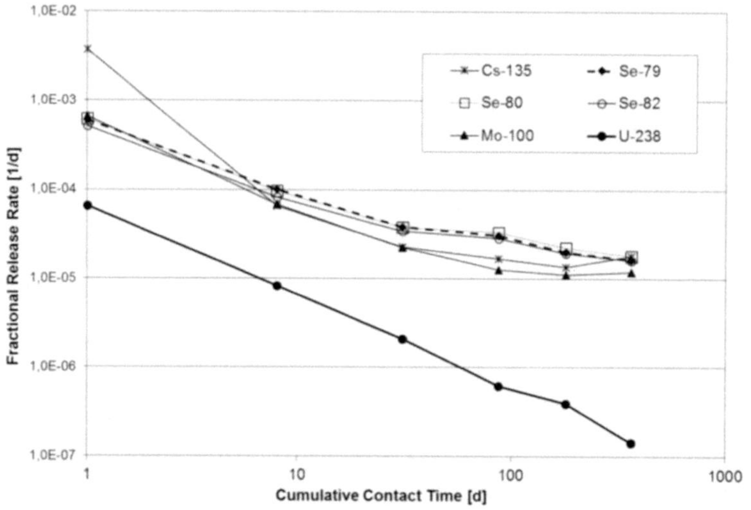

Figure 21.5: Release rates of ^{79}Se and selected radionuclides from the fragment leaching of sample C1.

Table 21.3 shows the spread in cumulative ^{79}Se release fractions from the two leaching experiments 5A2 and C1 with an estimated ^{79}Se analytical error of ±0.1 ppb.

Table 21.3: Cumulative ^{79}Se release fractions from the two leaching experiments 5A2 and C1 with an estimated ^{79}Se analytical error of ±0.1 ppb.

Cumulative Contact Time [d]	1	8	31	87	178	360
5A2 ^{79}Se high	0.10%	0.17%	0.28%	0.39%	0.50%	0.66%
5A2 ^{79}Se low	0.05%	0.08%	0.13%	0.19%	0.26%	0.37%
C1 ^{79}Se high	0.09%	0.18%	0.30%	0.50%	0.71%	1.03%
C1 ^{79}Se low	0.03%	0.08%	0.14%	0.28%	0.44%	0.71%

21.5 Discussion

The Se findings are supported by the good agreement of the analysed Se-isotopic composition in the samples and that of the calculated fuel inventory, which has a very different isotopic abundance relative to naturally occurring Se. The Se isotopic ratios also agree with radial Se profiles obtained by Laser Ablation ICP-MS on fuel cross sections from the studied fuels (Puranen et al., 2014). The results from the standard UO_2 sample (5A2) indicate a cumulative ^{79}Se release fraction evolution over 0-360 days of leaching (based on six samplings) that approximately follows that of Mo (Figure 21.2). The results are in fairly good agreement with the <0.22% ^{79}Se release fraction reported by Johnson et al., (2012) for 98 days of leaching of spent fuel with a similar burnup (this study, ~0.29±0.1% cumulative ^{79}Se release fraction at 87 days, Table 21.3), especially if potentially important factors such as surface to volume ratios, inventory related uncertainties and procedural differences are considered. The initial Se release from the standard UO_2 sample is slower than rapid release elements such as Cs, but higher than the matrix dissolution rate of U (Figure 21.3). It is interesting to note that the initial Se release fractions from the Al/Cr additive sample C1 up to 31 days of leaching also seem to follow the Mo release fraction (Figure 21.4). For longer leaching durations the Se release is however faster, approaching the Cs release fraction. It should be noted that the cumulative Cs release from the additive sample is lower than from the standard UO_2 sample. If these difference can be attributed to the effects of the additive such as Cr substitution in the UO_2 matrix, larger grain sizes or the presence of Al-rich secondary phases is hard to determine given the limited data set.

21.6 Conclusions

The similar or faster Se release fraction compared to the much more abundant fission product Mo, which is known to either dissolve in the UO_2 matrix or to aggregate in secondary phases such as the noble metal particles, depending on the local redox conditions, does suggests that Se might also be heterogeneously distributed in the fuel. This conclusion is supported by the intermediate nature of the Se release rates, given the lack of a very high initial Se release, such as that displayed by Cs, on the other hand, the up to one year Se release rate clearly remain much higher than the U matrix dissolution rate.

21.7 Acknowledgement

The research leading to these results has received funding from the European Union's European Atomic Energy Community's (Euratom) Seventh Framework Programme FP7/2007-2011 under grant agreement n° 295722 (FIRST-Nuclides project).

The Swedish Nuclear Fuel and Waste Management Co (SKB) is gratefully acknowledged for financial support.

21.8 References

Johnson, L., Günther-Leopold, I., Kobler Waldis, J., Linder, H.P., Low, J., Cui, D., Ekeroth, E., Spahiu, K., Evins, L.Z. (2012). Rapid Aqueous Release of Fission Products from High Burn-Up LWR Fuel: Experimental Results and Correlations with Fission Gas Release. Journal of Nuclear Materials, 420, 54-62.

Jörg, G., Bühnemann, R., Hollas, S., Kivel, N., Kossert, K., Van Winckel, S., Gostomski, C.L.V. (2010). Preparation of Radiochemically Pure ^{79}Se and Highly Precise Determination of Its Half-Life. Applied Radiation and Isotopes, 68, 2339.

Puranen, A., Granfors, M., Roth, O. (2014). Laser Ablation Study of Irradiated Standard UO_2 Fuel and Al/Cr Doped UO_2 Fuel. Final Workshop Proceedings of the 7th EC FP CP FIRST-Nuclides Project (eds. Kienzler et al.).

Roth, O., Puranen, A. (2012). Selection of Materials, Preparations and Experimental Set-Up. 1st Annual Workshop Proceedings of the 7th EC FP CP FIRST-Nuclides project (eds. Kienzler et al.). KIT Scientific Reports; 7639, 137-144.

Roth, O., Puranen, A., Low, J., Granfors, M., Cui, D., Askeljung, C. (2013). Spent Fuel Leaching Experiments and Laser Ablation Studies Performed in Studsvik - Status and Preliminary Results. 2nd Annual Workshop Proceedings of the 7th EC FP CP FIRST-Nuclides project (eds. Kienzler et al.). KIT Scientific Reports; 7676, 163-174.

Roth, O., Puranen, A., Askeljung, C., Cui, D., (2014). Selection of Materials and Characterization of Samples Used in Spent Fuel Leaching and Laser Ablation Studies. Final Workshop Proceedings of the 7th EC FP CP FIRST-Nuclides Project (eds. Kienzler et al.).

Sloth, J., Larsen, E. (2000). The Application of Inductively Coupled Plasma Dynamic Reaction Cell Mass Spectrometry for Measurement of Selenium Isotopes, Isotope Ratios and Chromatographic Detection of Selenoamino Acids. Journal of Analytical Atomic Spectrometry, 15, 669-672.

Wang, J., Ren, T., Lu, H., Zhou, T., Zhao, M., (2011). Absolute Isotopic Composition and Atomic Weight of Selenium Using Multi-Collector Inductively Coupled Plasma Mass Spectrometry. International Journal of Mass Spectrometry, 308, 65-70.

22 Leaching of high burn up spent fuel with and without matrix dopants

Olivia Roth, Charlotta Askeljung, Anders Puranen, Michael Granfors, Daqing Cui and Jeanett Low*

Studsvik Nuclear AB (SE)

** Corresponding author: olivia.roth@studsvik.se*

22.1 Abstract

This paper presents results from experiments performed at Studsvik Nuclear AB within the EURATOM FP7 Collaborative Project "Fast / Instant Release of Safety Relevant Radionuclides from Spent Nuclear Fuel (CP FIRST-Nuclides)".

The main focus of the investigations is to explore the effects of additives and dopants on the fast/instant release of fission products such as caesium and iodine. These effects are studied in spent fuel leaching experiments. The results indicate that Al/Cr doped fuel have a lower release of caesium and iodine than standard UO_2 fuel, this effect can probably be attributed to the larger grain size of the Al/Cr doped fuel.

Furthermore, analysis tests on previously obtained leaching solutions have also been performed to investigate the feasibility of measuring fast/instant release of ^{14}C during spent fuel leaching. However, further validation of the method is needed in order to obtain quantitative results.

22.2 Introduction

This paper describes ongoing efforts at Studsvik Nuclear AB within the EURATOM FP7 Collaborative Project "Fast / Instant Release of Safety Relevant Radionuclides from Spent Nuclear Fuel (CP FIRST-Nuclides)".

The aim of the project is to study the fraction of fission and activation products that are fast/instantly released from spent nuclear fuel upon contact with aqueous media. The fraction consists of readily soluble phases in the gap between fuel and cladding, cracks and grain boundaries. Some of these fission and activation products have a long half-life and are for this reason important for the safety assessment of deep repositories for spent nuclear fuel.

At the Studsvik Hot Cell laboratory, spent fuel leaching studies have been conducted since around 1980. During 1990-1996 a comprehensive research program was initiated aiming at mapping the most important parameters influencing the stability of spent nuclear fuel in water (Forsyth, 1997). Since then the program has been extended with leaching experiments of high burn-up fuel and instant release experiments (Johnson et al., 2012; Zwicky et al., 2011).

In these previous experiments, standard UO_2 fuel has been used. Today new fuel types with additives and dopants are taken into use in commercial reactors. The additives and dopants affects properties such as grain size and fission gas release which in turn may affect the instant release behavior of the fuel. The main objective of this study is to investigate how these changes in the fuel matrix affect the instant release process.

Four of the rods have standard UO_2 pellets, one has Al/Cr doped UO_2 pellets and one has Gd-doped pellets. This selection was made in particular to assess the effect of doping on the instant release behavior. One of the standard fuel samples was leached by a simultaneous grinding and leaching method to specifically assess the release of the grain boundary inventory.

The solutions from two of the samples described here (5A2 and C1) have been used to assess the aqueous leaching of ^{79}Se presented in Puranen et al. (2014).

Furthermore, in order to investigate the feasibility of measuring fast/instant release of ^{14}C during spent fuel leaching analysis tests on previously obtained leaching solutions have also been performed and are presented here.

22.3 Sample selection and preparation

The sample selection and preparation are described in (Roth et al., 2014). The six high burn-up fuels chosen for the studies are listed in Table 22.1. The FGR values in Table 22.1 are based on rod puncturing and gas analysis and the calculated BU values are based on core calculations. The methods are described in (Roth and Puranen, 2012).

Table 22.1: *Fuels selected for investigation at Studsvik.*

Sample name	Reactor type	Fuel type	FGR[1] [%]	Calculated BU (rod average) [MWd/kg_U]	Sample type
D07	BWR	Std UO_2	~1.6	50.2	cladded segments
L04	BWR	Std UO_2	~3.1	54.8	cladded segments
5A2	BWR	Std UO_2	~2.4	57.1	fragments + sep. cladding
C1	BWR	Al/Cr doped UO_2	~1.4	59.1	fragments + sep. cladding
VG81	PWR	Gd doped UO_2	~2.2	54.4	fragments + sep. cladding
AM2K12	PWR	Std UO_2	~4.9	70.2	powder

[1] Based on rod puncturing and gas analysis as described in Roth and Puranen (2012).

22.4 Sample preparation

Spent fuel leaching studies have been performed using samples from 6 different fuel rods (see Table 22.1). The samples are leached as cladded fuel segments, fuel fragments + separated cladding or fuel powder.

The preparation of cladded fuel segments, fuel fragments + separated cladding has been described in Roth et al. (2013) and includes gamma scanning, cutting and, for fuel fragments + separated cladding, decladding by longitudinal sawing. Each sample consists of approximately 2 fuel pellets including cladding.

Sample C1 and 5A2 have been decladded by longitudinal sawing, whereas VG81 was decladded using the crushing method described in Roth et al., (2013).

Powdered specimens are leached using a simultaneous grinding and leaching technique similar to that described by Stroes-Gascoyne et al., (1995). Further details of this method are given below. The objective of this method is to expose the grain boundaries by grinding the fuel down to the same size range as the individual fuel grains. This would ideally make the entire grain boundary inventory available for leaching. By combining the grinding and leaching into one wet grinding step surface oxidation and temperature effects (from the friction of grinding) can be minimized.

The fuel leached as powdered specimen has been leached in 10 mM NaCl + 2 mM $NaHCO_3$ for 4 years prior to the simultaneous grinding and leaching.

22.5 Experimental Setup

Spent fuel leaching studies

Samples D07, L07, 5A2, C1 and VG81 were leached as cladded fuel segments and fuel fragments + separated cladding respectively. As previously described (Roth and Puranen, 2012; Roth et al., 2013) the leaching of cladded fuel segments and fuel fragments + separated cladding is performed in glass flasks. Each flask contains 200 mL leaching solution (10 mM NaCl + 2 mM $NaHCO_3$). Cladded fuel segments are placed in a 'basket' made of platinum wire. Fuel fragments + separated cladding are placed in glass baskets with filter bottom. The baskets are attached to stop cocks and immersed in the leaching solution. The experiments are performed under aerated, stagnant conditions (no stirring).

The leaching is performed in a cumulative way in which the leaching solution is completely exchanged at every sampling interval. After 1 d, 7 d, 21 d, 63 d, 92 d and ~182 d (resulting in a cumulative leaching time of ~365 d), the stopcock and fuel basket is transferred to a new leaching flask containing a fresh leaching solution. Samples of the old leaching solutions are withdrawn from the used leaching flasks and analysed by ICP-MS and gamma analysis (^{137}Cs). In order to calculate the leached fraction of the fuel inventory up to a given leaching time the analysed concentrations are thus added to the results of the prior samplings.

All aqueous samples from the leaching studies are removed from the hot cell and transported to the chemistry laboratory where the samples are centrifuged (and filtered if necessary in the case of powdered samples) and analysed for Cs and I with ICP-MS.

Sample AM2K12 was leached as powdered sample using simultaneous grinding and leaching performed in a Fritsch Pulverisette 6 planetary ball mill. The ball mill is equipped with a tungsten carbide milling bowl and balls, see **Fehler! Verweisquelle konnte nicht gefunden werden.**.

Grinding tests using non-irradiated UO_2 were performed to obtain the best grinding conditions to reduce the spent fuel to grain size (approx. 10 µm) powder. The experiment was set up in a fume hood behind lead shielding. Some 7 - 8 fragments with a total weight of 1.8 gram, were chosen from a previously leached batch of fragments (Roth et al., 2014). The fuel fragments was grinded for 2 min at 400 rpm together with 30 mL of the leaching solution.

Figure 22.1: Tungsten carbide milling bowl and balls.

The leaching solution used was simplified ground water, 10 mM NaCl + 2 mM $NaHCO_3$. Directly after grinding another 20 mL leaching solution was added to the milling bowl. After 30 min of static leaching in the milling bowl, 4 liquid samples were taken from the bowl with a pipette. The samples were immediately centrifuged and the super natant was removed and preserved for ICP-MS analysis. In addition a sample of the slurry from the bottom of the bowl was taken. This was also centrifuged and the centrifuge rest was dried and prepared for SEM analysis to determine the obtained particle size distribution. Parallel to the leaching of grinded fuel a corresponding portion of the fuel was leached without grinding. This experiment was performed in order to assess the leaching from the fragment surface (as opposed to grain boundaries) caused by any potential pre-oxidation of the fuel fragments. Fragments (1.8 g) were removed from the same lot (previous leaching experiment) as the fragments used for grinding. The fragments were placed in 50 mL leaching solution in a bottle in Hot Cell. After 30 min 4 samples were taken for ICP-MS analysis.

^{14}C analysis

For the ^{14}C analysis, CO_2 separation by acidification and slight heating was employed, as described in Figure 22.2. As shown in the figure, the leaching solution is acidified and gently heated while argon gas is flowing the sample vessel. A washing bottle with acid captures interfering nuclides. Finally the gas flow is passing a bottle with Carbon Trap (a commercially available organic solution). The Carbon Trap is mixed with scintillation cocktail and beta measurement is performed by liquid scintillation.

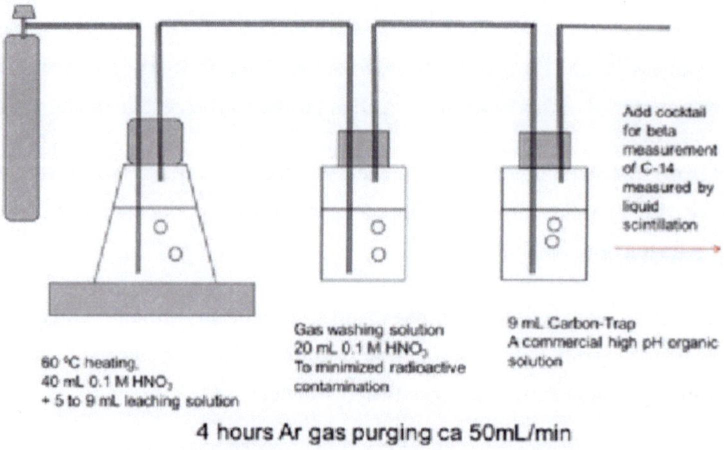

Figure 22.2: Schematic picture of CO_2 separation method.

The separation method was tested using solutions with known amount of ^{14}C. Four spent LWR fuel leaching solutions from previously performed studies (outside the scope of the present study) was analysed for ^{14}C using the method described above.

22.6 Results and Discussion

Spent fuel leaching studies

The results from the leaching studies are presented in Appendix A (samples D07 and L04), Appendix B (samples 5A2 and C1), Appendix C (sample VG81) and Appendix D (sample AM2K12).

The uncertainties in the data presented here is a combination of uncertainties related to the sample preparation and sampling, uncertainties in the analysis and uncertainties in the inventories used for calculation of the release fractions. The uncertainties related to the ICP-MS analysis varies depending on the nuclide and the sample, but is considered to be minor in this type of experiment. The main uncertainties are considered to be related to the weighing of the fuel sample (in particular in the case of fuel segments where the weight of the cladding needs to be estimated) and uncertainties in the inventory data. It is judged that the overall uncertainty of the data presented here is around 10 - 15%.

Preliminary results from samples 5A2 and C1 have been published in Roth et al. (2012) and Roth et al. (2013). Due to the unexpectedly low release of iodine measured for sample 5A2, a parallel study was initiated using a different sample preparation method (Roth et al., 2013). This parallel study showed that the low iodine release for the standard fuel (5A2) probably can be explained by loss of iodine during e.g. during sample preparation. For this reason the iodine results presented in table B2 should be used with care.

For the C1 fuel the difference in iodine release between the two sample preparation methods was minor relative to the difference observed for the 5A2 fuel. This leads to the conclusion that the loss of iodine from the 5A2 sample can be regarded as a single event and not something that will necessarily affect all samples prepared by the same method. Also, for this reason, the ^{129}I results obtained for sample C1 in the present study are considered reliable.

In Figure 22.3 to Figure 22.5 (and Appendix A-C) release rates of ^{137}Cs and ^{129}I (omitted for 5A2 as discussed above) are presented.

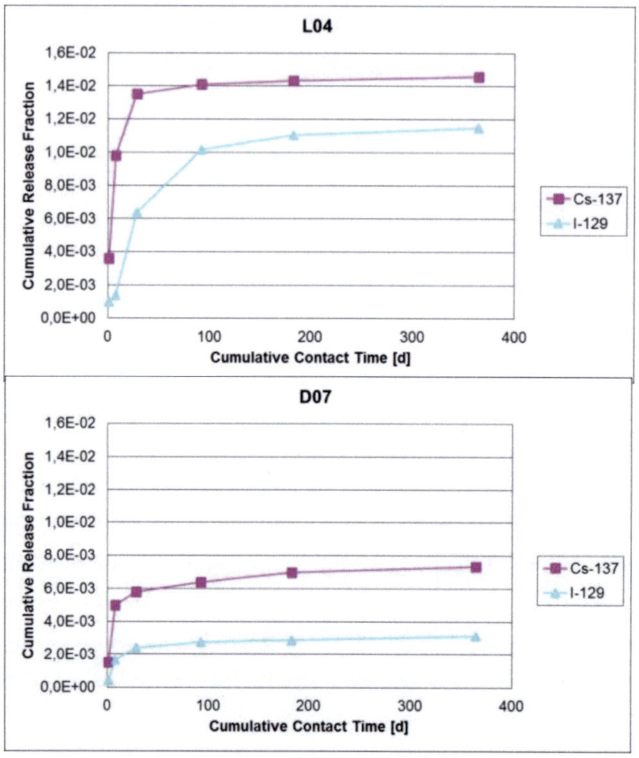

Figure 22.3: Release rates of ^{137}Cs and ^{129}I sample L01 and D07.

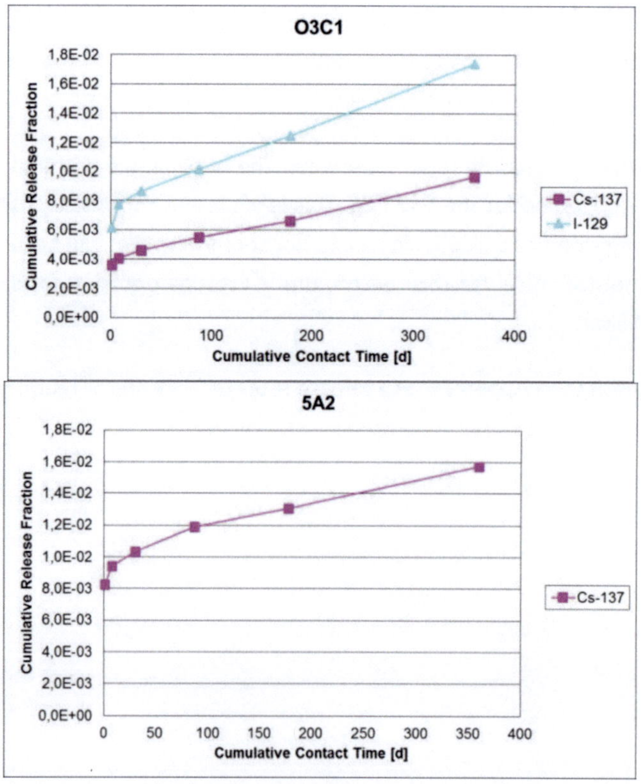

Figure 22.4: Release rates of 137Cs and 129I sample C1 and 5A2.

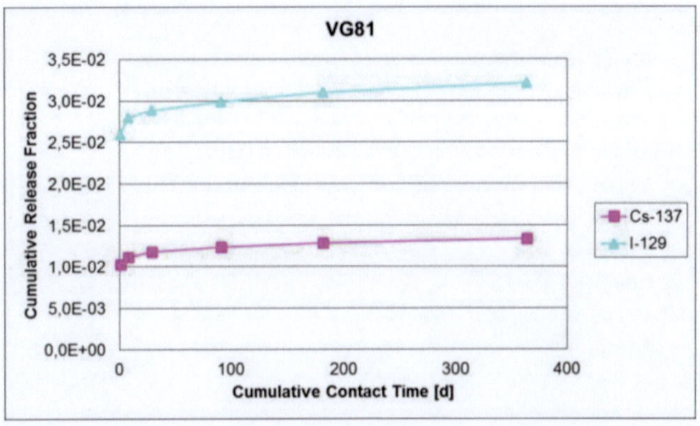

Figure 22.5: Release rates of ^{137}Cs and ^{129}I sample VG81.

Comparing the samples leached as cladded segment (L04 and D07) to the results presented in Zwicky et al. (2011) and Ekeroth et al. (2012), it is seen that the caesium and iodine releases measured in this study are within the expected order of magnitude. The results for L04 seem to fall in the higher range and the D07 results in the lower, which is expected due their different fission gas releases.

Results from the leached fragments (5A2, C1 and VG81) can be compared with data in Ekeroth et al. (2012). All data are within the expected order of magnitude for caesium and iodine release. It can be noted that the caesium and iodine release from sample C1 (Al/Cr doped) is lower than for both 5A2 and VG81 and also lower than for the samples studied in Ekeroth et al. (2012). This is expected from its lower FGR, which can be explained by the doping. This behaviour is further discussed in Roth et al. (2013) and Nilsson et al. (2014). It is however interesting to note that the release of caesium and iodine from sample C1 and 5A2 seem to continue at a relatively fast rate even though the iodine release from sample C1 has exceeded the FGR level. It is also interesting to note that all samples seems to have reached a rather low steady-state dissolution rate of caesium and iodine, except for the C1 and 5A2 samples. This difference in behaviour aggravates the comparison between the samples.

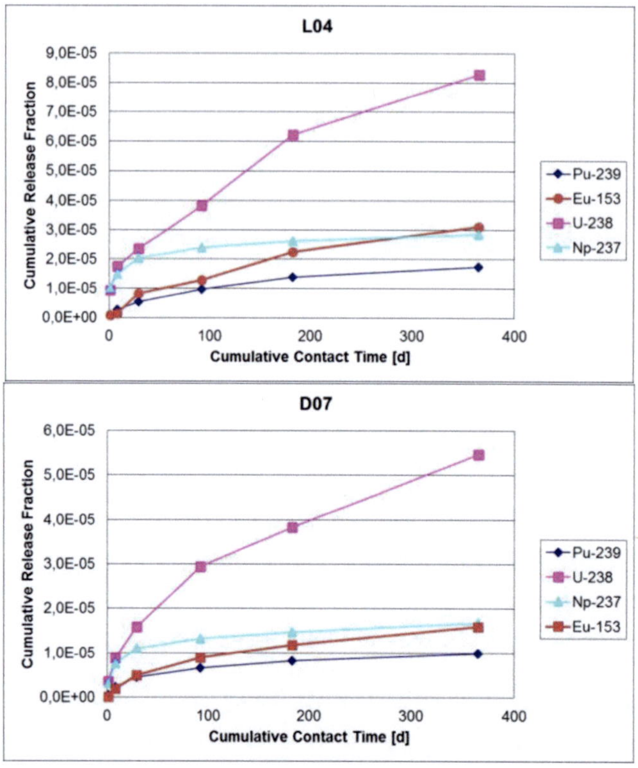

Figure 22.6: Release rates of 239Pu, 153Eu, 238U and 237Np sample L01 and D07.

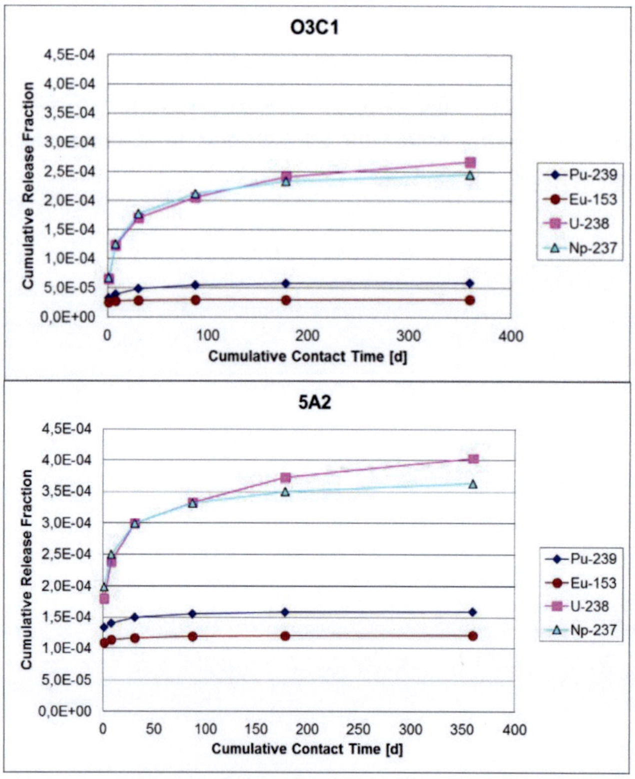

Figure 22.7: Release rates of ^{239}Pu, ^{153}Eu, ^{238}U and ^{237}Np sample C1 and 5A2.

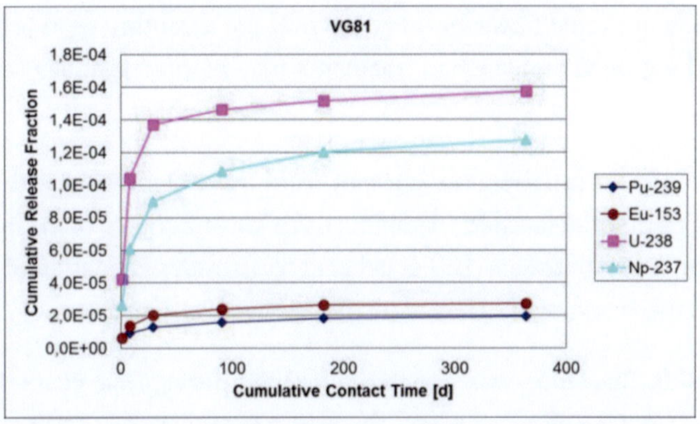

Figure 22.8: Release rates of ^{239}Pu, ^{153}Eu, ^{238}U and ^{237}Np sample VG81.

In Figure 22.6 to 8 (and in Appendices A-C), the release rates of ^{238}U, ^{239}Pu, ^{153}Eu and ^{237}Np are compared. These release rates are all within the expected range as compared to previous experiments (Zwicky et al., 2011). In all cases ^{238}U displays the highest release rate, generally followed by ^{237}Np. It is interesting to note that for the data set presented there seem to be a difference in neptunium leaching behaviour depending on the sample type; for the leached fuel segments (D07 and L04), neptunium seem to follow the behaviour of plutonium and europium, whereas for the leached decladded fragments (O3C1, 5A2 and VG81) neptunium follows the behaviour of uranium. However, previous data (Zwicky et al., 2011) obtained for leached fuel segment in some cases also show the correlation between uranium and neptunium behaviour leading to the conclusion that this behaviour cannot be explained by the sample type alone.

The effect of doping on the release of those nuclides seem to be insignificant. It has previously been noted that the release of uranium is slightly lower for the Al/Cr doped sample (O3C1) compared to the standard sample irradiated under similar conditions (5A2) (Roth et al., 2013). Considering the experimental uncertainties this difference can however not be considered significant.

As in the case of caesium and iodine release, the samples show different behaviours when it comes to reaching steady-state dissolution rates. The VG81 sample seem to have reached steady-state, and also for 5A2 and C1 a steady-state is approached. This is not the case for L04 and D07. In this case (e.g. for uranium and matrix dissolution) the behaviour can be explained by the sample preparation method. The decladded samples reach a steady-state faster due to facilitated water penetration. The VG81 sample (decladded by crushing) the water penetration is probably faster than for the 5A2 and C1 samples due to smaller fragment size.

The results from the 30 minutes grinding and leaching experiment (AM2K12) are shown in Appendix D. During this experiment, the leaching of caesium and iodine was measured to about 1.5% of the total inventory, and the release of uranium was about 0.04%. The parallel leaching (without grinding) of fragments from the same batch showed a release of Cs, I and U in the order of 0.01%. From this data it can be concluded that the iodine and caesium released during the 30 minutes grinding and leaching originates from freshly exposed grain boundaries and/or surfaces of secondary phases and not from matrix dissolution and/or dissolution of pre-oxidized UO_2-phases.

The 1.5% release of iodine and caesium correspond to a significantly higher release rate than measured for any of the other samples in this study. This fuel has significantly higher burn-up and FGR than the other samples used in this study. It should however be noted that the inventory used is an average over the full pellet radius, whereas the ground and leached fragments may be preferentially taken from rim or centre positions.

A comparison of the $P^{240}u/^{242}Pu$ ratio for the leachate from the experiments with the same ratio for the nuclide inventory indicates that the leached fragments have lower burn-up than the pellet average, i.e. are preferentially taken from centre positions. This could lead to an underestimation of the calculated radionuclide release fractions as these number are based on the 75 MWd/kg_U inventory.

The fuel fragments used in this study was previously leached during four years. In order to allow a full evaluation of the powder leaching data and assess the extent of additionally accessed grain boundaries, this needs to be compared to the data from the four year leaching period (which is not available yet).

Looking at data obtained for AM2K12 samples previously (Ekeroth et al., 2012), it can be seen that the present release rate (i.e. ~1.5% during 30 minutes leaching) is in line with the release rates measured for fresh (not pre-leached) samples. In Ekeroth et al. (2012), it is shown that the release of caesium and iodine approaches a steady-state after about 200 days leaching time, the steady-state release fraction obtained for decladded fragment is around 6% for iodine and 3% for caesium. The present results suggest that the exposure of grain boundaries would add around minimum 1.5% fractional release to this data.

^{14}C analysis

The initial tests with solutions of known ^{14}C content the yield in the separation step was approximately 80%. ^{14}C was detected in all four samples from previous spent fuel leaching studies, however further validation of the method is needed in order to obtain quantitative results. Further tests and validation is ongoing outside the scope of the present study.

22.7 Conclusions

The results indicate that Al/Cr doped fuel have a lower release of caesium and iodine than standard UO_2 fuel, this effect can probably be attributed to the larger grain size of the Al/Cr doped fuel. No effect by Al/Cr doping on matrix dissolution could be observed. The Gd-doped fuel showed a release behaviour similar to standard fuel of similar burn-up and hence, no effect of the doping could be observed. The results from the leaching of powdered fuel showed that the release of caesium and iodine increased significantly when exposing fresh grain boundaries and/or surfaces of secondary phases to the aqueous phase.

The initial tests with solutions of known ^{14}C content the yield in the separation step was approximately 80%. However, further validation of the method is needed in order to obtain quantitative results.

22.8 Acknowledgement

This project is funded by The Swedish Nuclear Fuel and Waste Management Co. (SKB) , Posiva Oy and the European Union's European Atomic Energy Community's (Euratom) Seventh Framework Programme FP7/2007-2011 under grant agreement n° 295722 (FIRST-Nuclides project).

22.9 References

Ekeroth, E., Cui, D., Low, J., Granfors, M., Zwicky, H.-U., Spahiu, K., Zetterström Evins, L. (2012). Instant Release Fractions from Corrosion Studies with High Burnup LWR Fuel. MRS Proceedings, 1475.

Forsyth, R. (1997). The SKB Spent Fuel Corrosion Program – An Evaluation of Results from the Experimental Program Performed in the Studsvik Hot Cell Laboratory. SKB Technical Report, TR 97-25.

Granfors M., Puranen, A., Zwicky, H.U. (2012). Radial Profiling and Isotopic Inventory Analysis of Irradiated Nuclear Fuel Using Laser Ablation ICP-MS. Proceedings of Top Fuel 2012 Transactions – Eurpean Nuclear Society.

Jochum, K.P., Weis, U., Stoll, B., Kuzmin, D., Yang, Q., Raczek, I., Jacob, D.E., Stracke, A., Birbaum, K., Frick, D.A., Günther, D., Enzweiler, J. (2011). Determination of Reference Values for NIST SRM 610–617 Glasses Following ISO Guidelines. Geostandards & Geoanalytical Research, 35 (4), 397-429.

Johnson, L., Günther-Leopold, I., Kobler Waldis, J., Linder, H.P., Low J., Cui, D., Ekeroth, E., Spahiu, K., Evins, L.Z. (2012). Rapid Aqueous Release of Fission Products from High Burn-Up LWR Fuel: Experimental Results and Correlations with Fission Gas Release. Journal of Nuclear Materials, 420, 54-62.

Nilsson, K., Roth, O., Jonsson, M. (2014). Oxidative Dissolution of ADOPT Compared to Standard UO_2 Fuel. Submitted to Journal of Nuclear Materials.

Pearce, N., Perkins, W., Westgate, J., Gorton, M., Jackson, S., Neal, C., Chenery, S. (1997). A Compilation of New and Published Major and Trace Element Data for NIST SRM 610 and NIST SRM 612 Glass Reference Materials. The Journal of Geostandards and Geoanalyses, 21, 115-144.

Puranen, A., Granfors, M., Roth, O. (2014). Aqueous Leaching of ^{79}Se from Spent Nuclear Fuel. Final Workshop Proceedings of the 7th EC FP CP FIRST-Nuclides Project (eds. Kienzler et al.).

Roth, O., Puranen, A., Askeljung, C., Cui, D., (2014). Selection of Materials and Characterization of Samples Used in Spent Fuel Leaching and Laser Ablation Studies. Final Workshop Proceedings of the 7th EC FP CP FIRST-Nuclides Project (eds. Kienzler et al.).

Roth, O., Low, J., Granfors, M., Spahiu, K. (2013). Effects of Matrix Composition on Instant Release Fractions from High Burn-Up Nuclear Fuel. MRS Proceedings, 1518, 145-150.

Roth, O., Low, J., Spahiu, K. (2013). Effects of Matrix Composition and Sample Preparation on Instant Release Fractions from High Burnup Nuclear Fuel. Submitted to MRS Proceedings.

Roth, O., Puranen, A. (2012). Selection of Materials, Preparations and Experimental Set-Up. 1st Annual Workshop Proceedings of the 7th EC FP CP FIRST-Nuclides project (eds. Kienzler et al.). Scientific Reports; 7639, 137-144

Stroes-Gascoyne, S., Moir, D.L., Kolar, M., Porth, R.J., McConnell, J.L., Kerr, A.H., (1995). Measurement of Gap and Grain-Boundary Inventories Of ^{129}I in Used CANDU Fuels. MRS Proceedings, 353, 625-631.

Zwicky, H.-U., Low, J., Ekeroth, E. (2011). Corrosion Studies with High Burnup Light Water Reactor Fuel; Release of Nuclides into Simulated Groundwater During Accumulated Contact Time of Up to Two Years. SKB Technical Report, TR-11-03.

23 Diffusion of fission products in fuel rods

Bernhard Kienzler[1], Volker Metz[1], Ernesto Gonzáles-Robles[1], Michel Herm[1] and Olivia Roth[2]

[1]*Karlsruhe Institute of Technology (KIT) (DE)*
[2]*Studsvik Nuclear AB (SE)*
** Corresponding author: bernhard.keinzler@kit.edu*

23.1 Abstract

The behaviour of volatile fission products in fuel rods is simulated in order to describe potential accumulations within the fuel rods and to which extend visible cracks in the fuel contribute to the instant release fraction. It is shown that the diffusion processes depend strongly on the temperature. The radial distribution of I and Cs in a spent fuel pellet is described by a temperature-dependent sorption parameter and the simulated concentration distributions correspond well to Studsvik's measurements by Laser Ablation Techniques. Furthermore, the calculated release of volatile radionuclides into the gap regions corroborate with the fission gas release (FGR) reported for the fuel rod under investigation.

23.2 Introduction

The added value of the CP FIRST-Nuclides consists in a direct answer to the inadequately supported question on fast/instant release of radionuclides for license applications related to disposal of high burn-up UO_2 fuel. The knowledge and data obtained within the project shall reduce the uncertainties and provide for realistic data on the relevant radionuclide release for the Safety Case. Due to the high dose rates, the experimental work can be only done with small samples. The up-scaling from the analytical and modelling micro-scale to the experimental bulk observations and to the release on a fuel-rod scale is a key challenge of FIRST-Nuclides. The experimentally investigated size range from UO_2 grains, in the range of several µm, to segment/pellet and to fuel rods. Furthermore, the temperature ranges from above 1,200°C during the irradiation in a reactor to some 300°C during the interim storage period to room temperature during the experiments. The modelling work within the project has to clarify which of the individual contributors with their different geometry and scales and/or temperature regimes will primarily contribute to the fast/instant release fraction (IRF). Aims concern (1) the migration of gaseous and non-gaseous fission products in micro-structures of spent fuel pellets (grains, grain boundaries, gas bubbles or pores) which are characterized by their specific diffusion coefficients, and (2) the development of improved models to predict the fission gas release (FGR) and release of non-gaseous fission products on the fuel rod scale. Modelling the distribution of radionuclides which belong to the instant release fraction is the primary objective of this study and will give indications on potential accumulations within the fuel rods and to which extend visible cracks in the fuel contribute to the instant release fraction.

23.3 Background

The boiling point of some fission products such as of selenium is 684.9 ± 1.0 °C (Brooks, 1952), of caesium 678.4 °C, of antimony 630.0 and of iodine 184.0 °C (Bentor, 2012) in elemental form. Diffusion within the UO_2 grains is controlled mainly by the temperature. Outside the grains, a gas phase exists which allows gas phase transport of the volatile elements. However, in regions of the fuel rods where the temperature is below the boiling temperature, the elements may accumulate. This is not the case for the fission gases Kr and Xe. The diffusion coefficients show a huge dependence on the temperature. For the potential mobilization of Se and Cs, the temperature ranges of the fuel and the related diffusion properties need to be considered.

During the 2nd Annual Workshop of the project, investigations by laser ablation were presented by Studsvik (Kienzler et al., 2014) studying the radial distribution of I, Xe and Cs in a spent fuel pellet and to explore correlation to the fission gas release and instant release leach rates of the corresponding fuel samples. In this study, it was concluded that the caesium and iodine profiles follow each other. Pronounced caesium and iodine peaks are typically associated with visible cracks in the fuel. The caesium, iodine and xenon profiles are rather flat in the interior of the pellet and showed increases towards the outer boundary of the pellet.

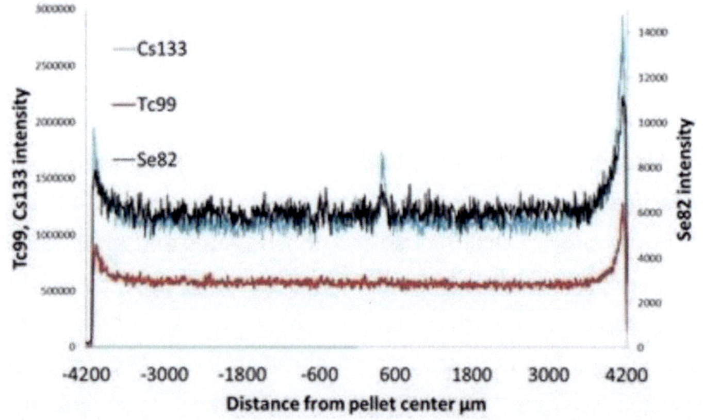

Initial Enrichment 4.1% 235U

Burn-up: 59.1 MWd/t_{HM}

Fission gas release: 1.3% Kr
1.5% Xe

Figure 23.1: Studsvik profiles of Cs, Tc and Se in a Al/Cr doped UO_2 rod C1 (Tentative Se-identification C1).

Figure 23.1 shows a significant increase up to a factor 2 of ^{99}Tc, ^{133}Cs and ^{82}Se concentrations towards the rim zone of the pellet. In the rim zone, a higher concentration of plutonium exists influencing the actual concentrations of fission products in the rim zone. According to Koo et al. (2001), the rim zone burn-up exceeds the average burn-up by 33% and the fission yields are shifted to some extend due to the increased fission of plutonium in this region. For example, the cumulative fission yield of ^{82}Se originating from plutonium is about 40% lower than the yields resulting from ^{235}U fission. In the case of ^{99}Tc, the thermal yields from ^{239}Pu and ^{235}U are almost equal, whereas for ^{133}Cs(stable) the cumulative yield from ^{239}Pu is only 6% above the yield from ^{235}U. (Nucleonica GmbH, 2011).

The observed increase of ^{82}Se, ^{99}Tc, and ^{133}Cs cannot be explained solely by the burn-up in the rim zone.

23.4 Modelling the temperature profiles and diffusion processes in UO_2 fuel under irradiation

Some data for modelling the boundary conditions for the temperature history of a 50.4 GWd/t_{HM} fuel rod segment have been provided in (Kienzler et al., 2013). In this paper, the central temperature of a fuel rod was calculated using heat conductivity as a function of the burn-up (Lucuta et al., 1996) and the heat capacity of UO_2 fuel taken from IAEA (1997) and Popov et al. (2000). For calculating the diffusion and distribution of fission products in UO_2 fuel, Popov recommends Lucuta's equation for the heat conductivity λ of UO_2, see Equation 23.1 (Kienzler et al., 2013):

$$\lambda(T,B,\rho,x)=\lambda_0 * FD * FP * FM * FR \quad (23.1)$$

where *B* is the burn-up (in atom %); *t* is the temperature as T (in K)/1,000; *p* is the porosity and x is the radius of the fuel pellet. λ_0 is the heat conductivity (in W/m/K) of UO_2; the *FD* factor represents the effect of dissolved fission products; the *FP* factor represents the effect of precipitated fission products; the *FM* factor corresponds to the effect of porosity; and *FR* corresponds to the effect of radiation damage.

For the thermal conductivity of 100% dense solid UO_2 the Equation 23.2 is the recommended one:

$$\lambda_0(T)=\frac{115.8}{7.5408+17.692\cdot t+3.6142\cdot t^2}+7410.5\cdot t^{-\frac{5}{2}}\cdot \exp(-\frac{16.35}{t}) \quad (23.2$$

The effect of the dissolved fission products is reflected by a burn-up and temperature- dependent factor FD given by Equation 23.3 (Kienzler et al., 2013):

$$FD=\omega^* \arctan\left(\frac{1}{\omega}\right) \quad (23.3)$$

$$\text{with } \omega=1.09\cdot B^{3.265}+0.0643\cdot \left(\frac{T}{B}\right)^{1/2}$$

The effect of the precipitated fission products is also reflected by a burn-up- and temperature-dependent factor FD expressed in Equation 23.4 (Kienzler et al., 2013):

$$FP = 1+\frac{0.019\cdot B}{3-0.019\cdot B}\cdot (1+\exp\left(-\frac{T-1200}{100}\right) \quad (23.4)$$

The effect of porosity is accounted for by the well-known Maxwell-Eucken factor, FM, following Equation 23.5:

$$FM=\frac{1\text{-}p}{1+2\cdot p} \quad (23.5)$$

Kienzler et al. (Nucleonica GmbH, 2011) used relations for calculating the rim burn-up and the rim porosity and derived a porosity of 3.8% at the relative radius of a pellet r/r_0 = 0.9 and 17.7% at r/r_0 = 1. The average rim porosity ($0.9 \leq r/r_0 \leq 1$) was calculated to 6.2%.

The radiation effect is given by the factor FR in Equation 23.6 (Kienzler et al., 2013):

$$FR=1-\frac{0.2}{1+\exp\left(\frac{T-900}{80}\right)} \quad (23.6)$$

The fuel density and the specific heat capacity of UO_2 fuel are also temperature dependent. The fuel density ρ, is defined by Equation 23.7:

$$\rho(T)=\rho_0\cdot(0.99734+9.802\times10^{-6}\cdot T-2.705\times10^{-10}\cdot T^2+4.391\times10^{-13}\cdot T^3)^{-3} \quad (23.7)$$

where ρ_0 is the UO_2 density at room temperature.

The specific heat capacity, C_p, is defined by Equation .23.8 (in J/kg/K) (IAEA, 1997):

$$C_p=C_1\cdot\left(\frac{\theta}{T}\right)^2\cdot\frac{\exp\frac{\theta}{T}}{\left[\exp\frac{\theta}{T}-1\right]^2}+2\cdot C_2\cdot T+C_3\cdot E_a\cdot\frac{\exp\left(\frac{-E_a}{T}\right)}{T^2} \quad (23.8)$$

where T is the temperature (K); θ is the Einstein temperature (K); and E_a is the electron activation energy divided by the Boltzman constant (K). The value of these constants are: C_1 = 302.27 (in J/kg/K); C_2= 8.463×10^{-3} (in J/kg/K^2); C_3 = 8.741×10^7 (in J/kg); θ = 548.68 (in K); E_a = 18 531.7 in (K).

Diffusion of elements in irradiated UO_2 has been investigated by several authors, e.g. (Matzke, 1983; Prussin et al., 1988). Prussin et al. (1988) reports diffusion data of xenon as function of the temperature for a variety of UO_2 samples. The investigations show that below 800 °C, the diffusion coefficient is in the range of solid state diffusion coefficients. The results of Prussin's work is shown in Figure 23.2 and 3.

In this paper, Arrhenius data are reported $D = D_0\cdot\exp(A/R\cdot T)$ (R: gas constant). Unfortunately, the measurements covered very high temperatures 1,500 K ≤ T ≤ 2,500 K. The derived A values are A = 99 ± 6 kcal/mol for Cs and 156 ± 23 kcal/mol for Xe. For the present study, an Arrhenius function with A = 90 kcal/mol was applied.

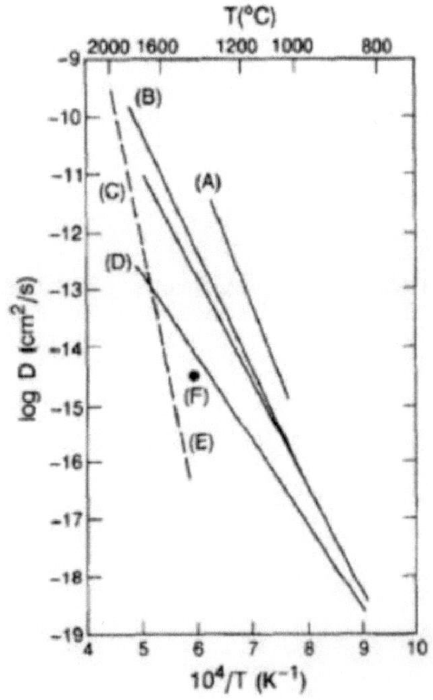

Figure 23.2: Diffusion coefficient for Xe as function of the temperature.

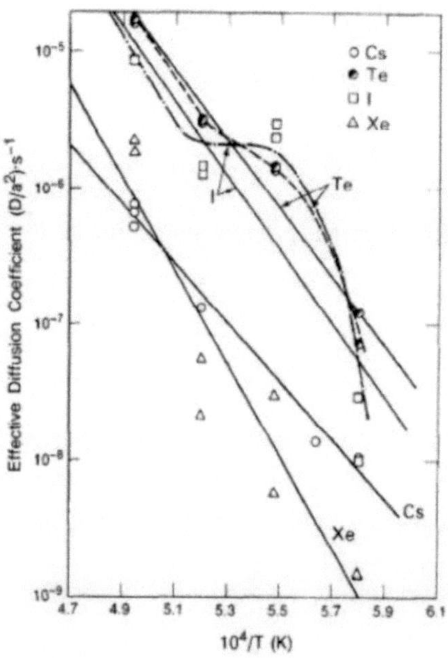

Figure 23.3: Effective diffusion coefficients for the volatile fission products in trace-irradiated sintered UO_2. Systematic counting uncertainties range from about 3% (Te) to 20% (Xe).

For the following calculations, the FlexPDE 6.35 code was used, which is a flexible solution system for partial differential equations. A standard geometry was applied for the calculations: Fuel zone: 0.00 - 4.40 mm, rim zone: 4.40 - 4.60 mm, gap: 4.60 - 4.65 mm, cladding: 4.65 - 5.375 mm. The calculations are performed in R, Z cylinder coordinates. The initial grid is shown in Figure 23.4.

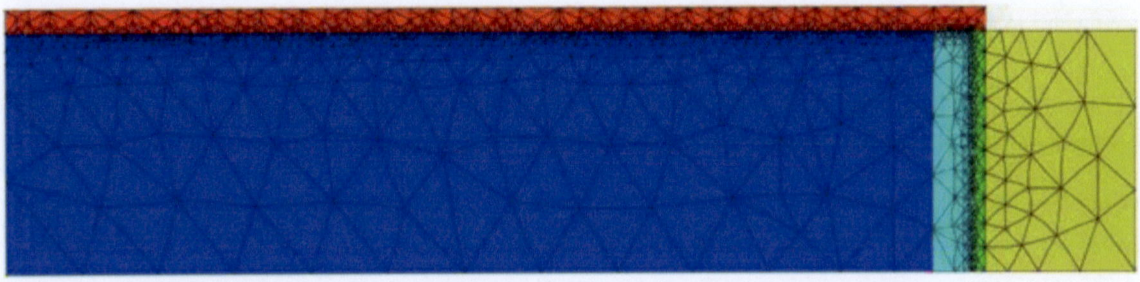

Figure 23.4: Finite element discretization (R, Z cylinder geometry) of the rod slice. (blue: fuel; turquois: rim zone; dark green: gap; light green: cladding; red: gap between pellets).

23.5 Results

To cover the burn-up and power rate ranges of fuels investigated in FIRST-Nuclides, the modelling is performed for an average burn-up of 50 GWd/t_{HM} and power rates of 180 W/cm (KKL), 260 W/cm (KKG) and 340 W/cm (e.g. power ramps or fuel investigations by SCK·CEN from PWR Tihange-1). A coolant's temperature of 325°C is a fixed boundary condition. Temperature calculations are performed in steady state mode.

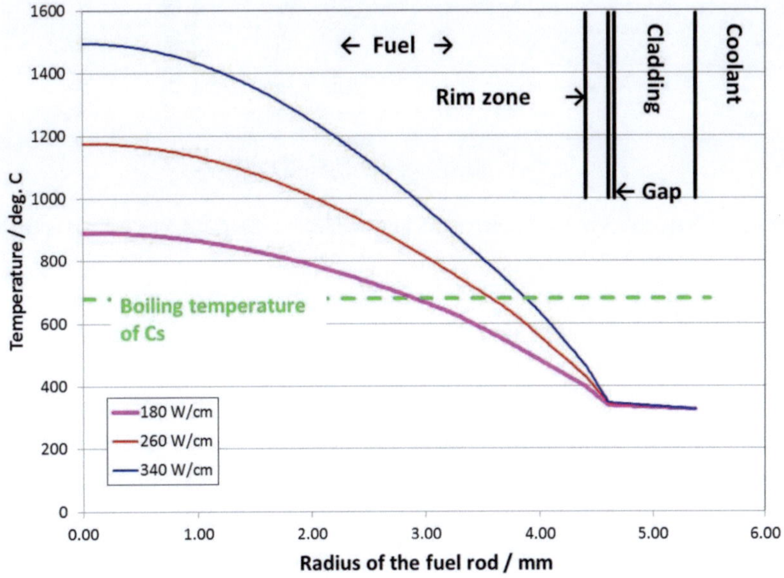

Figure 23.5: Calculated temperature profiles of a UO_2 pellet in a fuel rod under irradiation.

Figure 23.5 shows the temperature profiles obtained in the fuel during its use in a power plant. The complete fuel rod shows temperatures above the boiling temperature of pure iodine. The figure shows that in the gap between cladding and fuel (here 50 µm) as well as in the rim zone the temperature remains always below the boiling temperature of Cs. Precipitation of either a liquid or solid would lead to an accumulation of the respective element in this region. Based on the calculated temperature profiles, the diffusion coefficient is calculated by applying an A value of 90 kcal/mol. The results are shown in Figure 23.6. A lower limit of 10^{-20} mm^2/s is defined.

Figure 23.6 shows relatively high diffusion coefficients in the case of high temperatures. The diffusion coefficient in the centerline of the pellet under consideration varies from 10^{-17} for the 180 W/cm power rate to approx. 10^{-11} in the case of 340 W/cm power rate. In the gap region, a diffusion coefficient is selected in the calculations which exceeds the coefficient in the fuel by several orders of magnitude. This selection provides for a homogeneous distribution of the elements in the gap zone.

Figure 23.6 also shows clearly that diffusion takes place significantly only in the inner part of the fuel pellet, close to the rim zone, diffusion and mass transport is strongly retarded.

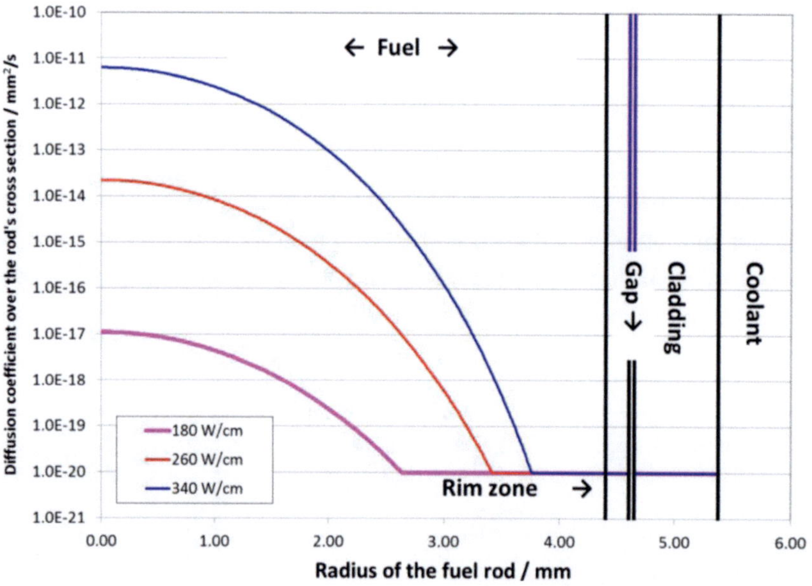

Figure 23.6: Calculated temperature depending diffusion coefficient profiles in a UO_2 pellet for a fuel rod under irradiation.

23.6 Application to other fuel rods

The same procedure was applied to Studsvik's fuel nominated as C1 (see Metz, 2013). In order to simulate the observed accumulation of radionuclides close to the outer surface of a pellet (at temperatures below 680°C), a phase transition of Cs and Se from gaseous state is assumed. This phase transition is modeled by assuming a temperature dependent condensation, which is modeled a specific coefficient similar to a sorption process. The geometry of a C1 fuel pellet is used as follows: height: 5 mm, gap between two pellets 50 µm, radius of the fuel zone: 4.0 mm, rim zone 100 µm, gap: 20 µm, cladding 0.69 mm. Initially, a homogeneous distribution of the radionuclides in the fuel and rim zones is assumed. The diffusion coefficient is chosen according to the previous chapter. Based on the fact that the boiling point of caesium is in the range of 680°C, a temperature function of a condensation coefficient is assumed: $k_d = 0$ for $T \geq 680$, $k_d = 2$ for $T \leq 500$°C and a linear interpolation between. The results of the calculations are shown in Figure 23.7, Figure 23.8 and Figure 23.9.

For the C1 fuel pellet, a center temperature of 1,000°C was computed, decreasing towards the cladding. The locations of cladding, gap, rim zone are shown in Figure 23.7. This figure shows also the regions where different condensation coefficients are used. In the following Figure 23.8 and Figure 23.99 the contour of the calculated temperatures as well as the calculated concentration distribution of the volatile elements in the pellet are presented for the C1 rod. The calculated total concentration in the colder region corresponds to Studsvik's measurements. Figure 23.9 demonstrates that the accumulation of volatile elements can be described by a temperature dependent condensation process. It also shows that retention of these elements in the gap zones between the pellet and between pellet and cladding are less relevant for the element distribution. After one year in the reactor, the calculated release of the volatile elements account to 1.3%, a number in the range of the fission gas release.

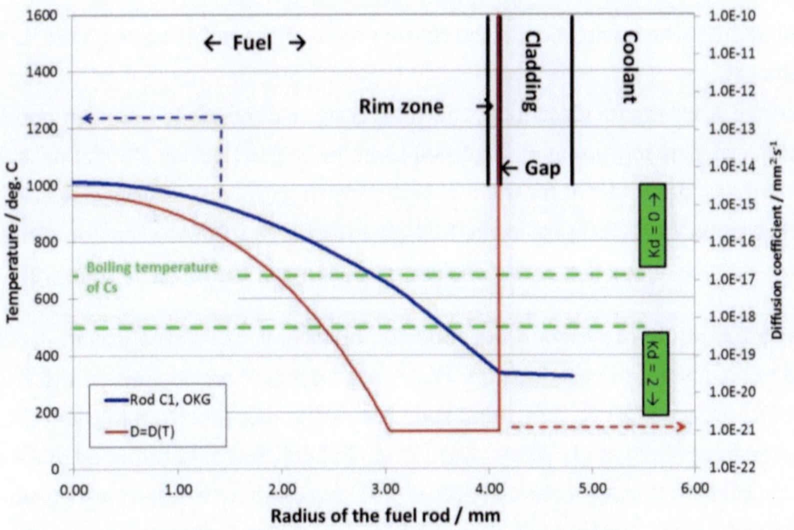

Figure 23.7: Profiles of temperature and diffusion coefficient over the radius of the C1 pellet.

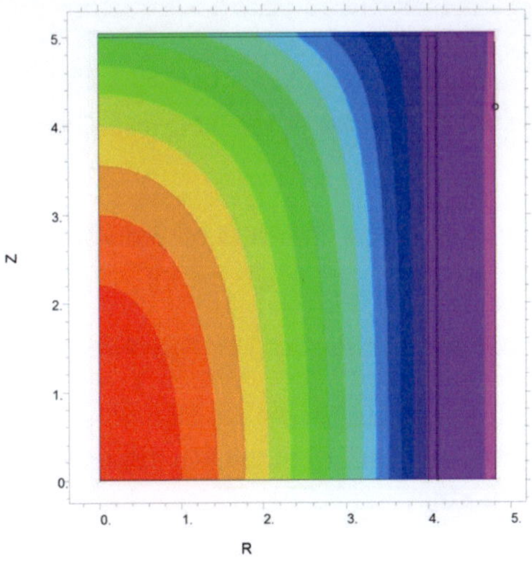

Figure 23.8: Contour of the calculated temperature in the C1 rod (red: highest, magenta lowest). File "rod C1 Studsvik_b.pde".

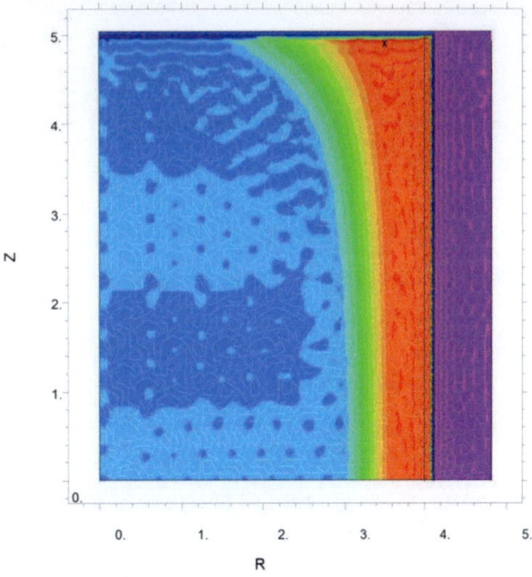

Figure 23.9: Concentration distribution of the volatile elements in the pellet (red highest concentrations, blue: initial concentration, magenta: zero concentration).

23.7 Discussion and Conclusions

A model is presented which allows to calculate the temperature dependent diffusion and retention of volatile radionuclides within a fuel pellet during the irradiation period. This model does not compete with reactor codes such as FRAPCON, FRAPTRAN, and MATPRO (Luscher and Geelhood, 2011) used by the utilities, it provides for a method of relating results obtained from small samples of irradiated fuel investigated within the CP FIRST-Nuclides to complete fuel rods. Following statements can be drawn:

- Diffusion processes within a pellet depend on the temperature.
- Small gaps between the pellets or between pellet and cladding effect the temperature significantly.
- The radial distribution of volatile elements such as I, Xe, and Cs in a spent fuel pellet can be modelled by assuming a temperature-dependent condensation parameter.
- The calculated total concentration in the colder region of the pellet corresponds to Studsvik's measurements.
- The published A values for describing the temperature dependent diffusion coefficients in a pure UO_2 matrix according to Prussin et al. (1988) seem to be too high for simulating spent fuel pellets in a temperature ranges relevant for the fuel under investigation in FIRST-Nuclides.
- Due to the temperature decrease in the fuel after discharge from the reactor, diffusion processes of volatiles do not play any role due to diffusion coefficients in the range of solid state diffusion.

It was intended by this study to describe the diffusion behaviour of volatile fission products in fuel rods which contribute to the instant release fraction. The model approach works well for the FPs I and Cs. For Se, however, it has been shown by X-ray absorption spectroscopy in UO_2 fuel from KKL and OKG-3 that Se is not present in elemental form (Curti et al., 2014a; Curti et al., 2014b). The data obtained in the X-ray absorption spectroscopy study indicate that a substantial part of ^{79}Se and other Se isotopes are present in UO_2 SNF in a homogeneous chemical form, probably as dispersed Se(-II) replacing oxygen sites in the UO_2 lattice. This Se(-II) sites in the UO_2 lattice do certainly not diffuse similar to the volatile elements.

23.8 References

Brooks, L.S. (1952). The Vapor Pressures of Tellurium and Selenium. Journal of the American Chemical Society, 74(1), 227-229.

Bentor, Y. (2012). Chemical Elements (Formerly "The Periodic Table of the Elements on the Internet"). Available from: http://www.chemicalelements.com/index.html.

Kienzler, B., et al., *2nd Annual Workshop Proceedings of the Collaborative Project 'FIRST-Nuclides'*, 2014, Karlsruhe Institute of Technology (KIT): Karlsruhe.

Koo, Y.-H., Lee, B.-H., Cheon, J.-S., Sohn, D.-S. (2001). Pore Pressure and Swelling in the Rim Region of LWR High Burnup UO_2 Fuel. Journal of Nuclear Materials, 295(2–3), 213-220.

Nucleonica GmbH (2011). Nucleonica Nuclear Science Portal (www.nucleonica.com). Version 3.0.11, Karlsruhe.

Kienzler, B., et al., *1st Annual Workshop Proceedings of the Collaborative Project 'FIRST-Nuclides'*, 2013, Karlsruhe Institute of Technology (KIT): Karlsruhe.

Lucuta, P.G., Matzke, H., Hastings, I.J. (1996). A Pragmatic Approach to Modelling Thermal Conductivity of Irradiated UO_2 Fuel: Review and Recommendations. Journal of Nuclear Materials, 232(2–3), 166-180.

IAEA (1997). Thermophysical Properties of Materials for Water Cooled Reactors. International Atomic Energy Agency, Vienna.

Popov, S.G., Carbajo, J.J., Ivanov, V.K., Yoder, G.L. (2000). Thermophysical Properties of MOX and UO_2 Fuels Including the Effect of Irradiation. Oak Ridge National Laboratory (ORNL).

Matzke, H., (1983). Radiation Enhanced Diffusion in UO_2 and $(U, Pu)O_2$. Radiation Effects, 75(1-4), 317-325.

Prussin, S.G., Olander, D.R., Lau, W.K., Hansson, L. (1988). Release of Fission Products (Xe, I, Te, Cs, Mo and Tc) from Polycrystalline UO_2. Journal of Nuclear Materials, 154(1), 25-37.

Metz, V., *Deliverable No: 1.2 "Characterisation of spent nuclear fuel samples and description of methodologies and tools to be applied in FIRST-Nuclides"* 2013, KIT-INE.

Luscher, W.G., Geelhood, K.J. (2011). Material Property Correlations: Comparisons between FRAPCON-3.4, FRAPTRAN 1.4, and MATPRO. Office of Nuclear Regulatory Research.

Curti, E., Froideval-Zumbiehl, A., Günther-Leopold, I., Martin, M., Bullemera, A., Lindera, H., Borcab, C. N., Grolimund, D. (2014a). Selenium Redox Speciation and Coordination in High-Burnup UO_2 Fuel: Consequences for the Release of ^{79}Se in a Deep Underground Repository. Journal of Nuclear Materials, 453, 98-106.

Curti, E., et al. (2014b). X-Ray Absorption Spectroscopy of Selenium in High Burnup UO_2 Spent Fuel from the Leibstadt and Oskarshamn-3 Reactors. Final Workshop Proceedings of the 7th EC FP CP FIRST-Nuclides Project (eds. Kienzler et al.).

24 Modelling of spent fuel saturation with water – implications for the instant release fraction (IRF)

Marek Pękala, Andrés Idiart and Lara Duro*

AMPHOS 21 Consulting (ES)
** Corresponding author: marek.pekala@amphos21.com*

24.1 Abstract

Radionuclide Instant Release Fraction can impact the performance of an underground repository for Spent Nuclear Fuel. In this contribution we propose that Instant Release Fraction can, for some radionuclides, be controlled by wetting by water of the surfaces of cracks in the Spent Nuclear Fuel. In order to test this hypothesis we have developed a numerical model that simulates water flow and tracer transport within the interconnected network of cracks. The model allows calculate the cumulative mass of tracer that leaves the fuel and accumulates in an external reservoir. Modelling results can be directly compared with laboratory data on Spent Fuel leaching and indicate that the proposed model is capable of capturing radionuclide release trends in a semi-quantitative manner.

24.2 Introduction

Disposal in a deep underground repository is currently considered by several countries as the preferred option for the long-term management of spent nuclear fuel (SNF). For regulatory reasons the safety of such a disposal facility must be demonstrated quantitatively through Performance Assessment (PA) calculations. One of the factors that have the potential to significantly influence the results of PA is the radionuclide release rate from the waste upon contact with groundwater. In current PA calculations the radionuclide release from the SNF is typically represented by a two-stage process: (1) an instantaneous release of a fraction of the inventory immediately upon contact with water (the *instant release fraction* – IRF), followed by (2) a relatively slow release rate of the inventory confined in the SNF matrix over a long period of time. PA calculations indicate that instant release is important for the overall safety assessment of the facility, but there are significant uncertainties about the nature of this phenomenon and how it should be represented in PA.

In the context of IRF, the knowledge of the rate of SF saturation is important as the release of some radionuclides is conditioned by wetting of the fuel surfaces by water. Therefore, the instant release rate of these radionuclides depends on the rate of SNF saturation with water. An improved understanding of the rate of saturation will help elucidate radionuclide instant release both under repository conditions and with regard to laboratory experiments conducted on the SNF under near-room conditions.

24.3 Background and Motivation

Laboratory leaching experiments are performed on SNF in order to shed light on fuel behaviour under repository conditions, when spent fuel might be contacted by groundwater (e.g.: Johnson and Tait, 1997; Johnson et al., 2002; Johnson et al., 2004; Gonzáles-Robles, 2011; Johnson et al., 2012; Casas et al., 2012). The experiments are performed on powdered, fragmented or complete SNF pellet material (fully cladded, partially cladded or un-cladded), where the concentration of selected radionuclides is monitored with time in the surrounding solution that remains in contact with the SNF. Batch leaching experiments indicate that the release of radionuclides from SNF into the solution follows an asymptotic curve, with the initial (hours to days) release rate relatively high, followed (over weeks to months) by gradually decreasing release rates. This behaviour is typically explained in terms of varying dissolution rates of the SNF (González-Robles 2011; Casas et al., 2012). However, as was previously postulated by Pękala et al. (2012), an alternative/ complementary explanation might exist, where the observed radionuclide release rates can be accounted for by the rate of wetting of internal surfaces of cracks in the SNF, desorption of radionuclides from the surface into water, and simultaneous diffusion of the radionuclides dissolved in water towards an external solution. In this contribution we elaborate the model of saturation of SNF with water and follow up with modelling of diffusive transport of dissolved radionuclides. The objective of this study is to demonstrate whether the water saturation-based approach can quantitatively explain experimental results obtained from SNF leaching in the laboratory. In doing so, this work attempts to contribute to better understanding of the mechanisms responsible for IRF release of radionuclides from SNF under both laboratory and repository conditions.

24.4 Overview of the Conceptual Model

As explained in Pękala et al. (2012), the conceptual model assumes that the rate of release for some radionuclides from a SNF pellet (or a fragment of it) in contact with water is controlled by the wetting of surfaces of cracks present in the pellet. Cracks are formed in the SNF pellets due to thermo-mechanical forcing induced by very steep temperature gradients across the radius of the pellet while in the reactor under high burn-up conditions (e.g. around 60 kW/m). The cracks vary largely in their aperture, geometry and interconnectedness, depending on the specific properties of the fuel pellet and irradiation conditions. It is expected that on contact with water this interconnected network of cracks will be gradually filled with water. The model presented attempts to quantify this process. It is considered that wetting of crack surfaces will liberate (desorb) some of the radionuclides present there and allow their diffusive transport towards an external reservoir (e.g. the batch experimental solution). In a general case, the processes of radionuclide release and transport within the inter-connected crack system could be accompanied by additional simultaneous surface and dissolution/precipitation reactions. Calculations were carried out to quantify the mass (moles) of radionuclides released from the crack surfaces and accumulated in the external reservoir as a function of time. This allows direct comparison with data available from laboratory leaching experiments. The contribution is divided into three main parts: at first the **Saturation** model is briefly reviewed, then **Transport** processes are considered, and finally a **Comparison with Experimental Results** is shown.

24.5 The Model for SNF Saturation with Water

Main features of the model were previously presented in Pękala et al. (2012). Here, only a condensed description is given.

Interconnected Voidage Available for Saturation with Water

Fuel pellets suffer a varying degree of cracking due to thermo-mechanical processes while irradiated in the reactor (e.g. Faya, 1981). This is due to a sharp thermal gradient that is established radially within a pellet and the resulting differential expansion. The degree of cracking (aperture geometry and crack patterns) depends on the reactor operating conditions and can be somewhat correlated with the degree of burn-up (Oguma, 1983; Williford, 1984). Although a significant increase of porosity occurs in the SNF pellet rim as a result of recrystallization processes during irradiation in the reactor, these are assumed to be non-connected, and therefore were excluded from the model (Spino et al., 1996). In this work, saturation is understood as the physical process of filling up of the interconnected network of cracks within the SNF pellet. In the saturation calculations presented here only cracks are assumed to provide continuous passage (connected "porosity") for moving water, and are therefore the only pathways included in the model.

Conceptualisation of the Pellet

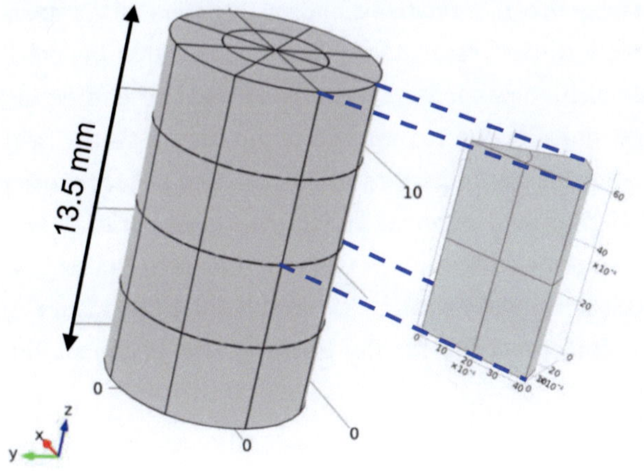

Figure 24.1: The Reference Pellet with idealised crack pattern (left) and a 1/16th section selected (based on symmetry criteria) for calculations (right). Pellet dimensions are height = 13.5 mm, diameter = 8.19 mm.

The aperture, geometry and inter-connectedness of crack networks vary significantly from pellet to pellet. For modelling purposes, the complete crack network was subdivided into two groups: "macro cracks" (the larger cracks with a mean aperture assumed to be equal 20 µm) and "micro cracks" (small cracks with a mean aperture of 0.1 µm). The two crack types were represented in the model differently: 1) "macro cracks" were represented as discrete fractures (with a defined geometry), while 2) "micro cracks" were represented using the equivalent porous medium approach. The mean apertures of the cracks were defined based on visual inspection of available microscope photographs (e.g. Serna et al., 2006). For modelling purposes an idealised Reference Pellet representative of relatively high linear power and burn-up conditions (around 60 kW/m) was defined (Figure 24.1). The "macro cracks" characteristics of the Reference Pellet were defined based on literature sources relating mean pellet crack aperture (20 µm) and crack pattern (regularly distributed cracks: 4 axial, 3 radial and 1 circumferential) with power applied to rod in the reactor (Oguma, 1983; Williford, 1984). The external edges of "macro cracks" are indicated in Figure 24.1by black lines (except for the upper and lower pellet edges).

The "micro cracks" are not included in the model explicitly, but are represented by a porous medium (the so called equivalent porous medium approach). The calculations of saturation with water for the two crack populations were de-coupled: at first the saturation of "macro cracks" was calculated, followed by the calculation for "micro cracks". This approach was based on results of preliminary calculations (Pękala et al., 2012), which indicated largely different time scales for the saturation with water of the two crack systems.

Mathematical Description of Water Flow (Saturation with Water)

Modelling of saturation with water involves solving equations describing variably-saturated flow of water. The simplest mathematical model for such a problem is based on the Richards equation (Bear and Cheng, 2010). The solution of the Richards equation must however be conducted simultaneously with a set of constitutive algebraic equations describing water retention and relative permeability properties of the medium. In this study the commonly used van Genuchten model (van Genuchten, 1980) is used in conjunction with the Richards flow equation.

Parameterisation of the Model

Two parameters (n and α) are required for the van Genuchten model. These were estimated based on a model derived from statistical considerations of cracks apertures (Fredlund and Xing, 1994; Zhang and Fredlund, 2003). The retention curve was calculated assuming a mean value of crack aperture (20 µm and 0.1 µm for the "macro cracks" and "micro cracks", respectively), a relative standard deviation of the apertures (100% and 70%, for the "macro cracks" and "micro cracks", respectively) and assuming that the crack apertures follow a log-normal distribution. The van Genuchten parameters were then fitted to the retention curves by means of non-linear optimisation. The pellet porosity due to presence of the "macro cracks" was calculated from crack and pellet geometry (assuming mean crack aperture of 20 µm). The pellet porosity due to "micro cracks" was assumed to be 0.03. The saturated permeability of the "macro cracks" was estimated according to the cubic law, which relates the permeability with the cube of the aperture (Marsily, 1986), while that for the "micro cracks" (porous equivalent) was calculated from the cubic law assuming averaged in-series and in-parallel connections. Zero compressibility for the fractures was assumed. The residual water content was also assumed to be zero.

Numerical Implementation

The model was implemented in the finite element code *Comsol Multiphysics* using the Fracture Flow Interface (for flow in the "macro cracks") and the Richards Interface (for flow in the "micro cracks"). The Fracture Flow Interface was modified by manually implementing a storage-retention term and the van Genuchten constitutive equations so as to account for variably-saturated water flow within the fractures. Advantage was taken of symmetry of the crack network of the Reference Pellet to downsize the complete pellet geometry sixteen times (see Figure 24.1 and Figure 24.2), which allowed to reduce computation times significantly.

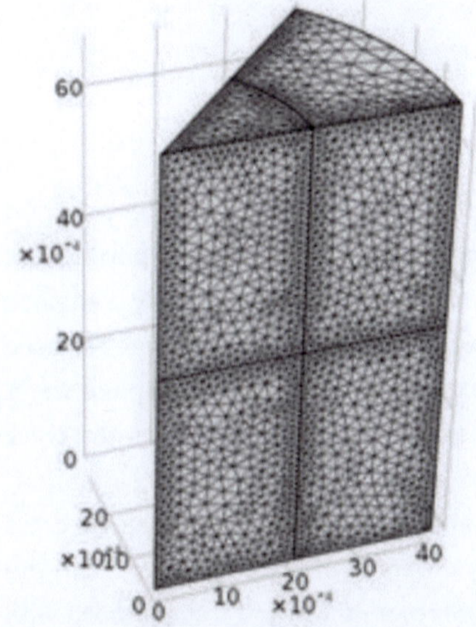

Figure 24.2: The geometry and finite element mesh of the 1/16th fragment of the Reference Pellet for which calculations were performed. Dimensions are in metres.

Initial and Boundary Conditions

For the Richard's equation to be solved appropriate initial and boundary conditions must be defined. Unfortunately, the initial (time zero) saturation state of the pellet is not known precisely. Consequently, a small initial saturation of 10^{-5} was assumed. This is expected to represent very dry initial conditions. As a boundary condition, a small and constant pressure (Dirichlet type boundary), corresponding to 5 cm water column, was prescribed to the external edges of the cracks. This represents the boundary water pressure that might be expected in a typical batch leaching experiment.

Results of Saturation Calculations

The results of saturation calculations for the "macro cracks" and "micro cracks" networks are shown in **Fehler! Verweisquelle konnte nicht gefunden werden.**. The results indicate that saturation of the "macro cracks" occurs rapidly, over a period of about 1 day. On the other hand, the saturation of "micro cracks" requires a much longer period of time (over 50 days).

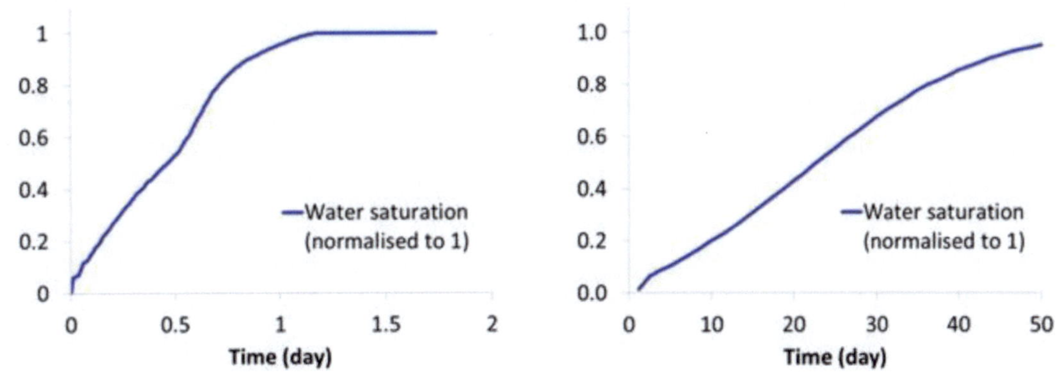

Figure 24.3: Total water content (surface integral over all "macro cracks" and volume integral over the pellet volume for "macro" and "micro" cracks, respectively), normalised to unity (complete saturation), as a function of time. "Macro cracks" – left, "micro cracks" – right.

24.6 Radionuclide Transport

Scoping Calculations

Preliminary 1D calculations of variably-saturated water flow (saturation) and diffusive-advective transport of a tracer from a single crack were performed in order to better understand the relative importance of flow and transport processes for the considered problem. The crack was 5 mm long and initially contained adsorbed radionuclides, but was devoid of water. As water invaded the crack, the tracer was released instantaneously after surface wetting and diffused against the direction of water flow. After complete water saturation, diffusion could continue for some time until all gradients of concentration between the crack and the outside reservoir were eliminated. The exercise considered two limiting cases: 1) "slow saturation" (a single crack of 5 mm completely water saturated within 27 days), and 2) "fast saturation" (a single crack of 5 mm completely water saturated within 2 hours). In both cases, the mass of tracer (arriving from the crack as it was being saturated with water) in the external reservoir was monitored, and a comparison with pure diffusion (assuming instant water saturation) was made. The results of this exercise indicate that if saturation with water is slower than a few hours, the release of a tracer from the pellet should be controlled by the rate of saturation with water. The consequence of this is that in cases of slow rate of water saturation (hours or longer) transport processes within the crack network are of negligible importance for the release of a tracer (and need not be modelled explicitly).

Coupled Saturation and Transport calculation for the "Macro Cracks"

In order to further verify that transport of a tracer is "instantaneous" relative to saturation with water for the complete interconnected network of "macro cracks" given the time scales considered, an additional calculation was performed. In the transport calculation, the initial condition considered a tracer concentration of unity inside all "macro cracks", while the boundary condition was prescribed as constant null concentration along the external crack edges. The boundary condition may be thought of to represent a large and well-mixed external reservoir, where tracer concentrations are maintained close to zero. Figure 24.4 shows a snapshot of tracer concentration (left) and saturation with water (right) in the "macro cracks" at simulation time of approx. 3 hours. Note that concentrations are shown on the log10 scale. Inspecting the figure it can be seen that the fronts of water saturation and tracer concentration coincide closely. This is expected for a problem where water saturation constitutes a limiting factor for transport. This behaviour is additionally illustrated in

Figure 24.5, which shows a comparison between the degree of water saturation of "macro cracks" (surface integral of water content over all "macro cracks" as a function of time) and the total accumulated mass of a tracer released from the "macro cracks" network and transported to an external reservoir.

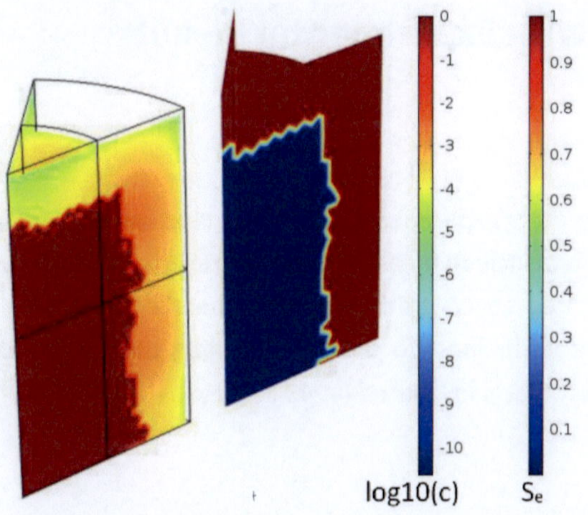

Figure 24.4: Tracer concentration (log10[mol/L] – left) and saturation with water (S_e – right) calculated at time 3 hours in "macro cracks" of the (1/16th)fragment of the pellet.

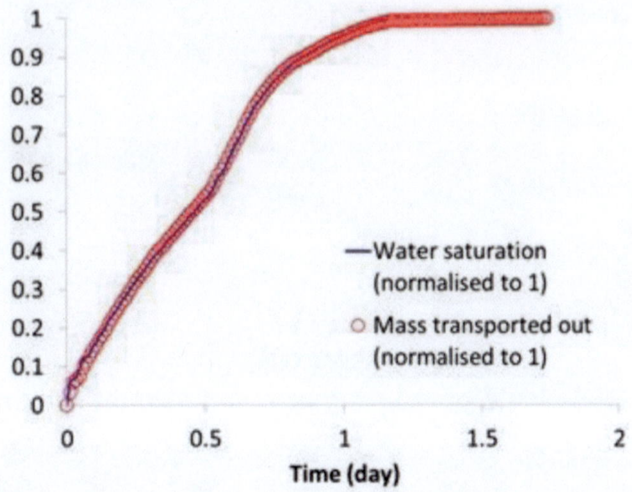

Figure 24.5: Comparison of the degree of saturation of "macro cracks" (surface integral of water content over all "macro cracks" as a function of time) and the total accumulated mass of a tracer released from the "macro cracks" network and transported to an external reservoir. Both variables are normalised to unity (complete saturation and all radionuclide mass outside the cracks, respectively) to help comparison.

It can be seen from Figure 24.5 that the two curves closely overlap. This indicates that, indeed, the system is saturation-limited and transport can be considered to be very fast or instantaneous. Importantly, this is even more so for the "micro cracks", where water saturation rates are still slower. Consequently, modelling of the release of a tracer from both the "macro" and "micro" cracks does not require transport processes to be accounted for explicitly – the normalised total mass of the tracer that has left the crack system in time may be accurately represented by the curve of total water content in the crack system.

24.7 Comparison with Experimental Results

Experimental Data

The model was applied to laboratory experimental data from leaching of high burn-up (54 kW/m) SNF pellet (González-Robles, 2011). The data include measured concentrations of Rb, Sr, Mo, Tc, Te, Cs and U. To allow data inter-comparison (integrated concentrations vary by nearly five orders of magnitude between the radionuclides), the data were normalised to unity by dividing time series concentration data for each radionuclide by the final concentration measured (at 49.17 days).

Data Interpretation

Inspection of the experimental data indicates a change of release rate after about one day. An example of this behaviour (for Cs) is shown in Figure 24.6. We interpret this to indicate a change in release mechanism and hypothesise that it marks the time of complete saturation of the "macro cracks" and the onset of the saturation of the "micro cracks".

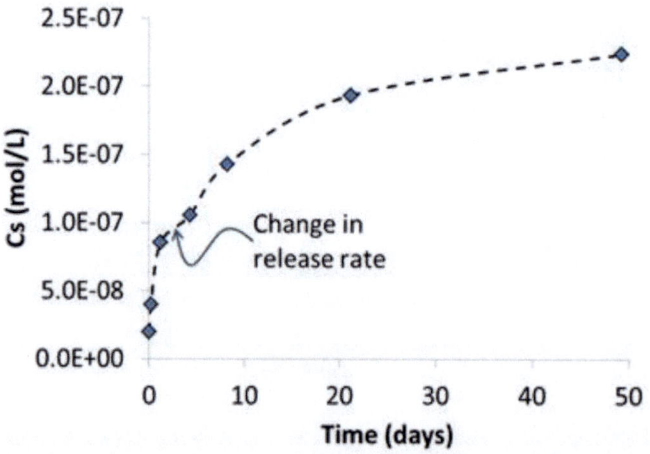

Figure 24.6: Experimental laboratory data on Cs release from SNF pellet (data of González-Robles (2011)).

Calculation Approach

The proposed model enables to calculate total mass of a radionuclide tracer (one that does not undergo complexation/exchange reactions with the pellet surfaces or precipitation/dissolution reactions) accumulating in an external reservoir during SNF pellet leaching. Given the results of scoping calculations presented above, for the problem considered, tracer transport within the interconnected crack network need not be represented explicitly. The total radionuclide mass accumulated in the outside reservoir can be calculated from the curve representing total degree of saturation of the pellet with water (water content integrated over the total pellet volume or crack surface area for the "micro" and "macro" cracks, respectively) as a function of time.

Results

The saturation curves shown in Figure 24.4 can be used directly to calculate the total accumulated mass of a tracer released from the crack systems. However, this requires that both curves be added up and scaled to unity. For this it is necessary to know the partitioning of the tracer mass between the two crack systems. This information is not directly available, but it can be inferred from experimental data based on the assumption that the change in radionuclide release rate coincides with the transition between the release from "macro cracks" and "micro cracks". The partitioning is expected to vary depending on the specific radionuclide considered. The experimental data suggest values in the range between 15 and 65% of the total mass being associated with the "macro cracks" (Figure 24.7).

In this work we intend to represent the overall behaviour of the system rather than focus on a specific radionuclide, for this reason an average value of 40% wt. was selected. The calculated model prediction of radionuclide mass accumulated in the external reservoir as a function of time is shown in Figure 24.7.

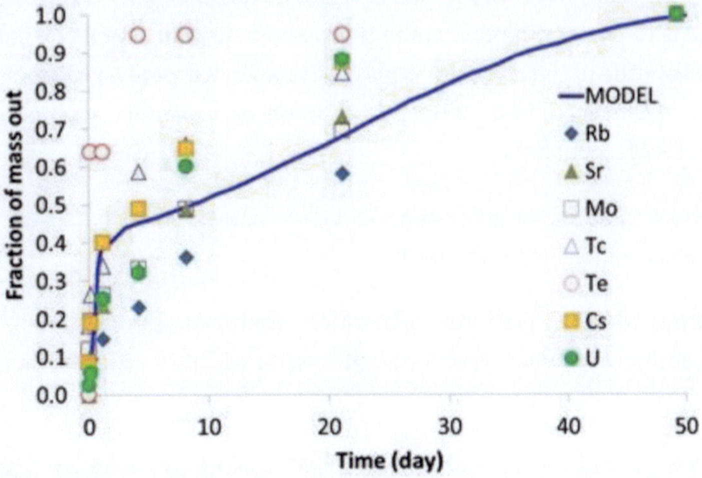

Figure 24.7: Experimental laboratory data (González-Robles, 2011) of radionuclide release from SNF pellet (normalised to unity) and model prediction assuming that 40% of total tracer mass is initially associated with "macro cracks".

Discussion

The model can reproduce experimental results in terms of their main two features: 1) fast initial release on the scale of a few days, followed by 2) slower rate of release during the remaining 50 days or so. Although the calculated results do not provide unique evidence that the proposed explanation is correct, it clearly demonstrates its feasibility.

From Figure 24.7 it is evident that the release rates are radionuclide-specific. This is expected given that different radionuclides associate with distinct parts of the SNF pellet (e.g. concentration gradients between the core and rim) and because of their distinct chemical properties. For example, different radial distribution of radionuclides in combination with irregular and/or radially varying crack system density would results in different fractions of radionuclide inventories being associated with the "macro" and "micro" cracks. A question that might arise is why, for some radionuclides (notably U), the model under-predicts the total mass released (release shouldn't be faster than surface wetting according to the model proposed). This could be explained by either or a combination of (at least) two factors. One is that the modelling results shown assume homogenous tracer distribution, while it is known that the rim zone of a pellet is typically enriched compared with the core (e.g. Manzel and Walker, 2002). Such heterogeneous distribution would have the effect to make the curve corresponding to saturation of the "micro cracks" more strongly concave upwards. Another point worth keeping in mind is that the normalisation of experimental results to unity may be somewhat confounding (in some cases) as it assumes that the last point in time of measurement (to which all earlier data points are normalised) represents 100% mass of the radionuclide accumulated in the experimental solution. This is probably a fairly accurate assumption in some cases (e.g. Cs or U), but somewhat erroneous for other radionuclides (such as Rb, Sr, Mo). If data points collected at later times were available, these effects would not occur. Nonetheless, for those radionuclides which did not approach steady state concentrations sufficiently closely, re-normalising the data points to a higher concentration value (acquired at a later time) would cause the intermediate points to be shifted somewhat downwards.

Finally, it is important to keep in mind that the underlying saturation model relies on a number of important assumptions and employs many parameters, whose values are generally not known. Specifically, the underlying constitutive model for water retention in the crack system (e.g. the statistical geometric properties of the cracks and the initial state of the cracks in terms of saturation with water) is poorly constrained. For this reason the modelled results should be thought of as more an illustrative example than a deterministic calculation based on reliable input data.

For the proposed model to be radionuclide-specific and to possess more predictive power the following knowledge gaps should be eliminated or reduced:

The statistical properties (density, apertures, orientation, inter-connectedness, roughness, etc.) of cracks should be known quantitatively, which would permit a more accurate estimation of the water retention model

The initial concentrations of radionuclides of interest associated with the different cracks should be quantified – information on how different radionuclide concentrations (surface densities) correlate with crack properties and vary in space

24.8 Conclusions

Modelling results presented in this contribution indicate that wetting of crack surfaces present in a SNF pellet may quantitatively explain radionuclide release patterns observed in laboratory leaching experiments performed on SNF pellets or their fragments. Specifically, the time scales of initial fast radionuclide release and the later slower releases are on a scale comparable with the times calculated for complete saturation of the "macro" and "micro" crack systems. The proposed model is capable of capturing these general trends in a semi-quantitative manner. In order to increase the predictive capacity of the model further data on the statistical properties of cracks and on the distribution of radionuclides on the surfaces of cracks is needed.

24.9 References

Bear, J., Cheng, A. H-D. (2010). Modelling Groundwater Flow and Contaminant Transport. Theory and Applications of Transport in Porous Media, Volume 23, Springer Science+Business Media B.V.

Casas, I., Espriu, A., Serrano-Purroy, D., Martínez-Esparza, A., de Pablo J. (2012). IRF Modelling from High Burn-Up Spent Fuel Leaching Experiments. 1st Annual Workshop Proceedings of the 7th EC FP CP FIRST-Nuclides Project (eds. Kienzler et al.). KITScientific Reports 7639, 31-39.

Faya, S.C.S (1981). A Survey on Fuel Pellet and Healing Phenomena in Reactor Operation. Instituto de Pesquisas Energéticas e Nucleares, Centro de Engenharia Nuclear, Report OUTUBRO/1981.

Fredlund, D.G., Xing, A. (1994). Equations for the Soil-Water Characteristic Curve. Canadian Geotechnical Journal, 31, 521-532.

Van Genuchten, M.Th. (1980). A Closed-Form Equation for Predicting the Hydraulic Conductivity of Unsaturated Soils. Soil Science Society of America Journal, 44, 892–898.

Gonzáles-Robles, E. (2011). Study of Radionuclide Reléase in Comercial UO_2 Spent Nuclear Fuels. Effect of Burn-Up and High Burn-Up Structure. Ph. D. Thesis, Universitat Politècnica de Catalunya.

Johnson, L., Günther-Leopold, I, Kobler Waldis, J., Linder, H.P., Low, J., Cui, D., Ekeroth, E., Spahiu, K., Evins, L.Z. (2012). Rapid Aqueous Release of Fission Products from High Burn-Up LWR Fuel: Experimental Results and Correlations with Fission Gas Release. Journal of Nuclear Materials 420, 54-62.

Johnson, L., Poinssot, C., Ferry, C., Lovera, P. (2004). Estimates of the Instant Release Fraction for UO_2 and MOX Fuel at t=0, in: A Report of the Spent Fuel Stability (SFS). Project of the 5th Euratom Framework Program, NAGRA Technical Report 04-08.

Johnson, L.H., Tait, J.C. (1997). Release of Segregated Nuclides from Spent Fuel. SKB Technical Report, TR-97-18.

Johnson, L.H., McGinnes, D.F. (2002). Partitioning of Radionuclides in Swiss Power Reactor Fuels. Nagra Technical Report, TR-02-07.

Manzel, R., Walker, C.T. (2002). EPMA and SEM Fuel Samples from PWR Rods with an Average Burnup of around 100 MWd/kg_{HM}. Journal of Nuclear Materials, 301, 170.

Oguma, M. (1983). Cracking and Relocation Behaviour of Nuclear Fuel Pellets During Rise To Power. Nuclear Engineering and Design, 76, 35.

Serna, J.J., Tolonen, P., Abeta, S., Watanabe, S., Kosaka, Y., Sendo, T., Gonzales, P. (2006). Experimental Observations on Fuel Pellet Performance at High Burnup. Journal of Nuclear Science and Technology, 43(9), 1-9.

Spino, J., Vennix, K., Coquerelle, M. (1996). Detailed Characterisation of the Rim Microstructure in PWR fuels in the Burn-Up Range 40-67 GWd/t_{HM}. Journal of Nuclear Materials, 231, 179-190.

Zhang, L., Fredlund, D.G. (2003). Characteristics of Water Retention Curves for Unsaturated Fractured Rocks. Second Asian Conference on Unsaturated Soils. UNSAT-ASIA 2003, Osaka, Japan.

25 Brief summary of selenium chemistry update

Emily Moore, Ondřej Beneš and Rudy Konings*

European Commission, JRC, ITU, P.O. Box 2340, D-76125 Karlsruhe, Germany
Corresponding author: ondrej.benes@ec.europa.eu

25.1 Abstract

The review of chemical thermodynamic data of selenium to assess the current state of available data concerning the long lived fission product ($T_{1/2}$ = 3.77 x 10^5 years) is still ongoing. The solid and liquid forms of elementary selenium have been evaluated in terms of their relevant thermodynamic properties such as standard entropy, heat capacity and phase transition data. With that set of data it is possible to calculate the Gibbs energy functions of relevant phases that are expected to be formed during irradiation within the fuel matrix. Since Gibbs energy defines the chemical stability of a studied system its knowledge allows to assess the chemical form of the fission product that is likely to be present in the fuel. The form present in the fuel is not only important with respect to the vaporization behaviour, and consequently for determination of gas release upon an accidental - high temperature event, but also for the determination of instant release of selenium from spent nuclear fuel upon e.g. interaction with surrounding water and other environmental components at lower temperatures.

Apart of the data of elementary forms of selenium, data of selenium-containing gaseous species plus the data of oxide compounds have been critically reviewed and introduced in a thermochemical database for the purpose of modelling long term fission product retention. With the complete database have predicted the selenium chemistry with respect to external variables, such as temperature, concentration of selenium (or partial pressure of selenium) or oxygen potential so that the environmental implications can be accounted for in the event of container failure during long term disposal of nuclear waste.

Figure 25.1 - 3 show the main outcome of this study highlighting the predominance phase diagrams of the Se-U-O system calculated as function of oxygen and selenium pressure for 800, 1,000 and 1,500 K, respectively. The temperatures were selected as average temperatures of LWR fuel during irradiation upon the periphery of the fuel pin (800 K), in the intermediate zone (1,000 K) and in the hottest part, in the middle of the fuel pin (1,500 K). The pressure of oxygen (i.e. oxygen potential) which significantly influences the fuel chemistry is defined by the oxygen stoichiometry of the UO_{2-x} matrix which is kept slightly hypo-stoichiometric for LWR fuels. In the figures the oxygen potential range that is expected in the fuel corresponds to the O/M ratio between 1.98 and 2.00 highlighted by the bold horizontal lines.

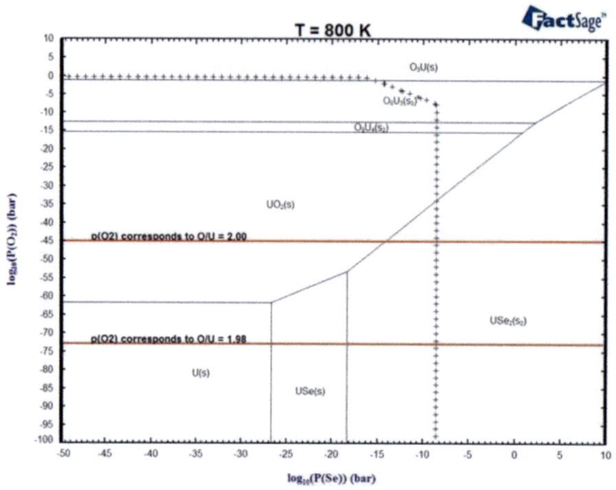

Figure 25.1: Predominance phase diagram of the Se-U-O system calculated for 800 K.

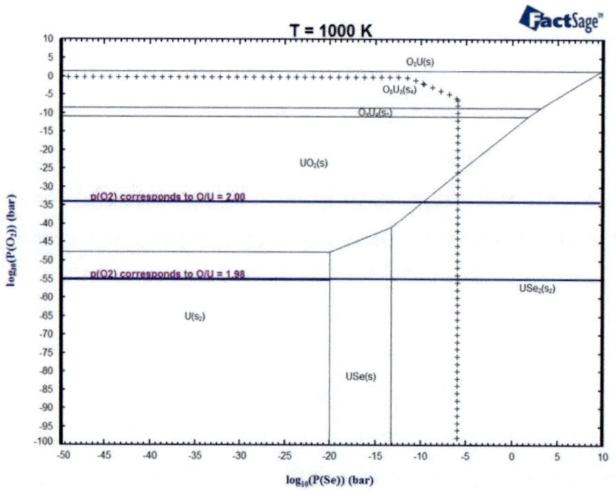

Figure 25.2: Predominance phase diagram of the Se-U-O system calculated for 1000 K.

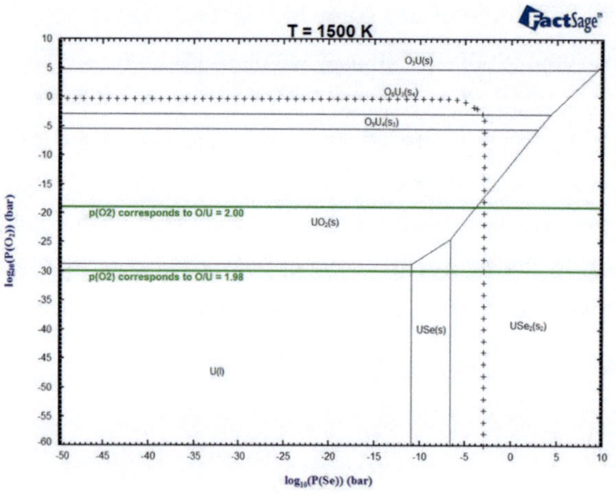

Figure 25.3: Predominance phase diagram of the Se-U-O system calculated for 1500 K.

As depicted in the figures above, there is an absence of the any selenium oxide compounds for the given temperatures and oxygen partial pressure ranges (stoichiometry). Even at high oxygen potentials, the selenide phases are less stable than their uranium counterparts. At operating temperatures of the LWR reactor and at low or moderate partial pressures of produced selenium the selenium species are present in the gas phase rather than in the solid state, as the sublimation point of the binary oxide is quite low and the monoxide exists exclusively in the gaseous form. With further increase of selenium pressure, which can be caused by accumulation of this fission product trapped within the UO_2 fuel matrix, selenium will react with uranium matrix forming selenides, first USe, followed by formation of USe_2. Furthermore, the calculation showed that oxidation of selenium in equilibrium with the fuel matrix is only feasible at unreasonably high partial pressures of oxygen, not indicative of operating conditions. The findings indicate that even during extreme conditions, the majority species will consist of uranium oxides or uranium selenides.

25.2 Acknowledgement

The research leading to these results has received funding from the European Union's European Atomic Energy Community's (Euratom) Seventh Framework Programme FP7/2007-2011 under grant agreement n° 295722 (FIRST-Nuclides project).

26 IRF modelling from spent fuel leaching experiments

Ignasi Casas[1], Alexandra Espriu [1], Daniel Serrano-Purroy[3], Albert Martínez Torrents[2], Aurora Martínez-Esparza[4] and Joan de Pablo[1,2]*

[1] *Chemical Engineering Department, UPC-Barcelona Tech (ES)*

[2] *Environmental Technology Area, Fundació CTM (ES)*

[3] *European Commission, JRC, ITU (EC)*

[4] *ENRESA (ES)*

** Corresponding author: ignasi.casas@upc.edu*

26.1 Abstract

The present work focuses on the Instant Release Fraction modeling of several leaching experiments performed in the framework of the First Nuclides Project. As a comparison, some other results obtained at the ITU are included as well. The fuels studied in this report are from both BWR (burnups 42, 53 and 54 Mwd/kg_U) and PWR (burnups 48, 52 and 60 Mwd/kg_U). Experimental results from powder samples from the outer part of the fuel pellets and cladded fuel segment samples are evaluated. Additional data obtained from samples taken at the inner part of the fuel as well as data obtained from dynamic experiments are not included in the present report.

U, Cs, Rb, Mo, Sr and Tc releases have been evaluated by using a kinetic model already described in previous reports. In the present work we focus on the very first release obtained from the different series of experiments. Results are presented in terms of normalized to inventory release rate and percentage of RN dissolved.

26.2 Introduction

From a theoretical point of view, radionuclides not located in the UO_2-matrix can release immediately when water contacts the fuel. Different estimations of the RN percentage corresponding to this instant release fraction (IRF) have been proposed by different authors. A good discussion of this can be found in Johnson et al. (2004) where for high burn-up (HBU) fuels best and pessimistic estimated values vary more than 50%. The contribution to IRF from gap and grain boundary is also unclear as it is pointed out in the same report.

These inconsistencies promoted an important experimental program on IRF from High Burn-up Fuel which started during the NF-Pro (Clarens et al., 2007) and follows with the present First Nuclides project. To obtain information about the location of the different RN's, powders from the center and the periphery of the fuel and cladded segments were leached in different media (González-Robles, 2011).

In a recent paper (Serrano-Purroy et al., 2012), an attempt to identify the RN's release contribution from the external and the internal part of a HBU pellet was discussed, in an effort to elucidate the effect of the high-burnup structure of the periphery. In the same sense, Roudil et al. (2007) give the contribution of the gap and grain boundaries to the IRF.

As it is conceptually summarized in Figure 26.1, the experimentally obtained availability of the different radionuclides to be released depends on the way the studied sample has been pretreated, because the accessibility to water will be different in each case. According to that, separate contributions could be differentiated, in a broad sense, based on the sample pretreatment. In the real case, the three situations may take place simultaneously as the water is reaching the different areas and the release of the RN's might show it. However, the time at which this may happen is something basically impossible to predict a priori.

In addition to the different contributions presented in Figure 26.1, a matrix contribution is going to be experimentally obtained. However, this fraction should not be considered in the evaluation of the experimental IRF. It is believed that values obtained at very short experimental times, like the ones presented in this work, have a negligible matrix contribution.

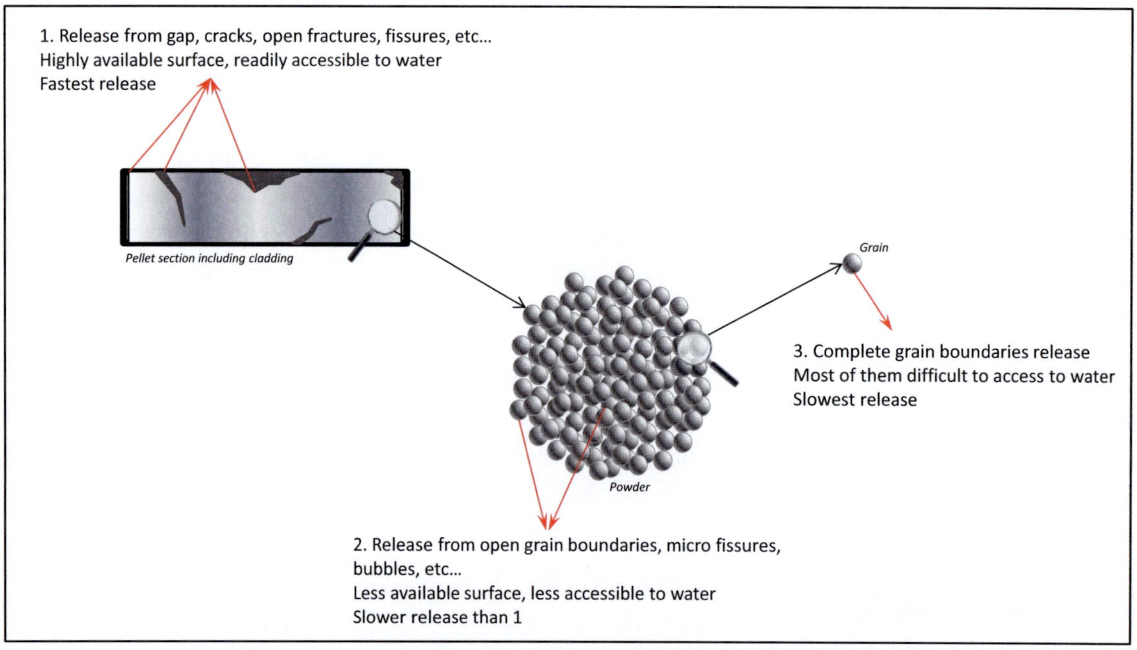

Figure 26.1: Conceptual description of the experimentally determined release of different radionuclides segregated from the matrix based on sample location, availability to water and sample pretreatment.

26.3 Kinetic model

The kinetic model for the instant SNF dissolution is developed as a sum of the contributions from different fuel phases. The dissolution of each different phase is considered as a first order kinetics.

$$m_{RN}(t)= \sum_{i=1}^{N} m_{RN,i\infty}\cdot(1-e^{-k_i\cdot t}) \qquad (26.1)$$

The main parameters of the equation are: $m_{RN}(t)$ which accounts for the total measured/predicted cumulative moles of the correspondent fission product released at a time *t*; $m_{RN,i\infty}$ which gives the total moles measured/predicted for the specified dissolving phase of the fission product and, finally, k_i which is the kinetic constant for the dissolution of that specific phase.

The model has a number of implicit assumptions and limitations. We assume the homogeneity of the different samples; in the case of powders this means that every single particle has the same size, composition and radionuclide distribution. In the case of cladded fuel samples, it is assumed that the fuel surface is regular. It is evident that these assumptions are not correct, although they may constitute a valid approximation for a short experimental time period.

The model includes a term that accounts for matrix dissolution, since its dissolution rate will determine the rate at which the different radionuclides located inside the matrix will be released. However, since the interest of the present work relays on the very initial dissolution, the matrix contribution is not considered in the results presented.

In Figure 26.2, an example of the result of the data fitting obtained by using Equation 26.1 is shown. The Cs data presented in the figure were obtained from Serrano-Purroy et al. (2012). In the example presented, three different contributions were considered. The one called "Model 1" is mostly assumed to account for a very rapid initial dissolution from oxidized phases and/or fine particles. The so-called "Model 3" would be related to the matrix contribution. Finally, "Model 2" is identified as the fraction of Cs that is being dissolved independently of U. This is the contribution to be considered as rapid or instant release fraction and it is the one that is presented in the data summarized below in the results section. The full line "Cs model" is the sum of the three contributions.

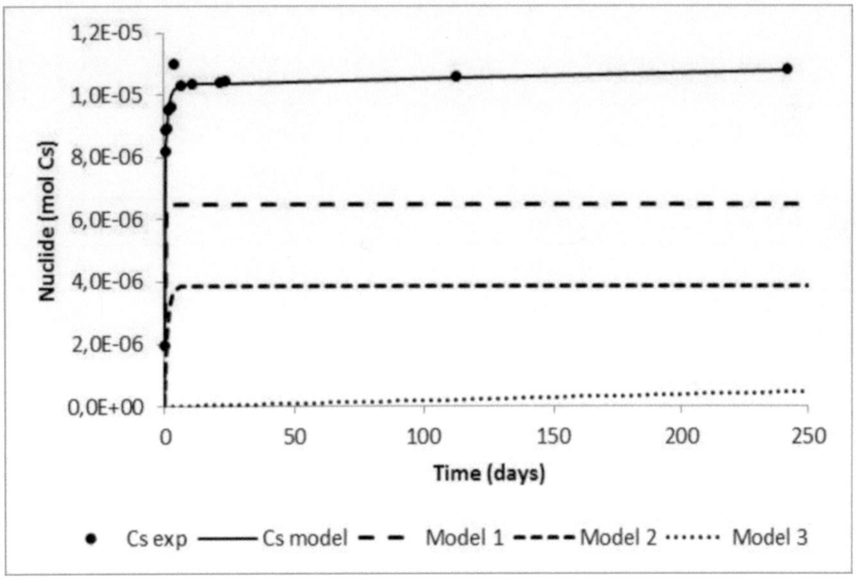

Figure 26.2: Example of Cesium fitting from static experiments using a cladded fuel segment (González-Robles, 2011). See explanations of the different models in the text.

26.4 Spent Fuel samples

The characteristics and preparation of the spent fuel samples that we have used in the modeling exercise are given in detail elsewhere (González-Robles, 2011) as well as in previous reports (Clarens et al., 2007). In the present work, results obtained from cladded pellet samples and from powder samples from the periphery of the pellet (OUT) are included (Serrano-Purroy et al., 2012 and data not yet published).

26.5 Results

A modelling of different experimentally obtained radionuclide's release was performed in a similar way as it is explained above for Figure 26.2. The results obtained for the selected radionuclides and for the "Model 2" contributionare summarized in **Table 26.1** to 6. In these tables, results are presented for each fuel sample studied and data from cladded pellet samples as well as from OUT powder samples are given.

The radionuclides presented are a selection based on radionuclides that are of interest for the determination of the IRF. Uranium is included as a comparison. The dissolution rates are normalized with respect to the sample inventory and are given in mol·day^{-1}. FIAP data refer to the total inventory dissolved at the end of the experimental time, which in all cases refers to the initial days (see Figure 26.2).

Tables 26.1-26.6: Summary of selected radionuclide determinations for the fuel samples studied. Normalized rate values account for the initial dissolution rate normalized to the radionuclide content in the inventory, whereas FIAP accounts for the total percentage of radionuclide dissolved in the corresponding experimental time. Pellet refers to cladded fuel samples whereas Powder (OUT) refers to the experiments performed with powdered samples from the pellet periphery.

Table 26.1	**Pellet 60 BURNUP PWR**		**Powder (OUT) 60 BURNUP PWR**	
	Normalized rate	FIAP	Normalized rate	FIAP
U	4.6E-07	0.003	9.2E-07	0.8
Tc	2.5E-06	0.014	4.7E-07	1.0
Mo	2.7E-06	0.014	1.5E-06	1.1
Sr	1.3E-05	0.064	1.2E-06	2.1
Rb	3.6E-05	0.21	1.1E-06	3.2
Cs	7.7E-04	4.3	1.6E-06	4.5
Table 26.2	**Pellet 54 BURNUP BWR**		**Powder (OUT) 54 BURNUP BWR**	
	Normalized rate	FIAP	Normalized rate	FIAP
U	5.7E-09	0.0008	1.2E-08	0.1
Tc	3.4E-10	0.00001	4.5E-08	0.3
Mo	3.4E-07	0.013	4.7E-08	0.3
Sr	2.4E-07	0.006	6.7E-08	0.4
Rb	1.6E-07	0.004	7.7E-08	0.5
Cs	7.1E-06	0.2	8.4E-08	0.6

Table 26.3	Pellet 53 BURNUP BWR		Powder (OUT) 53 BURNUP BWR	
	Normalized rate	FIAP	Normalized rate	FIAP
U	3.1E-08	0.002	4.9E-07	0.3
Tc	3.8E-08	0.003	7.1E-08	0.1
Mo	6.1E-08	0.006	4.9E-07	1.2
Sr	4.1E-07	0.02	3.6E-07	1.0
Rb	6.7E-07	0.05	3.4E-06	3.3
Cs	4.1E-06	0.4	2.5E-06	4.7

Table 26.4	Pellet 52 BURNUP PWR		Powder (OUT) 52 BURNUP PWR	
	Normalized rate	FIAP	Normalized rate	FIAP
U	6.1E-07	0.02	3.2E-07	0.2
Tc	3.1E-07	0.01	1.2E-07	0.1
Mo	1.2E-06	0.05	2.4E-07	0.3
Sr	2.5E-06	0.08	2.1E-06	1.7
Rb	9.2E-06	0.9	2.2E-06	2.2
Cs	5.7E-05	3.2	1.7E-06	2.8

Table 26.5	Pellet 48 BURNUP PWR		Powder (OUT) 48 BURNUP PWR	
	Normalized rate	FIAP	Normalized rate	FIAP
U	9.9E-07	0.007	5.0E-07	0.4
Tc	2.8E-06	0.02	3.9E-07	0.3
Mo	1.5E-06	0.008	4.8E-07	0.5
Sr	5.8E-06	0.04	1.0E-06	0.8
Rb	3.3E-04	1.8	1.0E-06	1.8
Cs	1.4E-03	7.0	1.3E-06	4.8

Table 26.6	Pellet 42 BURNUP BWR		Powder (OUT) 42 BURNUP BWR	
	Normalized rate	FIAP	Normalized rate	FIAP
U	1.1E-08	0.003	2.0E-07	0.8
Tc	9.5E-10	0.001	3.0E-07	1.3
Mo	2.6E-07	0.05	3.6E-07	1.5
Sr	7.5E-08	0.01	3.6E-07	1.5
Rb	1.5E-07	0.03	3.0E-07	1.3
Cs	6.0E-07	0.1	4.6E-07	2.0

In Figure 26.3 to 8, a comparison is made for each radionuclide and for the different samples studied. Bars are used to represent the initial normalized release rates in mol/day (to be read on left X-axe) and dots correspond to the percentage dissolved in the corresponding experimental time (to be read on right X-axe). Full bars and dots correspond to pellet results (either for PWR or BWR) while open bars and dots refer to powder results.

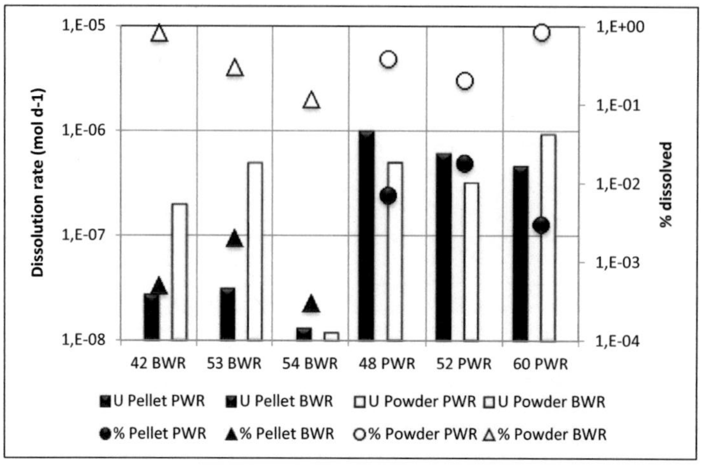

Figure 26.3: Results for U for the different fuel samples studied.

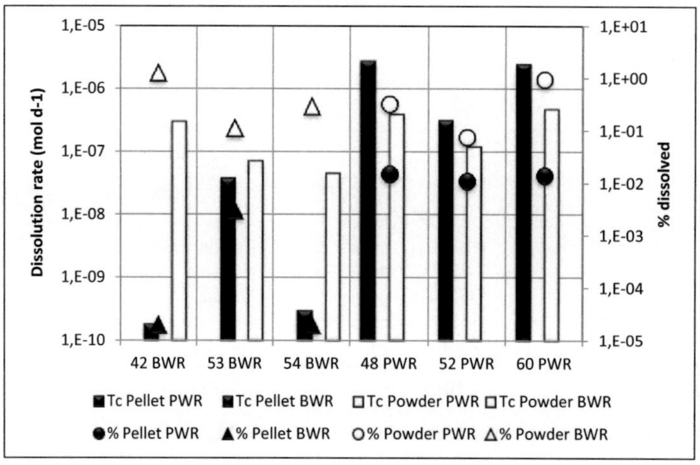

Figure 26.4: Results for Tc for the different fuel samples studied.

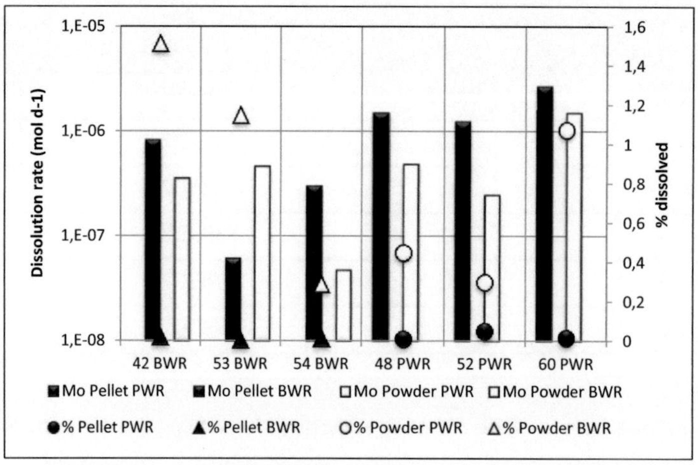

Figure 26.5: Results for Mo for the different fuel samples studied.

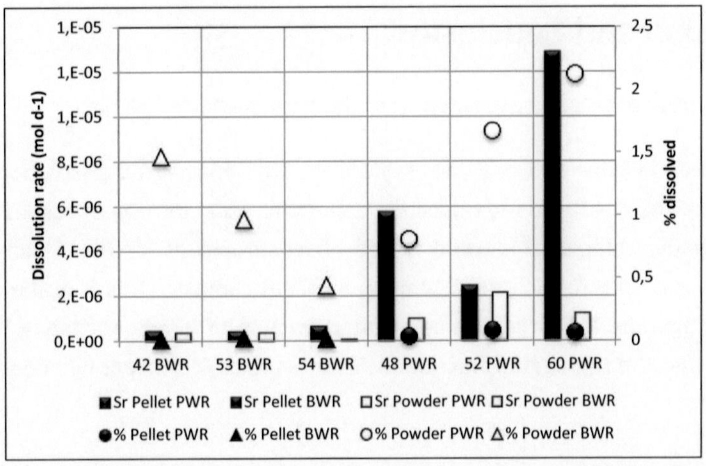

Figure 26.6: Results for Sr for the different fuel samples studied.

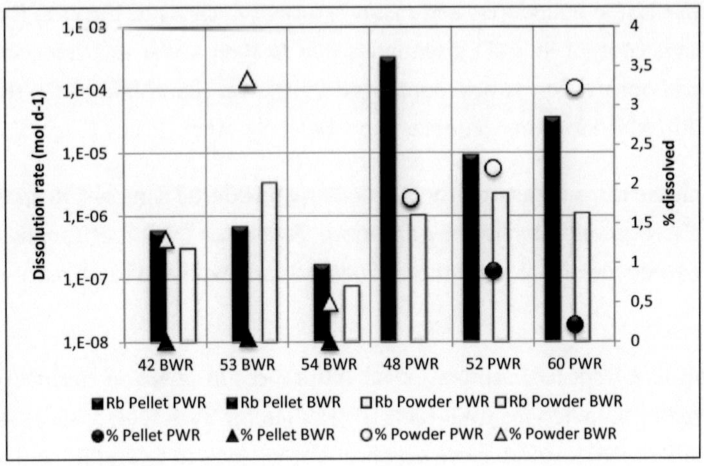

Figure 26.7: Results for Rb for the different fuel samples studied.

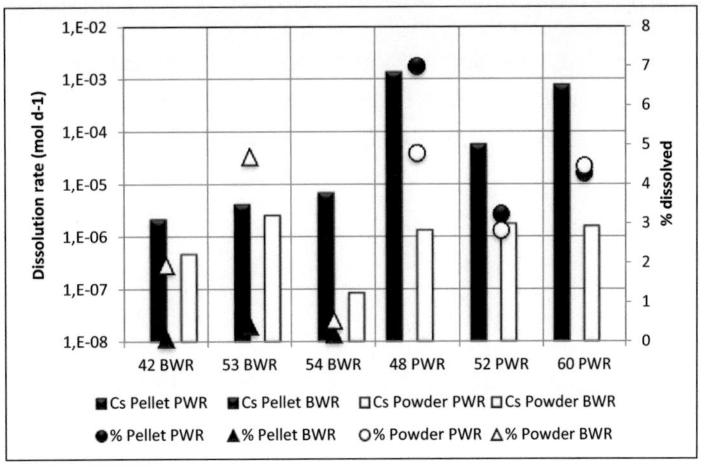

Figure 26.8: Results for Cs for the different fuel samples studied.

26.6 Discussion and Conclusions

Some interesting observations can be extracted from the data presented above.

From **Table 26.1** to 6 and Figure 26.3 to 8, it is evident that the normalized dissolution rate of Cs is always higher than for the rest of measured radionuclides, especially when looking at results obtained from cladded samples. For powder samples this trend is also observed except for Rb, which shows more similar release rates. This is also true for Sr rates obtained for PWR samples. This is assumed to be due to the degree of matrix segregation. The larger the segregated fraction of a radionuclide is, the larger is the total release. We assume that the segregated fraction is instantaneously released, when sufficient water gets in contact.

In general, the percentage of radionuclide released is higher for powdered than for cladded samples, except in the case of Cs in PWR samples. In those cases similar values are obtained. Considering the radionuclides distribution described in Figure 26.1, it can be assumed that for Cs its content in gaps and fractures is similar than its content in grain boundaries and micro fissures, which is not the case for the rest of radionuclides, that show greater content in grain boundaries due to their lower migration once segregated from the matrix. This general observation is also supported by the results obtained for the other high mobile radionuclide studied, Rb, which for some samples show behaviors that closely follow those of Cs.

With respect to the release rates measured for cladded and powdered samples, in general those values are larger for powdered PWR samples, while they are more similar for BWR fuel samples. Differences, again, become more evident when looking at the radionuclides that are expected to be more segregated (especially for Cs).

When comparing PWR and BWR fuel samples, clear differences in terms of normalized release rates are observed, being the values obtained for PWR fuels larger than for BWR fuels (clearly seen in Figure 26.3 to 8). Once again, such differences become more significant as we look to Cs and Rb and to some extent to Sr, which are the ones expected to be segregated from the matrix in higher degrees. When looking at the percentages dissolved, they seem to be higher for pellets (higher availability to water) than for powders, except for Cs and PWR fuels, as it was already commented above.

Being Cs identified, as expected, as an important contribution to the IRF, some interesting observations can be also obtained when we compare its behavior with respect to the rest of radionuclides. For this reason, we prepared Table 7, were the ratio between the normalized release rate of Cs to the rest of radionuclides is presented.

Table 26.7: Ratio between the initial normalized dissolution rates of Cs to the rest of radionuclides and for the different fuels and sample pretreatment.

PELLET					
	Cs/U	**Cs/Tc**	**Cs/Mo**	**Cs/Sr**	**Cs/Rb**
42 BWR	78	11657	2.6	4.9	3.6
48 PWR	1370	491	912	234	4.1
52 PWR	94	182	47	23	6.2
53 BWR	132	108	67	10	6.1
54 BWR	525	23120	23	11	41
60 PWR	1689	315	290	60	22
POWDER					
	Cs/U	**Cs/Tc**	**Cs/Mo**	**Cs/Sr**	**Cs/Rb**
42 BWR	2.3	1.5	1.3	1.3	1.5
48 PWR	2.7	3.4	2.7	1.3	1.3
52 PWR	5.4	14.5	7.1	0.8	0.8
53 BWR	5.1	35.4	5.5	7.0	0.7
54 BWR	7.1	1.9	1.8	1.3	1.1
60 PWR	1.7	3.3	1.0	1.3	1.4

Some clear tendencies that we observe in Table 7 are discussed as follows. Among radionuclides: The decrease of the ratio between the normalized release rates might indicate that Tc and Mo are clearly less segregated than Sr and Rb. This is especially evident for the results obtained from cladded samples and for all fuel samples, with a very few exceptions. It is important to notice here that the so-called 48 PWR sample show results that, in many cases, are very different than results shown by the rest of the fuels. This is also true for some of the tendencies described below. This behavior is attributed to a very high irradiation history of that sample, which gave a value for the fission release fraction of 15%, very high for such burnup.

When looking at the ratios between radionuclides obtained for the powdered fuel samples, the results are completely different. Although some tendencies can be observed, values are in general much lower than the values obtained for cladded samples. At the same time, they are more comparable among themselves.

We believe that, again, this is an indication of both, the degree of segregation of the radionuclides as well as their location. The lower ratios as well as the similarity among them, seems to indicate that for powdered samples, as expected, the segregated radionuclides are mostly found in grain boundaries, more difficult to access by the water. We believe that in this case what we measure is not indicative of their rate of dissolution. Instead, what we assume we measure is the rate of water penetration, which becomes the rate limiting step.

On the contrary, cladded samples have surfaces much more accessible to water (gap, fractures, fisures, etc...) so the fractions of radionuclides found there are released much more rapidly than in the case of powdered samples.

It is worth to notice that, with the exception of the fuel 48 PWR commented above, a clear tendency is observed for cladded samples with respect to burnup. There seems to be an increase of the values presented in Table VII as the burnup increases. The values corresponding to the ratio between Cs and U are perhaps the more evident, clearly indicating an increase of enrichment of Cs in the gap and macro fractures as the burnup is increased. Such a tendency is not so clearly observed for powdered samples. In the work of Roudil et al. (2007) performed with samples powdered to the size of grains, they reached similar conclusions, that is, that no clear dependence with burnup was observed.

Some of the more erratic values presented in this work are observed for Tc. We believe that this can be attributed to the fact that, even if this radionuclide segregates from the matrix, it has a large tendency to form insoluble phases, as it was already observed in experiments performed many years ago by Thomas and Guenther (1988) and Gray and Strachap (1991).

26.7 Acknowledgement

The research leading to these results has received funding from the European Union's European Atomic Energy Community's (Euratom) Seventh Framework Programme FP7/2007-2011 under grant agreement n° 295722 (FIRST-Nuclides project).

26.8 References

Clarens, F., Serrano-Purroy, D., Martínez-Esparza, A., Glatz, J-P., de Pablo, J., Wegen, D., Christiansen, B., Casas, I., González-Robles, E., Gago, J.A. (2007). RN Fractional Release of Spent Fuel as a Function of Burn-Up: Effect of RIM. NF-Pro Workshop, October, 2007.

González-Robles, E. (2011). Study of Radionuclide Release in Commercials UO2 Spent Nuclear Fuels. Effect of Burn-up and High Burn-up Structure. PhD Thesis, UPC-Barcelona Tech.

Gray, W. J. and Strachan, D. M. (1990). UO2. Matrix Dissolution Rates and Grain Boundary Inventories of Cs, Sr, and Tc in Spent Lwr Fuel. MRS Proceedings, 212, 205, doi:10.1557/PROC-212-205.

Johnson, L., Ferry, C., Poinssot, C., Lovera, P. (2004). Estimates of the Instant Release Fraction for UO2 and MOX fuels at t=0. Nagra Technical Report, TR 04-08.

Roudil, D., Jégou, C., Broudic, V., Muzeau, B., Peuget, S., Deschanels, X. (2007). Gap and Grain Boundaries Inventories from Pressurized Water Reactors Spent Fuels. Journal of Nuclear Materials, 362, 411-415.

Serrano-Purroy, D., Clarens, F., González-Robles, E., Glatz, J., Wegen, D., de Pablo, J., Casas, I., Giménez, J., Martínez-Esparza, A. (2012). Instant Release Fraction and Matrix Release of High Burn-Up UO2 Spent Nuclear Fuel: Effect of High Burn-Up Structure and Leaching Solution Comosition. Journal of Nuclear Materials, 427, 249-258.

Thomas, L. E. and Guenther, R. J. (1988). Characterization of Low-Gas-Release LHR Fuels by Transmission Electron Microscopy. MRS Proceedings, 127, 293, doi:10.1557/PROC-127-293.

Training and mobility measures

27 α-radiolysis under alkaline conditions in 0.05 and 5.0 molar nacl.

A. Martínez-Torrents, E. González-Robles, E. Bohnert, I. Casas, J. de Pablo, V. Metz

An improvement on the knowledge of the initial dissolution behaviour of high burn-up Spent Nuclear Fuel is one of the objectives of the FIRST-Nuclides collaborative project.

In workpackage 2, the effect of α-radiolysis on UO2(s) grain dissolution was simulated by the CNRS, using non- irradiated UO2 leached in pure water under 64.7 MeV 4He2+ irradiation between 0 and 600 Gy/h.

Following the same trend and complementing these studies our project studied the effect of alpha-radiolysis on UO2 grain /matrix dissolution, using 238Pu as alpha-emitter (5.5 MeV), under alkaline conditions (pH 12) in 0.05 and 5 molar NaCl.

The effects of alpha-radiolysis were determined, on one hand, through the generation of radiolytic products: H2, O2, HClO and H2O2, and on the other hand from the dissolution of both U and Pu. The studies were focused on the effect produced by different dose rates, different ionic strength as well as varying the location of the alpha- emitters (either into the pellets or dissolved in solution).

Regarding the O2 and H2 production neither the location of the alpha-emitters nor the ionic strength had any effect on the gas formation. The gas formation results were in well agreement with those found in the bibliography. The ionic strength is a key factor for the formation of HClO or H2O2. At high ionic strength only the HClO formation is observed, while at low ionic strength only the H2O2 formation is observed. Higher dose rates increases the formation of radiolysis products and the dissolution of U and Pu.

The experimental data regarding the formation of O2 and H2 were successfully fitted using the Macksima-Chemist software. The model proved to be sensitive to the changes in most of the G-values but robust to the changes in the kinetic constants.

28 First-nuclides mobility program summary report

Péter Szabó PhD student

Institution: MTA-EK, Budapest (Hungarian Academy of Sciences - Centre for Energy Research) Duration of Mobility measure: 8-19. July 2013
Host Institution: KIT-INE, Karlsruhe (Karlsruhe Institute of Technology - Institute for Nuclear Waste Disposal)
Supervisors: Dr. Volker Metz, Dr. Ernesto Gonzólez-Robles Corrales

28.1 Summary

During the internship at the KIT-INE Péter Szabó was instructed to work with spent nuclear fuel in KIT-INE's shielded box line. The training included among the others:

- handling of manipulators (in a mock manipulator station)
- Raman spectroscopy & optical microscopy of UO2 samples
- handling/working in inert gas glovebox in argon atmosphere (O2, CO2 < 1 ppm), to transfer samples into the glovebox by means of vacuum double sided locking system, instructions on bagging-out of samples
- sample preparation for 14C measurements by liquid scintillation counting
- uranyl solubility investigations' sample preparations for ICP-MS
- geochemical introduction to modeling
- gas sampling in a shielded box and sample preparation for gas mass spectrometry measurement
- laser fluorescence measurements
- visiting the ANKA abbreviation for "Angströmquelle Karlsruhe"), a synchrotron light source facility at the KIT
- lectures on vitrification, final disposal

29 SR sorption onto goethite and zirconium oxide in high ionic strength systems. experimental and preliminary modelling results

David García[1], *Johannes Lützenkirchen*[2], *Nicolas Finck*[2], *Maximilien Huguenel*[3], *Léa Calmels*[3], *Vladimir Petrov*[4], *Mireia Grivé*[1] *and Lara Duro*[1]

[1] *Amphos 21*
[2] *KIT-INE*
[3] *ECPM-Strasbourg*
[4] *Lomonosov University*

29.1 Abstract

In this work the sorption behaviour of Sr (Sr-90 is part of the elements within the high fast/Instant Release Fraction) onto zircalloy and/or iron materials (two of the most common materials studied for building up the canister and the cladding of the Spent Nuclear Fuel), has been studied. ZrO2(s) has been used as a zircalloy template, and goethite, FeOOH(s), as a representative of Fe3+ solid phases (iron corrosion phases). Both solid phases have been characterized by XRD, XPS and with respect to their specific surface areas. The two solids were subjected to fast acid-base titrations at different NaCl concentrations (from 0.1 to 5M). For each background electrolyte concentration, two series of Sr-uptake experiments were conducted. Sr sorption onto the two solids was studied at the same NaCl concentrations used for the acid-base titrations at two different levels 1·10-5 and 5·10-5 M Sr (Figure 29.1).

A Surface Complexation Model (SCM) was fitted to the experimental data, following a coupled Pitzer/surface complexation approach. The Pitzer model was applied to aqueous species. A Basic Stern Model was used for interfacial electrostatics of the system, which includes ion-specific effects (as does Pitzer); a simple 1-pKa model was applied (generic surface species, such as >SOH-1/2). Parameter fitting was done using the general parameter estimation software UCODE (Poeter and Hill, 1998) coupled to a modified version of FITEQL2 (Westall, 1982).The combined approach describes the full set of data (i.e. beyond those shown in Figure 29.1) reasonably well.

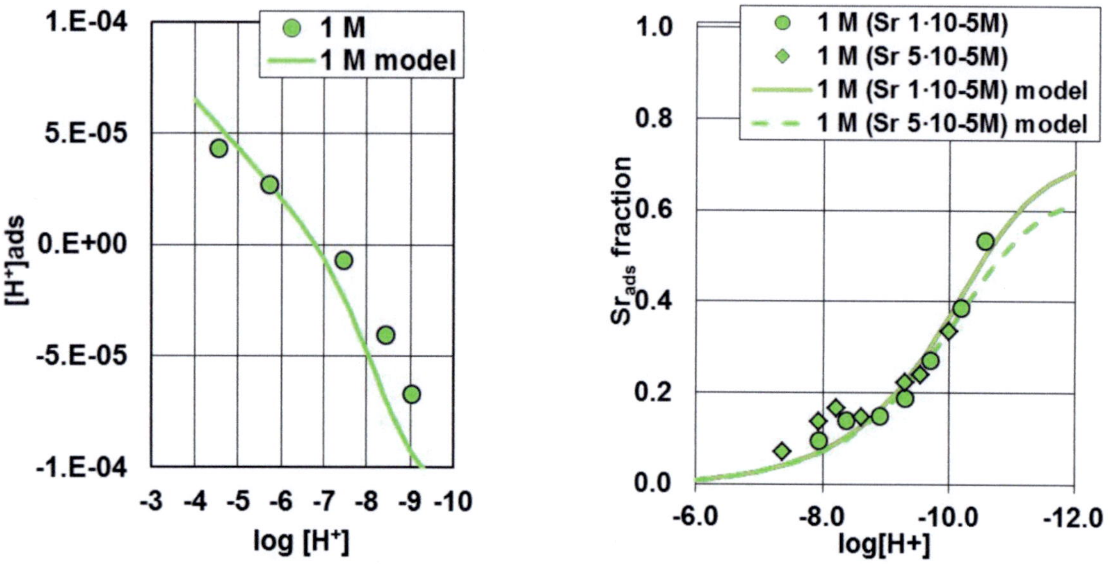

a) b)

Figure 29.1: a) H+ adsorption onto FeOOH(s) surface as a function of pH in 1M NaCl system, b) fraction of Sr adsorbed onto FeOOH(s) in 1M NaCl system (solid line 1·10-5 M Sr), dashed line 5·10-5 M Sr).

29.2 References

Poeter, E.P., Hill, M.C. (1998). Documentation of UCODE: A Computer Code for Universal Inverse Modeling (Vol. 98, No. 4080). DIANE Publishing.

Westall, J.C. (1982). FITEQL: A Computer Program for Determination of Chemical Equilibrium Constants from Experimental Data. Department of Chemistry, Oregon State University.

Additional presentations

30 XAFS investigation of a genuine HAWC glass fragment sampled from the Karlsruhe vitrification plant (VEK)

K. Dardenne[1], E. González-Robles[1], J. Rothe[1], N. Müller[1], G. Christill[2], B. Kienzler[1], V. Metz[1], G. Roth[1] and H. Geckeis[1]

[1]*Karlsruhe Institute of Technology, Institute for Nuclear Waste Disposal (KIT-INE) (DE)*
[2]*Sicherheit und Umwelt (KIT-SUM) (DE)*

30.1 Introduction

The Karlsruhe Reprocessing Plant (WAK) was operated from 1971 to 1991 as a pilot facility for reprocessing of spent nuclear fuels from German research reactors and commercial power plants. Reprocessing activities resulted in ~60 m3 of highly active waste concentrates (HAWC) stored on-site in liquid form. An important step in the current decommissioning and dismantling of the WAK was the HAWC vitrification in the Karlsruhe Vitrification Plant (VEK) constructed close to the HAWC storage facilities (Roth and Weisenburger, 2000; Fleisch et al., 2010). A total of 2.2 g from sections of genuine HAWC glass rods were retained during the vitrification process and transferred to the INE shielded box line for later glass product characterization.

30.2 Experimental

In 2013 a mm sized fragment with a contact dose rate of ~590 μSv/h (Figure 30.1, left) was selected and mounted in a specially designed sample holder (Figure 30.1, right) for pilot XAS/XRF investigations at the INE-Beamline at ANKA. The experiment aimed at elucidating the potential of direct actinide / radionuclide speciation (with an emphasis on the fission products Se and Tc, which are regarded as contributing to the initially mobile fraction of the radionuclide inventory in spent nuclear fuel and waste glasses following corrosive waste degradation in a geological repository) in highly active nuclear materials and assessing the possible influence of the γ-radiation field surrounding highly active samples on the XAS detection electronics.

Figure 30.1: HAWC glass fragment (arrow) selected for investigation at INE-Beamline (left, scale in mm); glass fragment inside a double containment mounted in the sample holder (right) prior to the transport to ANKA.

30.3 Results

While the influence of the γ-radiation field turned out to be negligible, initial radionuclide speciation studies by XAS were most promising. Exemplarily, Figure 30.2 depicts normalized Se K- (left) and Tc K-edge XANES measurements (right) of the HAWC glass fragment and corresponding Se and Tc reference samples, respectively. Edge position and simple spectral fingerprint analysis point to the presence of Se in the glass as selenite oxoanion (SeO_3) as in Na_2SeO_3. Pronounced dampening of the near-edge fine structure indicates dispersion of the SeO_3^{2-} oxoanions in the glass matrix, where the crystalline ordering such as in the Na_2SeO_3 reference is lost.

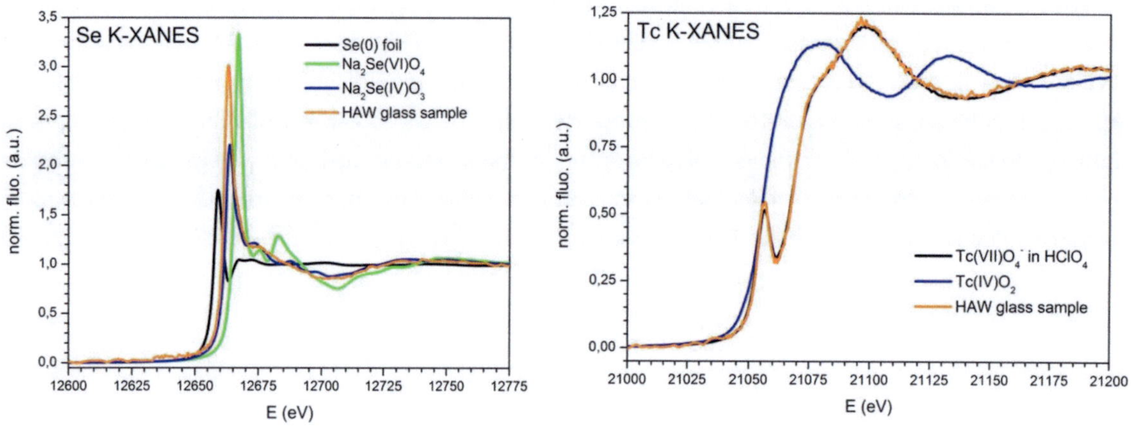

Figure 30.2: Normalized Se K- (left) and Tc K-XANES spectra of the HAWC glass fragment and corresponding reference spectra.

Tc is dispersed as pertechnetate oxoanion (TcO_4^-) in the glass matrix as in the aqueous Tc(VII)/$HClO_4$ reference sample, which is unequivocally proven by the edge shift relative to Tc(IV)O_2 and the pronounced pre-edge feature at ~21,056 eV indicative of tetrahedral oxygen coordination.

In addition to Se and Tc discussed above and the plethora of elements contained in the glass fragment, the focus of these initial XAFS investigations was on the possibility for direct actinide speciation by recording corresponding L3-edge XAFS data. Th was not expected to be contained in the glass and, hence, was not detectable in XAFS scans across the Th L3-edge. Only a very tiny absorption signature was detected by scanning the energy across the Cm L3-edge. This signal was not sufficient to obtain any useful XANES data. This is in full agreement with the known composition of the HAWC oxide residue [1], where the Cm content is expected to reach about 1/10 of the Pu content, which, in turn, amounts to ~1/30 of the U content of the HAWC glass. The concentrations of Am and Np are expected to reach twice the value obtained for Pu. For all actinide elements besides Th and Cm (U, Np, Pu, Am) as well as for Zr registration of high quality XANES data was possible - for some of these elements even reasonable EXAFS spectra were recorded. The XANES analysis reveals that uranium is present as U(VI), neptunium as Np(V), plutonium as Pu(IV), americium as Am(III) and zirconium as Zr(IV) in the HAWC glass, as expected from the oxidizing conditions during the HAWC vitrification process (Dardanne et al., 2014).

30.4 Acknowledgements

A. Bauer, Ch. Marquardt (both KIT-INE) and Th. Hoffmann (KIT-SUM) are acknowledged for invaluable technical and logistic support during preparation of these experiments. ANKA is acknowledged for the provision of beamtime.

30.5 References

Roth, G., Weisenburger, S., (2000). Vitrification of High-Level Liquid Waste: Glass Chemistry, Process Chemistry and Process Technology. Nuclear Engineering and Design, 202, 197-207.

Fleisch, F.J., Weishaupt, M., Roth, G., Grünewald, W., Tobie, W., Weisenburger, S. (2010). Verglasungsanlage VEK - Erfolgreiche heiße Inbetriebsetzung und erste Betriebserfahrungen. In: Jahrestagung Kerntechnik, Berlin, Germany.

Dardenne, K., González-Robles, E., Rothe, J., Müller, N., Christill, G., Lemmer, D., Praetorius, R., Kienzler, B., Metz, V., Roth, G., Geckeis, H. (2014). XAS and XRF Investigation of a Genuine HAWC Glass Fragment Obtained from The Karlsruhe Vitrification Plant (VEK). Submitted to Journal of Nuclear Materials.

31 First nuclides transport modelling from spent fuel to the biosphere and uncertainties propagation

[1]B. Pastina and [2]T. Hjerpe

[1] *Posiva Oy (FI)*
[2] *Facilia AB (SE)*

At the end of 2012, Posiva Oy presented the license application to construct a spent fuel repository in Finland. The license application comprises several components, including a safety case addressing the long-term safety of the repository after its closure (Posiva, 2012). One of the aims of the safety case is to identify the scenarios in which radionuclides could be released from the canisters and to model the transport of these radionuclides from the spent fuel into the surface environment. First-nuclides are used in the source term for the RN transport modelling chain. The modelling chain consists of a source term, a model for the radionuclide transport in the near field, one for the geosphere and one for the surface environment. The assessment endpoints are annual activity releases, annual doses to humans, absorbed dose rates to plants and animals and other complementary indicators. Key supporting models are the groundwater flow model and the terrain and ecosystem development model for the surface environment. Radionuclide releases are converted into a radiation dose (or radiological risk) to human and other biota. The doses/risks are then compared to the regulatory dose criteria to determine whether the spent fuel repository complies with long-term safety standards even in the unlikely scenarios of early radionuclide releases. The Finnish regulatory agency set a dose-based compliance criterion (0.1 mSv/year) and Posiva's assessment is mainly based on a deterministic approach based on scenarios and their analysis.

The source term for the radionuclide release model consists of four components: (1) the UO2 matrix, (2) the gaps and grain boundaries (i.e. the Instant Release Fraction, IRF), (3) the cladding and other Zr-based alloys, (4) metallic structures of the fuel assembly (Posiva, 2013). There are currently important uncertainties in the releases from the gap and grain boundaries that are managed by introducing a high level of conservatism of the source term. For example, the fraction of radionuclides susceptible to be located in the gaps and/or grain boundaries is set to be released instantaneously at the time of the initial canister release (hence the term Instant Release Fraction, IRF).

The three key IRF radionuclides that contribute to the doses to the biosphere are 14C, 36Cl and 129I. Other radionuclides are included in the IRF are: ^{135}Cs, ^{137}Cs, ^{10}Be, ^{107}Pd, ^{90}Sr, ^{99}Tc, ^{79}Se and ^{126}Sn.

The main results from the radionuclide modelling chain show that the total normalised release rates from the near field and to the biosphere are controlled mainly by 14C, 36Cl and 129I, three radionuclides that present weak or no sorption on buffer, backfill and rock matrix in most realisations. 14C (in non-sorbing, organic form) is the dominant radionuclide in the realisations with highest peak total normalised release rates. The IRFs in the fuel matrix have a significant influence on the peak release rates from the near field and to the biosphere only for 129I, 36Cl and 135Cs. This effect is much greater in the case of an enlarging defect that in the case of a constant defect (1 mm). The release locations locations (a geosphere/biosphere interface issue) play a great role in the results: for some scenarios, a 3 orders of magnitude difference in peak dose was observed depending on the assumption of release location. This is mainly due to the evolution of the terrain and ecosystems as a function of time and on the assumptions on land use. Given a fixed source term, the uncertainty totally dominating the variation in the assessment outcomes is the location of the geosphere releases into the surface environment, i.e. the assumption of which canister location in the repository the canister with an assumed undetected penetrating defect is placed in. The difference in peak doses across the three analysed release locations (BS-RC, BS-LOC1 and BS-LOC2) is in the order of three to four magnitudes (see for example Figures 9-1 and 9-2 in POSIVA, 2012-10). The assessment results also indicate that the surface environment, and its development, combined with different release locations may have a great impact on the doses. An important source of uncertainty are the ratios of activity in foodstuff and the soil or water (concentration ratios). However, outcomes from the analysed calculation cases (details in POSIVA, 2012-10, Chapter 6) show that the uncertainties in features, events and processes related to human diet, land use and radionuclide transport in soils has a rather minor impact on the assessment outcome, compared to the uncertainties in the source term

For a dynamic biosphere, the impact of an increased release from the fuel on the dose is not linear since the surface environment is rapidly changing at the time scale of the peak releases (especially for the IRF). The doses depend not only on the peak release rate but also the release rate curve (timing and shape).

Of particular interest to the safety case are the uncertainties in the following

- Speciation and partitioning of the key IRF nuclides activation products and 14C
- Quantification of the IRF and correlation with fission gas release (FGR) measurements
- Correlation between the irradiation history (e.g., burnup, linear power rating) and the IRF

Concerning the last bullet, the importance of correlating spent fuel data (e.g. irradiation history) with measurements of FGR or IRF has been recognized in the previous studies leading to FIRST-Nuclides.

Within the FIRST-Nuclides project, spent fuel from one of the Finnish nuclear power plants is undergoing leaching measurements at Studsvik. The results from these tests will be used in the next safety case in support of the operational license application, due in 2020. Gathering additional data (e.g. short-term leaching, FGR) on well characterised fuel is recommended to reduce, if possible, the conservativeness of IRF assumptions in the safety case.

31.1 Acknowledgment

The authors thank Posiva Oy for sponsoring their participation to the workshop.

31.2 References

Posiva (2012). Safety Case for the Disposal of Spent Nuclear Fuel at Olkiluoto - Synthesis 2012. Posiva Technical Report, POSIVA 2012-12.

Posiva (2013). Safety Case for the Disposal of Spent Nuclear Fuel at Olkiluoto - Models and Data for the Repository System 2012. Posiva Technical Report, POSIVA 2013-01.

32 Corrosion behaviour of agr simfuels

[1]Nadya Rauff-Nisthar, [1]*Colin Boxall, [2]David Hambley, [3]Zoltan Hiezl, [4]Cristiano Padovani and 1Richard Wilbraham

[1] Lancaster University (UK)
[2] National Nuclear Laboratory (UK)
[3] Imperial College London (UK)
[4] Nuclear Decommissioning Authority (UK)
* Corresponding author: c.boxall@lancaster.ac.uk

32.1 Abstract

Pure UO_2 pellets and, for the first time, AGR SIMFUEL pellets simulating 25 (pellet interior) and 43 GWd/tU burn-up (rim structure) have been fabricated. Using combined electrochemical and Raman spectroscopic studies, we have investigated the effect of the AGR SIMFUEL dopants on the structure of the UO_2 crystal matrix and studied the effect of H_2O_2 on the SIMFUEL surface as a simulant for the effect of α radiolysis of water within intruding groundwater in the near field. Preliminary results of electrochemical studies of AGR SIMFUEL samples held in direct contact with the AGR fuel cladding material, 20/25/Nb stainless steel, in modified simplified groundwater (10 mmol/dm-3 NaCl, 2 mmol dm-3 NaHCO3, pH 6.8) indicate that the open circuit potential of the coupled system is (i) negative of that adopted by the isolated SIMFUEL and (ii) positive of that adopted by the isolated cladding in electrolytes of the same composition. This implies that the SIMFUEL is protected against corrosion at the expense of the steel cladding, an observation that is currently the subject of further work in our laboratories.

32.2 Introduction

The direct disposal of spent nuclear fuel into a geological repository is part of the nuclear waste policy of several mature nuclear states (USA, Sweden, Spain and Finland are prominent examples). The vast majority of this fuel is from light-water moderated reactors (LWR) and a significant amount of research has been carried out to support the direct disposal concept in terms of the physical and aqueous durability of the container, the cladding and the irradiated UO_2-based fuel.

In the UK, the majority of spent nuclear fuel (SNF) is from indigenous Advanced Gas-cooled Reactors (AGRs). AGRs, whilst using UO_2-based fuel, employ CO_2 as coolant and are graphite moderated. Further, the fuel assembly cladding is comprised of 20/25/Nb stainless steel rather than zircalloy as is the case in Pressurised Water Reactors (PWRs). Consequently, AGR fuel has unique characteristics that need to be evaluated before disposal in an expected anoxic UK geological repository.

It is expected that, at some point after repository closure, the disposal canisters within which the SNF is sealed will fail, allowing for ingress of groundwater. This may come into contact with the 20/25/Nb cladding and the spent UO_2 fuel pellets, initiating corrosion processes on either or both. At this failure point, anticipated as being many thousands of years into the future, many of the fission products will have decayed away, as will the associated ß/γ radiation field, leaving an α-field derived from long-lived α-emitting actinides. Alpha radiolysis of the invading groundwater as it comes into contact with the UO_2 fuel pin will generate hydrogen peroxide as one of several end species (Sattonnay et al., 2001).

Hydrogen peroxide concentrations in the near-field are expected to be in the region of 10 μmol/dm3 (Buck et al., 2012). This locally oxidising environment may convert U(IV) to U(VI), so allowing U to pass into solution, with the possibility of re-precipitation of oxidized U(VI) back onto the fuel pellet surface, generating secondary mineral phases such as schoepite. The peroxide minerals studtite (UO_2)O2(H2O)4 and metastudtite $(UO_2)O_2(H_2O)$ are stable in H_2O_2-bearing environments even at low (H_2O_2). Studtites have recently been found in Chernobyl lavas (Burakov et al., 1997), formation possibly occurring at the expense of more common transformation phases. As such, studtite layers may form a barrier for retarding SNF corrosion and could retain radionuclides either within their structure or by surface adsorption. Whether more or less corrosion obtains in the presence of the studtites is a generic research question for all UO_2-based SNF, not just AGR SNF, which needs to be explored.

Further, the 20/25/Nb cladding of AGR fuel may also show susceptibility to near field corrosion. The direct contact of the steel to the fuel pellet surface means that these putative steel corrosion reactions may couple with processes occurring on the UO_2 surface, so altering the corrosion behavior of both cladding and fuel pellet. To our knowledge, this coupling has been unstudied hitherto; we therefore present some preliminary results pertaining to this system – the first from an AGR SIMFUEL-20/25/Nb steel system in direct contact.

32.3 Experimental

UO_2 pellets and SIMFUEL pellets simulating 25 and 43 GWd/tU burn-up were fabricated at the UK National Nuclear Laboratory (Springfields, Preston, UK). SIMFUEL pellets were prepared by sintering powder precursors containing the dopant compositions given in Table I. Compositions were calculated using FISPIN nuclear inventory calculation code (developed by the National Nuclear Laboratory UK and distributed by AMEC (Burstall, 1979)) and are representative of a fuel pellet that has aged for 100 years after removal from the reactor.

Both the UO_2 and SIMFUEL pellets were prepared by first formulating a 60 g blend of each composition. These were milled overnight with a ZrO2 milling medium. In order to pre-compact the powder into granulates, a pressure of 75 MPa was applied. 0.2 wt.% zinc stearate was added as the binder and the material was slowly mixed for 5 minutes using a rotary mixer. The granulates were then pressed into green pellets by applying a pressure of 400 MPa using a uniaxial machine. All pellets were sintered at a heating rate of 5 °C/min to 300 °C, and then 15 °C/min to 1,730 °C. The sintering time was 300 minutes which was followed by cooling, with a cooling rate of 15 °C/min until room temperature had been reached. The furnace (supplied by Cambridge Vacuum Engineering) atmosphere contained 99.5 vol% H2 and 0.5 vol% CO2. The finished pellets were cut into slices ranging from ~1–3 mm thick using a Struers Accutom-2 precision cut-off machine with a Struers MOD13 diamond cut-off wheel.

Table 32.1: Compositions of AGR SIMFUEL pellets of different burn-ups (Hiezl et al., 2015).

Dopant	wt% 25 GWd/tU	wt% in 43 GWd/tU
UO2	95.705	92.748
Nd2O3	0.761	1.284
ZrO2	0.793	1.276
MoO3	0.614	1.027
RuO2	0.512	0.892
BaCo3	0.328	0.576
CeO2	0.297	0.499
PdO	0.195	0.425
Rh2O3	0.080	0.115
La2O3	0.156	0.256
SrO	0.081	0.126
Y2O3	0.095	0.149
CsCO3	0.311	0.495
TeO2	0.073	0.130

32.4 Results & Discussion

Raman microscopy reveals an increase in the lattice damage (500 – 700 cm-1) with increasing burn-up, as the UO_2 cubic fluorite lattice structure becomes more distressed and moves towards a tetragonal structure (Figure 32.1). Cyclic voltammetric studies of the AGR SIMFUELs reveal them to be more susceptible to electrochemical oxidation than PWR SIMFUELs; in particular the peak associated with the in-grain UO_2 to UO_2+x oxidation (onset -0.2 V vs SCE, Ep = +0.14 V) is ~3 x larger for AGR SIMFUELs than PWR SIMFUELs of similar simulated burnup. Importantly, analogous studies on PWR SIMFUELs have shown that UO_2+x locations act as preferential corrosion sites.

At (H_2O_2) = 10 µmol/dm3, close to that encountered in an α dominated near field, H_2O_2 generates an oxidative stress of 0.05 V vs SCE on 43 GWd/tU AGR SIMFUEL in simplified modified groundwater. This corresponds to the near top of the voltammetric wave for the in-grain UO_2 to UO_{2+x} process (Figure 32.2), suggesting that near field peroxide may be capable of inducing a substantial population of anodically active UO_{2+x} sites on AGR SNF and consequent localised site corrosion at the spent fuel surface.

The open circuit potentials (EOC) for UO_2 and AGR SIMFUELs show a logarithmic dependence on (H_2O_2) in modified simplified groundwater (10 mmol/dm3 NaCl, 2 mmol/dm3 NaHCO3, pH 6.8) (Figure 32.3). In contrast, the EOC of PWR SIMFUELs of ~0.1 V vs SCE in aqueous alkaline solution is near independent of (H_2O_2), an observation attributed to surface-catalysed H_2O_2 decomposition by non-stoichiometric U(IV)/U(V) sites. This suggests that, in contrast to the non-complexing alkaline conditions, carbonate complexation of U(VI) in groundwater simulant removes the H_2O_2 decomposition sites from the UO_2 surface so resulting in a larger net (H_2O_2) and thus oxidative stress being induced at the SIMFUEL surface.

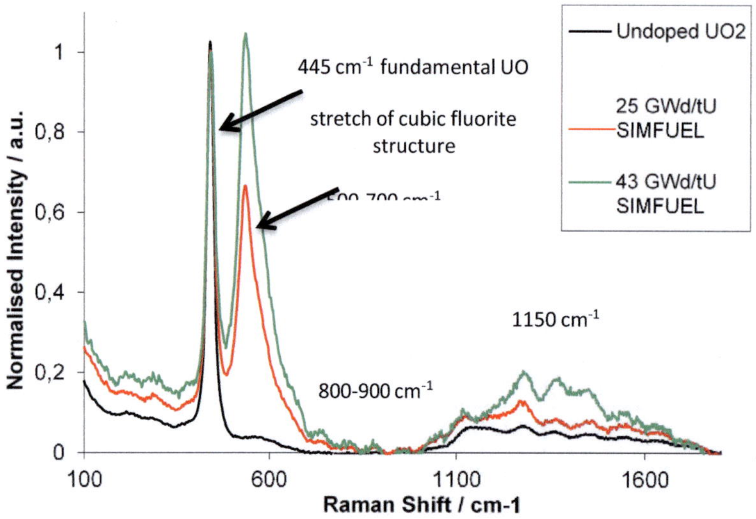

Figure 32.1: Effect of Fuel Burnup on Raman Spectra.

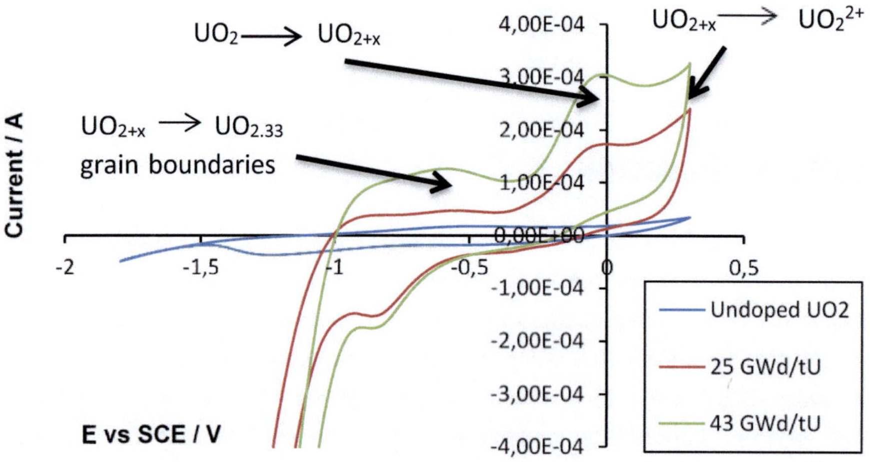

Figure 32.2: Cyclic Voltammetry of UO_2, 25 & 43 GWd /tU SIMFUELs, Ar sparged in 0.1 M Na2SO4.

Extended duration Raman studies on the oxidation of UO_2 on 25 GWd/tU SIMFUELs in 1 mmol/dm3 H_2O_2 in O_2-free deionised water (shown in part in Figure 32.4) reveal that, over a 1 week period, there is an increase in the intensity of the peak at 540 cm-1, associated with the formation of defect sites in the UO_2 fluorite lattice. The measured open circuit potential during this study has a value of +0.19 V vs SCE. The cyclic voltammetric data of Figure 32.2 indicates that this potential corresponds to the feature associated with in-grain oxidation of UO_2 to UO_{2+x}, an observation consistent with the evolution of the Raman spectra of Figure 32.4.

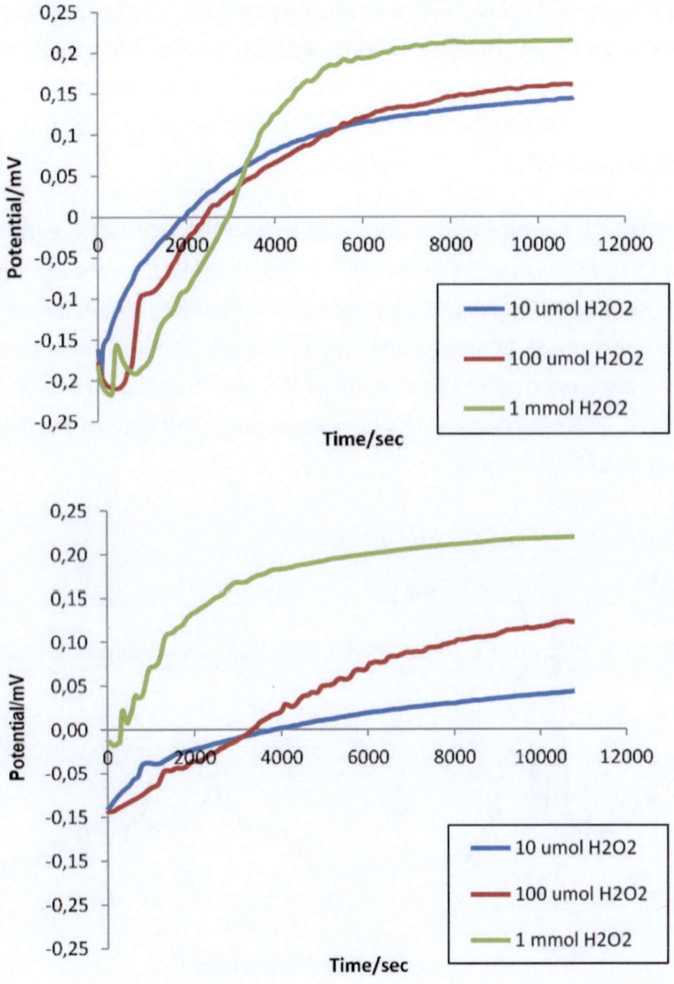

Figure 32.3: Open Circuit Potentials of UO_2 in anoxic modified simplified Groundwater (0.01 M NaCL & 0.002 M NaHCO3 and 43 GWd /tU SIMFUEL in anoxic modified simplified groundwater.

Shorter term studies on undoped UO_2 samples in 0.01 to 1 mmol/dm3 H_2O_2 solutions (not shown) also reveal the in-growth of a peak at ~610 cm-1, a feature that has been attributed to O sublattice distortions due to the generation of UO_{2+x} locations. This again indicates of the potential for near field concentration H_2O_2 to induce an increased susceptibility to corrosion for AGR SIMFUEL.

The ultimate product of H_2O_2 induced corrosion of either UO_2 or AGR SIMFUELs would be expected to be U(VI). The extended duration Raman studies on 25 GWd/tU SIMFUELs (Figure 32.4) show that a peak appears at 820 cm-1 after 5 hours exposure to 1 mmol/dm3 H_2O_2. This grows in intensity up to 50 hours exposure and has been assigned to metastudtite ($UO_2(O_2)$ $2H_2O$), a uranium peroxide secondary phase formed at the SIMFUEL surface as a result of the reaction between H_2O_2 and UO_2 - the latter necessarily generated as a result of above-expected H_2O_2–driven corrosion of the AGR SIMFUEL UO_2 matrix.

Peroxide aside, open circuit potential measurements on 43 GWd/tU simulated burnup AGR SIMFUELs samples in intimate contact with 20/25/Nb stainless steel and in modified simplified groundwater electrolyte show that they exhibit a mixed potential of ~-0.12 V vs SCE (Figure 32.5). To put this into context, the open circuit potentials for the steel and SIMFUEL in isolation from each other in electrolytes of the same composition were found to be -0.15 and -0.08 V vs SCE respectively. The former corresponds to a region of passivity in the 20/25/Nb linear sweep voltammogram, whilst the latter corresponds to the near peak of the oxidation wave for in-grain UO_2 to UO_{2+x} process in the associated voltammogram (Figure 32.2) – implying that AGR SIMFUEL corrosion may occur in the modified simplified groundwater even in the absence of a peroxide generated oxidative stress.

The mixed potential of -0.12 V recorded from the coupled SIMFUEL-20/25/Nb system then corresponds to decrease of 40 mV from the value recorded from the uncoupled SIMFUEL – indicating that the coupling may at least in part protect the UO_2 matrix against corrosion in the modified simplified groundwater. In contrast, -0.12 V corresponds to the onset of metastable pit formation in the linear sweep voltammogram of 20/25/Nb (not shown) – implying that the protection of the SIMFUEL is at the expense of the corrosion of the steel cladding material. This observation may have significant implications for the repository safety case and is currently the subject of further work.

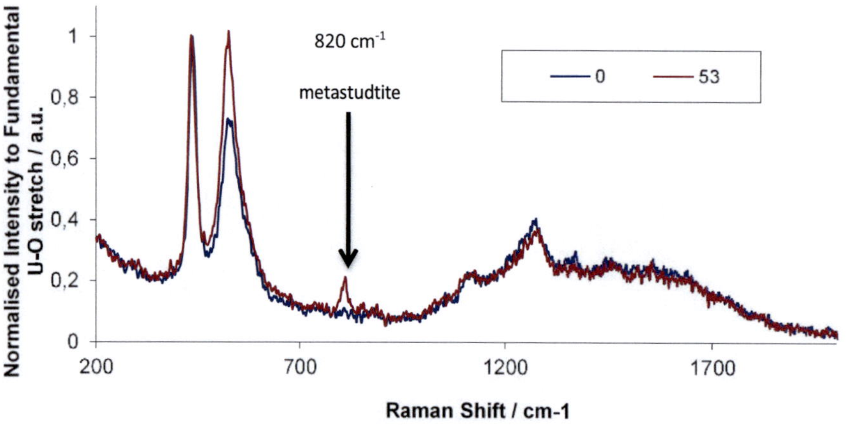

Figure 32.4: Raman Spectroscopy of a 25 GWd /tU SIMFUEL after 0 and 53 hours exposure to anoxic solution of 1mmol dm-3 H_2O_2, showing metastudtite formation at 820 cm-1 after 53 hrs exposure.

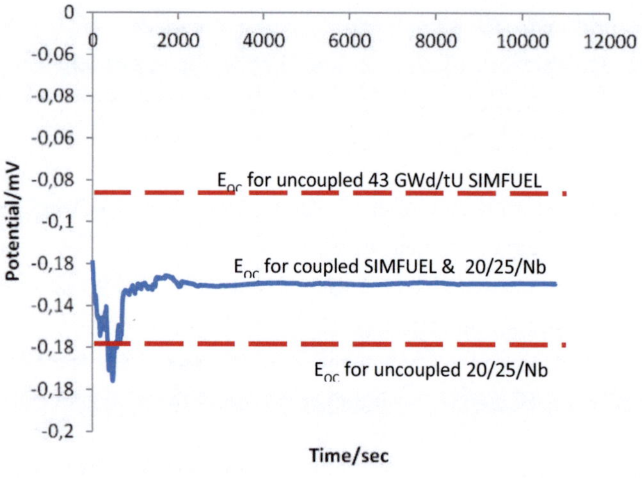

Figure 32.5: Open Circuit Potential of coupled 43 GWd/tU & 20/25/Nbvs time in anoxic modified simplified Groundwater.

32.5 Conclusions and Future work

AGR SIMFUELs have been prepared at a range of simulated burn-ups and, using combined electrochemical and Raman spectroscopic studies, have been studied for the effect of the SIMFUEL dopants on the UO_2 crystal structure. We have also studied the effect of H_2O_2 on the SIMFUEL surface as a simulant for the effect of α radiolysis on intruding groundwater in the near field. The intensity of the fundamental U-O stretch (445 cm-1) decreases as the amount of dopant increases in each SIMFUEL burn-up composition. A simultaneous increase in the lattice damage (500 – 700 cm-1) peak is observed as the UO_2 cubic fluorite lattice structure becomes more distressed and moves towards a tetragonal structure. Exposure to 100 µmol dm-3 H_2O_2 further decreases the fundamental U-O stretch and increases the lattice damage peak, suggesting that additional point defects are established as the concentration of interstitial oxygen is increased in the lattice via the H_2O_2-induced corrosion of the SIMFUEL.

Cyclic voltammetric studies of the AGR SIMFUELs reveal them to be more susceptible to electrochemical oxidation than PWR SIMFUELs. The open circuit potential for UO_2 and AGR SIMFUELs show a logarithmic dependence on (H_2O_2) in modified simplified groundwater. In contrast, the EOC of PWR SIMFUELs of ~0.1 V vs SCE in alkaline solution is near independent of (H_2O_2). The open circuit potential of -0.08 V vs SCE measured from 43 GWd/tU burnup SIMFUEL in modified simplified groundwater indicates that the SIMFUEL may corrode in the modified simplified groundwater even in the absence of a peroxide generated oxidative stress. Similar studies of the same SIMFUEL held in contact with samples of the 20/25/Nb steel cladding show that this coupled system exhibits a mixed potential of ~-0.12 V vs SCE. This potential is (i) negative of that adopted by the isolated SIMFUEL and (ii) positive of that adopted by the isolated cladding in electrolytes of the same composition, implying that the SIMFUEL is protected against corrosion at the expense of the steel cladding, an observation that is currently the subject of further work in our laboratories.

Finally, we recently have fabricated UO_2 coated QCM crystals which will be used in future works to study by electrochemical and thermal means the UO_{2+x} generation on UO2 that may occur upon exposure to H_2O_2. These crystals will also be used in the study of H_2O_2 driven UO2 dissolution and fission product uptake, namely Cs+ and Sr2+, from the aqueous phase and hydrogen uptake form the gas phase.

32.6 Acknowledgement

We thank the Nuclear Decommissioning Authority (NDA), the EPSRC (Award No EP/I036400/1) and The Lloyd's Register Foundation (LRF) for financial support. LRF, a UK registered charity and sole shareholder of Lloyd's Register Group Ltd, invests in science, engineering and technology for public benefit, worldwide.

32.7 References

Buck, E.C., Wittman, R.S., Skomurski, F.N., Cantrell, K.J., McNamara, B.K., Soderquist, C.Z. (2012). Radiolysis Process Modeling Results for Scenarios. PNNL Technical Report, FCRD-UFD-2012-000199 PNNL- 21554.

Burakov, B.E., Strykanova, E.E., Anderson, E.B. (1997). Secondary Uranium Minerals on the Surface of Chernobyl 'Lava'. MRS Symposium Proceedings, 465, 1309-1312.

Burstall, R.F. (1979). Fispin – A Computer Code for Nuclide Inventory Calculations. UK Atomic Energy Authority Report, ND-R-328 (R).

Corbel, C., Sattonnay, G., Guilbert, S., Garrido, F., Barthe, M.-F., Jegou, C. (2006). Addition Versus Radiolytic Production Effects of Hydrogen Peroxide on Aqueous Corrosion of UO2. Journal of Nuclear Materials, 348, 1-17.

Hiezl, Z., Hambley, D.I., Padovani, C., Lee, W.E. (2015). Processing and Microstructural Characterisation of a UO2-Based Ceramic for Disposal Studies on Spent AGR Fuel. Journal of Nuclear Materials, 456, 74–84.

Sattonnay, G., Ardois, C., Corbel, C., Lucchini, J.F., Barthe, M.-F., Garrido, F., Gosset, D. (2001). Alpha-Radiolysis Effects on UO2 Alteration in Water. Journal of Nucear Materials, 1, 11-19.

33 The characteristics and leaching behaviour of spent AGR fuel – scoping studies

Cristiano Padovani[1], C. Askeljung[2], A. Puranen[2], M. Granfors[2], O. Roth[2], M. Cowper[2] (AMEC)

[1] *Radioactive Waste Management Limited (UK)*
[2] *Studsvik Nuclear AB (SE)*
[3] *Amec (UK)*

33.1 Abstract

The presentation covered the results of pre-leaching chemical analysis of Advanced Gas-cooled Reactor (AGR) fuel and compared the results of leaching studies with data on LWR fuels. The different design of the AGR fuel elements was shown and the fuel types under investigation. The initial enrichment was 3.42 % ^{235}U, the burn-up ranged from 27 MWd/kgU (FGR 0.2%), 30 MWd/kgU (FGR 1.1 %) to 37 MWD/kgU (FGR 8.4 %). The leaching was performed in 'hot cells' for 2 years in naturally aerated 0.01 M NaCl / 0.002 M $NaHCO_3$ solution by static tests. The IRF and the matrix dissolution rates were determined.

The measured release rates of ^{129}I and ^{137}Cs from the three types of AGR fuels were compared to calculated data of LWR fuels according to IRF (%) = $0.024 \times e^{(0.08 \times BU)}$.

The results of the experiments could be summarized as follows:

- The chemical analysis of AGR fuel indicated (1) a loss of volatile elements such as Cs and I from pellet interior in regions associated with highest temperatures (bore in the pellets) and (2) the evidence of precipitates containing Cs, I and Te in cooler regions of the fuel pin at the bottom end cap. These precipitates might possibly be associated with macroscopic segregation of volatile chemical elements during irradiation.
- The leaching studies with AGR fuels of 'low/high' FGR (0.2-1.1%) and 'average' burn-up (~30 MWd/kgU) revealed similar results in comparison to LWR fuels under the same conditions. For fuels with 'very high' FGR (8.4%) and 'high' burn-up (~ 40 MWd/kgU), leaching had not reached the steady state. Brown-coloured precipitates (largely containing U and Fe) have been observed during leaching experiments in the presence of the (stainless steel) cladding.

34 Non-congruent release: What does it mean?

Brady Hanson, Edgar Buck, Edward Mausolf

Pacific Northwest National Laboratory

34.1 Abstract

Several examples of non-congruent release (solution and solids analyses) are presented. Static tests by Wilson in J-13 groundwater are described. The releases of radionuclides from bare fuel at 25°C and 85°C decreased with time. Other static tests have been performed for 19 yrs. by Stroes-Gascoyne et. al. MRS(465) 1997. The ANL "drip" tests in EJ-13 groundwater is shown and the PNNL single-pass flow-through tests (SPFT) using sieved samples. Preferential release of 137Cs and 90Sr was observed and the sensitivity to pH and radiolysis. Carbonate and peroxide dissolution were conducted to determine Tc release from ATM-106 fuel, crushed and sieved to 25 µm and 5 µm fractions and dissolved in an ammonium carbonate/hydrogen peroxide or nitric acid solutions. Up to 68% of the Tc remained in the undissolved solids, Pu was dissolved completely. This method allowed to determine the distribution of Tc between ε-phase and "free" in matrix.

Questions are raised concerning the Tc release which was above total gap and grain boundary inventory. Further questions cover the discrimination of the IRF from the following slow matrix dissolution and how to apply these data.

IRF depends on the distribution of nuclides between gap, grain boundary, and matrix, but also as distributed between "free"/dissolved and non-matrix phases (e.g., ε-phase). This is most likely a function of burnup and duty cycle. Further dependencies are as follows:

- Water chemistry affects the dissolution of nuclides differently.
- Radiolysis impacts dissolution rates, secondary phase formation and nuclide incorporation.
- Rapid changing of surface areas (opening grain boundaries, secondary phase precipitation, insoluble nuclides).
- Fuel chemistry effects especially after long times